SUSTAINABLE BIOFUEL AND BIOMASS

Advances and Impacts

SUSTAINABLE BIOFUEL AND BIOMASS

Advances and Impacts

Edited by

Arindam Kuila

Apple Academic Press Inc.
3333 Mistwell Crescent
Oakville, ON L6L 0A2
Canada USA

Apple Academic Press Inc.
1265 Goldenrod Circle NE
Palm Bay, Florida 32905
USA

First issued in paperback 2021

Exclusive worldwide distribution by CRC Press, a member of Taylor & Francis Group

ISBN 13: 978-1-77463-481-3 (pbk)
ISBN 13: 978-1-77188-807-3 (hbk)

Library and Archives Canada Cataloguing in Publication

Title: Sustainable biofuel and biomass : advances and impacts / edited by Arindam Kuila.

Names: Kuila, Arindam, editor.

Description: Includes bibliographical references and index

Identifiers: Canadiana (print) 20190095628 | Canadiana (ebook) 20190095652 | ISBN 9781771888073 (hardcover) | ISBN 9780429265099 (eBook)

Subjects: LCSH: Biomass energy. | LCSH: Lignocellulose.

Classification: LCC TP339 .S87 2019 | DDC 333.95/39—dc23

CIP data on file with US Library of Congress

Apple Academic Press also publishes its books in a variety of electronic formats. Some content that appears in print may not be available in electronic format. For information about Apple Academic Press products, visit our website at **www.appleacademicpress.com** and the CRC Press website at **www.crcpress.com**

About the Editor

Dr. Arindam Kuila, PhD

Arindam Kuila is currently working as Assistant Professor in the Department of Bioscience and Biotechnology at the university Banasthali Vidyapith in Rajasthan, India. Previously, he worked as a research associate at the Hindustan Petroleum Green R&D Centre, Bangalore, India. He earned his PhD from the Agricultural and Food Engineering Department at the Indian Institute of Technology in Kharagpur, India in 2013. His PhD research was focused on bioethanol production from lignocellulosic biomass. He was awarded a Petrotech Research Fellowship in 2008. He has co-authored 16 peer-reviewed papers, five review papers, two books, and six book chapters, and has filed five patents.

Contents

Contributors .... *ix*

Abbreviations .... *xiii*

Preface .... *xv*

1. **Biofuel Production from Lipase and Its Biotechnological Application** **1**
   Ritika Joshi and Arindam Kuila

2. **Biohydrogen Production from Dark Fermentation of Lignocellulosic Biomass** **19**
   Zumar M. A. Bundhoo

3. **Tailoring Triacylglycerol Biosynthetic Pathway in Plants for Biofuel Production** **41**
   Kshitija Sinha, Ranjeet Kaur, and Rupam Kumar Bhunia

4. **Industrial Technologies for Bioethanol Production from Lignocellulosic Biomass** **61**
   Amrita Saha, Soumyak Palei, Minhajul Abedin, and Bhaswati Uzir

5. **Production of Bio-Syngas for Biofuels and Chemicals** **73**
   Shritoma Sengupta, Debmallya Konar, Debalina Bhattacharya, and Mainak Mukhopadhyay

6. **Biofuel Cell from Biomass** **95**
   Rituparna Saha, Debalina Bhattacharya, and Mainak Mukhopadhyay

7. **Biofuel Production from Algal Biomass** **119**
   Rituparna Saha, Debalina Bhattacharya, and Mainak Mukhopadhyay

8. **Potential Feedstocks for Second-Generation Ethanol Production in Brazil** **145**
   Luiza Helena da Silva Martins, João Moreira Neto, Paulo Weslem Portal Gomes, Johnatt Allan Rocha de Oliveira, Eduardo Dellosso Penteado, and Andrea Komesu

9. **Perspective of Liquid and Gaseous Fuel Production from Aquatic Energy Crops** 167
Gunasekaran Rajeswari, Samuel Jacob, and Rintu Banerjee

10. **Metabolic Engineering for Liquid Biofuels Generations from Lignocellulosic Biomass** 183
Dipankar Ghosh

11. **Weed Biomass as Feedstock for Bioethanol Production: A Review** 213
Vinod Singh Gour, Ravneet Chug, and S. L. Kothari

12. **Implication of Anaerobic Digestion for Large-Scale Implementations** 225
Samuel Jacob and Rintu Banerjee

13. **Bioenergy Production: Biomass Sources and Applications** 245
Vartika Verma, Priya Singh, Gauri Singhal, and Nidhi Srivastava

14. **Third-Generation Biofuels: An Overview** 261
João Moreira Neto, Andrea Komesu, Luiza Helena da Silva Martins, Vinicius O.O. Gonçalves, Johnatt Allan Rocha de Oliveira, and Mahendra Rai

15. **Bioethanol Production from Different Lignocellulosic Biomass** 281
Saurabh Singh and Jay Prakash Verma

16. **Algal Biomass and Biodiesel Production** 301
Samakshi Verma and Arindam Kuila

17. **Bioenergy: Sources, Research, and Advances** 317
Sunanda Joshi, Monika Choudhary, Sameer Suresh Bhagyawant, and Nidhi Srivastava

Index *331*

Contributors

Minhajul Abedin
Department of Environmental Science, Amity Institute of Environmental Science,
Amity University Kolkata, West Bengal, India

Rintu Banerjee
Department of Agricultural and Food Engineering, Indian Institute of Technology,
Kharagpur 721302, West Bengal, India

Sameer Suresh Bhagyawant
School of Studies in Biotechnology, Jiwaji University, Gwalior, Madhya Pradesh, India

Debalina Bhattacharya
Department of Biotechnology, University of Calcutta, Kolkata, West Bengal, India

Rupam Kumar Bhunia
Plant Tissue Culture and Genetic Engineering, National Agri-Food Biotechnology Institute (NABI),
Mohali 140306, Punjab, India. E-mail: rupamb2005@gmail.com; rupamb@nabi.res.in

Zumar M. A. Bundhoo
Department of Chemical and Environmental Engineering, Faculty of Engineering,
University of Mauritius, Réduit, Mauritius. E-mail: zumar.bundhoo@gmail.com

Monika Choudhary
Department of Bioscience and Biotechnology, Banasthali Vidyapith, Rajasthan, India

Ravneet Chug
Amity Institute of Biotechnology, Amity University Rajasthan Jaipur, Kant Kalwar, Jaipur, India

Dipankar Ghosh
Department of Biotechnology, JIS University, 81 Nilgunj Road, Agarpara, Kolkata,
West Bengal 700109, India. E-mail: d.ghosh@jisuniversity.ac.in; dghosh.jisuniversity@gmail.com

Paulo Weslem Portal Gomes
Centro de Ciências Naturais e Tecnologia, StateUniversityof Pará (UEPA), Belém, PA, Brazil

Vinicius O.O. Gonçalves
Instituto de Química, Departamento de Físico-Química, Universidade Federal do Rio de Janeiro,
Rio de Janeiro, RJ, Brazil

Vinod Singh Gour
Amity Institute of Biotechnology, Amity University Rajasthan Jaipur, Kant Kalwar, Jaipur, India.
E-mail: vkgaur@jpr.amity.edu; vinodsingh2010@gmail.com

Samuel Jacob
Department of Biotechnology, School of Bioengineering, SRM Institute of Science and Technology,
Kattankulathur 603203, Tamil Nadu, India. E-mail: samueljacob.b@ktr.srmuniv.ac.in

Ritika Joshi
Department of Bioscience and Biotechnology, Banasthali Vdiyapith, Rajasthan 304022, India

Sunanda Joshi
Department of Bioscience and Biotechnology, Banasthali Vidyapith, Rajasthan, India

Ranjeet Kaur
Department of Biotechnology, Mangalmay Institute of Management and Technology (MIMT), Greater Noida 201306, Uttar Pradesh, India

Andrea Komesu
Departamento de Ciências do Mar (DCMar), Federal University of São Paulo (UNIFESP), Santos, SP, Brazil. E-mail: andrea_komesu@hotmail.com

Debmallya Konar
Department of Biotechnology, JIS University, Kolkata, West Bengal, India

S. L. Kothari
Amity Institute of Biotechnology, Amity University Rajasthan Jaipur, Kant Kalwar, Jaipur, India

Arindam Kuila
Department of Bioscience and Biotechnology, Banasthali Vdiyapith, Rajasthan 304022, India. E-mail: arindammcb@gmail.com

Luiza Helena da Silva Martins
Centro de Ciências Naturais e Tecnologia, StateUniversityof Pará (UEPA), Belém, PA, Brazil

Mainak Mukhopadhyay
Department of Biotechnology, JIS University, Kolkata, West Bengal, India. E-mail: m.mukhopadhyay85@gmail.com

João Moreira Neto
Departamento de Engenharia (DEG), University Federal of Lavras (UFLA), Lavras, MG, Brazil

Johnatt Allan Rocha de Oliveira
Instituto de Ciências da Saúde, Faculdade de Nutrição, Federal University of Pará (UFPA), Belém, PA, Brazil

Soumyak Palei
Department of Environmental Science, Amity Institute of Environmental Science, Amity University Kolkata, West Bengal, India

Eduardo Dellosso Penteado
Departamento de Ciências do Mar (DCMar), Federal Universityof São Paulo (UNIFESP), Santos, SP, Brazil

Mahendra Rai
Nanobiotechnology Lab, Department of Biotechnology, SGB Amravati University, Amravati, Maharashtra, India

Gunasekaran Rajeswari
Department of Biotechnology, School of Bioengineering, SRM Institute of Science and Technology, Kattankulathur 603203, Tamil Nadu, India

Amrita Saha
Department of Environmental Science, Amity Institute of Environmental Science, Amity University Kolkata, West Bengal, India. E-mail: asmicrobio@yahoo.in

Rituparna Saha
Department of Biotechnology, JIS University, Kolkata, West Bengal, India

Department of Biochemistry, University of Calcutta, Kolkata, West Bengal, India

Shritoma Sengupta
Department of Biotechnology, JIS University, Kolkata, West Bengal, India
Department of Biochemistry, University of Calcutta, Kolkata, West Bengal, India

Priya Singh
Department of Bioscience and Biotechnology, Banasthali Vidyapith, Rajasthan, India

Saurabh Singh
Institute of Environment and Sustainable Development, Banaras Hindu University, Varanasi 221005, India

Gauri Singhal
Department of Bioscience and Biotechnology, Banasthali Vidyapith, Rajasthan, India

Kshitija Sinha
Plant Tissue Culture and Genetic Engineering, National Agri-Food Biotechnology Institute (NABI), Mohali 140306, Punjab, India

Nidhi Srivastava
Department of Bioscience and Biotechnology, Banasthali Vidyapith, Rajasthan, India.
E-mail: nidhiscientist@gmail.com

Bhaswati Uzir
Department of Environmental Science, Amity Institute of Environmental Science, Amity University Kolkata, West Bengal, India

Jay Prakash Verma
Institute of Environment and Sustainable Development, Banaras Hindu University, Varanasi 221005, India
Hawkesbury Institute for the Environment, Hawkesbury Campus, Western Sydney University, Penrith, NSW 2750, Sydney, Australia.
E-mail: jpv.iesd@bhu.ac.in; verma_bhu@yahoo.co.in; j.verma@westernsydney.edu.au

Samakshi Verma
Department of Bioscience and Biotechnology, Banasthali Vidyapith, Rajasthan 304022, India

Vartika Verma
Department of Bioscience and Biotechnology, Banasthali Vidyapith, Rajasthan, India

Abbreviations

5-HMF	5-hydroxymethyl furfural
ABI	abscisic acid insensitive
ACX	acyl-CoA oxidase
acyl-CoA	acyl-Coenzyme A
ASTM	American Society for Testing and Materials
BGL	β-glucosidase
C:N	carbon to nitrogen ratio
CBM	carbohydrate binding module
CBP	consolidated bioprocessing
CHP	combined heat and power
COD	chemical oxygen demand
CTAB	cetyl trimethylammonium bromide
DAG	diacylglycerol
DAT	dilute acid treatment
DF	dark fermentation
DGAT	diacylglycerol acyltransferase
DGR	daily growth rate
DHAP	dihydroxyacetone phosphate
EEL	energy equivalent liter
ER	endoplasmic reticulum
FAD2	fatty acid desaturase 2
FFAs	free fatty acid
FVW	fruit and vegetable waste
G-3-P	*sn*-glycerol-3-phosphate
GA	gibberellic acid
GHGs	greenhouse gases
GPAT	glycerol-3-phosphate acyltransferase
H_C	H_2 yield from untreated substrates (control)
H_P	H_2 yield from pretreated substrates
HRT	hydraulic retention time
HTL	hydrothermal liquefaction
IEA	International Energy Agency

LB	lignocellulosic biomass
LEC1	leafy cotyledon-1
LEC2	leafy cotyledon-2
LEDs	light emitting diodes
LPAAT	lysophosphatidic acid acyltransferase enzyme
MSW	municipal solid waste
OB	oil bodies
OFMSW	organic fraction municipal solid wastes
OLR	organic loading rate
PAP	phosphatidate phosphatase
PDC	pyruvate decarboxylase
PEP	phosphoenolpyruvate
Pfl	pyruvate:formate lyase
Pfor	pyruvate:ferredoxin oxidoreductase
PXA1	peroxisomal ABC transporter-1
RBPDs	regional biomass processing depots
RSVs	reducing sugar values
SDP1	sugar-dependent-1
TAG	triacylglycerol
TS	total solids
VFA	volatile fatty acid
VPW	vegetable processing waste
VS	volatile solids

Preface

This book provides in-depth information on several types of biofuel from lignocellulosic biomass. Biofuel production from waste biomass is increasingly being focused on due to several advantages of lignocellulosic biomass—it is available in abundance from several sources, it is cost-effective, and it has little competition with food sources. The present book contains 17 chapters that deal with different aspects of biofuel production from lignocellulosic biomass. Chapter 1 deals with different lipase-mediated types of biofuel production. Chapter 2 describes biohydrogen production from lignocellulosic biomass. Chapter 3 describes triacylglycerol biosynthetic pathways in plants for biofuel application. Chapter 4 deals with the industrial prospects of lignocellulosic bioethanol production. Chapter 5 describes bio-syngas production from biomass. Chapter 6 describes biofuel cell production from lignocellulosic biomass. Chapter 7 describes biofuel production from algal biomass. Chapter 8 describes potential feedstock availability in Brazil for bioethanol production. Chapter 9 deals with the current status and future prospect of liquid and gaseous fuel production from aquatic weeds. Chapter 10 describes an metabolic engineering approach for biofuel production from lignocellulosic biomass. Chapter 11 deals with the prospect of weed biomass for biofuel production. Chapter 12 describes anaerobic digestion system for large-scale biogas production. Chapter 13 describes biomass sources and potential application for biofuel production. Chapter 14 deals with the current status and future prospect of third-generation biofuel. Chapter 15 gives detailed information about technological updates of lignocellulosic bioethanol production. Chapter 16 describes the prospect of algal biomass for biodiesel production. Chapter 17 gives detailed information about biomass sources, availability, and their utility for biofuel production. This book will be useful to senior undergraduate and graduate students, researchers, and others interested in the field of biofuel/bioenergy.

CHAPTER 1

Biofuel Production from Lipase and Its Biotechnological Application

RITIKA JOSHI and ARINDAM KUILA*

Department of Bioscience and Biotechnology, Banasthali Vidyapith, Rajasthan 304022, India

**Corresponding author. E-mail: arindammcb@gmail.com*

ABSTRACT

Biofuels have become progressively eye-catching in the recent years due to the growing attention on the depletion of environmental issues and fossil fuel resources. Biofuel production is potential and essential field of research because it is significant to the environmental advantages and increasing petrol price. In comparison to the conventional production process, this is very eco-friendly process. The feedstock plays a major role in the biofuel production process. Downstream processes largely decide the process efficiency. This chapter reviews the past and current developments of biofuels, together with the different types of biofuels, processing, and characteristics of biofuels industry. The biotechnological applications of biofuels are discussed as well.

1.1 INTRODUCTION

Currently, biofuels are the biodegradable and renewable sources of energy, obtained from different biomass. The invention of this technology is very important because of increasing distress of the effect that the petro fuels have on the habitat (Godbole and Dabhadkar, 2016). Fat and oils are the main feedstocks which represent supreme importance for lipase application. These materials comprises triglycerides (ester with saturated and unsaturated fatty acids) from microbial, plant, and animal origin (Ribeiro et al., 2011). Biofuels have various advantages such as high combustion efficiency, transport ability, low sulfur, high cetane number, biodegradability, and aromatic content (Demirbas, 2008).

In biotechnological applications, biocatalysts are receiving a great attention. Lipases (triacylglycerol acylhydrolases, EC 3.1.1.3) are flexible enzymes that are widely used as a natural catalyst. In addition to the breakdown of triglycerides, lipases can also accelerate the diverse variety of chemical reactions which comprise esterification, transesterification, acidolysis, and aminolysis. The production of biodiesel is one of the eye-catching applications of lipase catalysis (Mittelbach et al., 1998). Transesterification of lipase catalysis occurs in two steps, first is ester bond hydrolysis and second is substrate esterification. There are various sources by which lipase can be isolated such as pancreatic lipase (animal), papaya latex, castor seed, and oat seed lipase (plant) (Zhang et al., 2012). In industries, due to short generation time they prefer microorganisms for enzyme production (Villeneuve et al., 2000). The best features of microorganisms include great versatility in different conditions such as culture condition, high yield, and genetic manipulation (Ribeiro et al., 2011). *Rhizopus oryzae*, *Candida rugosa*, *Chromobacterium viscosum*, *Streptomyces* sp., *Aspergillus niger*, *Candida* sp., *Mucor miehei*, *Pseudomonas cepacia*, *Photobacterium lipolyticum*, *Pseudomonas fluorescens*, *Thermomyces lanuginosus*, and various Antarctic spp. are some fungal and bacterial species which are frequently used in biodiesel production (Yahya et al., 1998). Microorganism such as *Candida rugosa* isolated from yeast is used in the production of lipase. Current investigation on *Streptomyces* sp. was found as a powerful lipase producer as well as used in biodiesel production (Cho et al., 2012).

1.2 BIODIESEL

Biodiesels are monoalkyl esters of long-chain fatty acids and are considered as an alternative for the fossil fuels. They are originated from animal and plant fat as well as from natural oil. Biodiesel has enticed a broad awareness on the earth by virtue of its biodegradability, non-toxicity, renewability, and eco-friendly benefits (Alemayehu et al., 2015). Production of biodiesel requires vegetable oil and it holds a lot of environmental benefits. Large amount of vegetable oil generate at the time of cooking and it has disposable issues. The main benefit of using them for biodiesel production is low cost. Due to the short-reaction time, high alteration of pure triglyceride to fatty acid methyl ester, and availability of plant sources it is the excellent preliminary material for biodiesel production (Thirumarimurugan et al., 2012).

Biomass processing technologies have an extended history for improved production of biodiesel. Thermochemical and biochemical conversions are the two main categories of all biomass conversions. Liquefaction, gasification, and pyrolysis are the general thermochemical methods to make synthetic oil and biochemicals in that order from biomass. In contrast, production of bioethanol and biodiesel takes place in biochemical conversion (Hassan and Kalam, 2013) of wheat, potatoes, sugarcane, and maize that are the different sources used in hydrolysis or fermentation of bioethanol. Production of biodiesel is done by transesterification process. In the process of transesterification, breakdown of triglycerides to glycerol and free fatty acids takes place (Srivastava and Prasad, 2000). Bioethanol is used in spark-ignition engine because it is fungible to petrol; in the same way, biodiesel is used in compression-ignition engine as it can be used interchangeably with diesel.

E. Duffy and J. Patrick in 1853, first time conducted the process transesterification of triglycerides. In 1853, Rudolf Diesel a famous German scientist, who invented diesel engine in the same year when his paper got published on heat engine (Zhou and Thomson, 2009). Several researches are carried on the transesterification process with continuous improvements and several modifications to obtain the highest production of biodiesel for commercial purpose. This catalytic method is broadly used.

1.3 MAIN TYPES OF BIOFUELS

A selection of biofuels can be obtained from biomass (Table 1.1).

- Biodiesel and ethanol as biofuels are most commonly used in transportation today.
- In the future, Fischer–Tropsch and butanol fuels have the prospective to become liquid fuels.

- Gasification of wood by gas synthesis is used chiefly for electricity generation.
- Biogas and fuel wood formed by anaerobic digestion of animal and plant wastes are used for heating and cooking at the domestic level.

TABLE 1.1 Different Sources and Applicable Methods of Biodiesel Production.

S. no.	Methods	Sources	References
1.	Lipase-catalyzed interesterification	Wheat	Kafuku and Mbarawa, 2011
2.	Pyrolysis and transesterification	Soybeans	Shahid and Jamal, 2011
3.	Catalytic pyrolysis	Sunflower	Singh et al., 2010
4.	Lipase-catalyzed interesterification	Rice bran oil	Lin et al., 2011
5.	Hydrolysis and saponification	Peanut oil and corn	Shahid and Jamal, 2011
6.	Transesterification	*Jatropha*	Kibazohi and Sangwan, 2011
7.	Catalytic pyrolysis	Sunflower	Karmakar et al., 2010
8.	Enzymatic transesterification	Rapeseed	Ahmad et al., 2011

1.4 BIOFUEL FEEDSTOCKS AND CONSUMPTION

Biodiesel and bioethanol are the two types of biofuels which are commonly used. These biofuels are derived primarily from vegetable oils, lignocelluloses, and seeds. Biodiesel can be used as an alternative to diesel, whereas bioethanol is the replacement of petrol. Plant oil such as soybean, palm rapeseed, sunflower, and also nonedible oil such as neem, *Jatropha*, karanja, and mahua are some common biodiesel feedstocks. After refining, used animal fat (i.e., beef tallow) and waste cooking oil can also be used for of the production of biodiesel. At the same time, algae is considered as a new source of third-generation biofuel (Hassan and Kalam, 2013). Biodiesel contain no petroleum so it is blended at every ratio with No. 2 diesel fuel to be used in diesel engines with few or no changes. Fuels ranked biodiesels are bent around the transesterification method consistent to precise designation such as ASTM D6751 in command to assure good quality and operation. As an alternative, wheat, sugarcane, potatoes, maize, and corn are used in the fermentation methods for bioethanol production.

Brazil has invented fuel-flex vehicle which is made to order for the use of bioethanol as well as bioethanol–petrol blends which has effectively cut down their assurance on petrol.

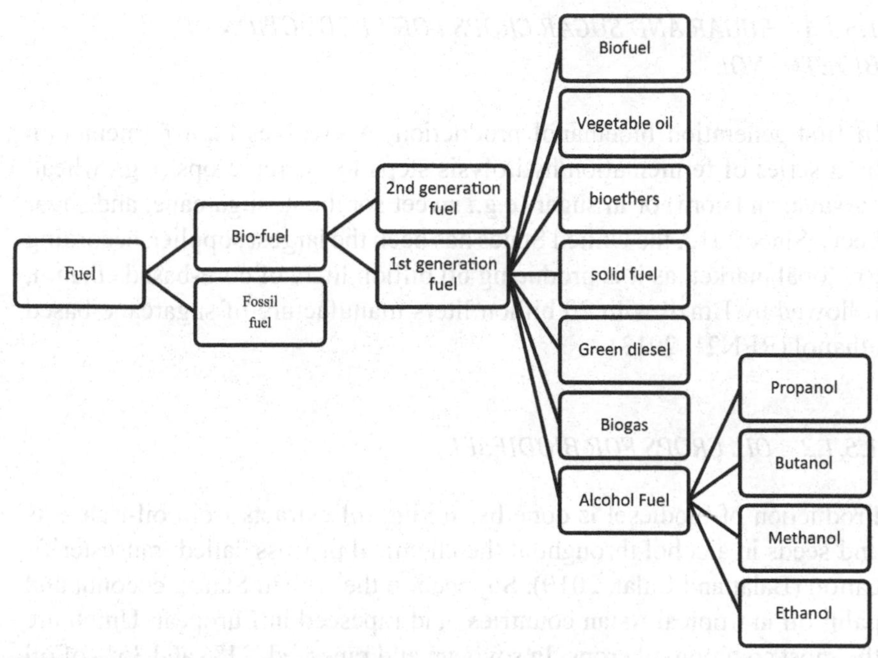

FIGURE 1.1 Classification of biofuels.

1.5 BIOMASS RESOURCES AND THEIR BIOENERGY PROSPECTIVE

1.5.1 FIRST-GENERATION FEEDSTOCK BIOFUELS

In recent year, production of biofuels has rapidly increased and presently 3.4% of global transportation is in fuel supplies, with 21% significantly contributed to Brazil, 3% in European Union, and greater than ever contributed to the United State (4%) (IEA, 2013). In biofuel production such as biogas, bioethanol, biodiesel, arable food crops are used as bioenergy crops, approximately 40 million gross hectares (FAOSTAT, 2011). Oil seeds are used for biodiesel and starch and sugar are used for bioethanol production; these are traditional food crops of first generation of biofuels production.

1.5.1.1 SUGAR AND SUGAR CROPS FOR PRODUCTION OF BIOETHANOL

In first-generation bioethanol production, it involves high fermentation by a series of fermentation/hydrolysis steps for starch crops (e.g., wheat, cassava, and corn) or in sugar (e.g., sweet sorghum, sugarcane, and sugar beet). Since 2012, the United States has been the largest supplier, according to global market, as it is producing 60 billion liters of corn-based ethanol, followed by Brazil with 20 billion liters manufacture of sugarcane-based ethanol (REN21, 2013).

1.5.1.2 OIL CROPS FOR BIODIESEL

Production of biodiesel is done by mixing oil extracts from oil-rich nuts and seeds in alcohol throughout the chemical process called transesterification (Balat and Balat, 2010). Soybean in the United States, coconut and palm oil in tropical Asian countries, and rapeseed in European Union are the most common oil crops. In soybean and rapeseed, 21% and 35% of oil content is found, respectively (Ramos et al., 2009). As compared to other seed, palm oil has the highest oil content—40% (Balat and Balat, 2010). Moreover, for the conversion of feedstocks for biodiesel production, it requires conversion of cooking oil and beef tallow (REN21, 2013). The main difference between oil feedstocks is determined by its molecular structure, different form of fatty acids attached in the triacylglycerol (TAG), and the degree of unsaturation/saturation (Ramos et al., 2009). All these factors, sequentially, quality, expenditure, and affect production processes of the biodiesel products (Ramos et al., 2009).

The transesterification is a step-by-step reaction of conversion of TAG with an alcohol to form glycerol and esters in the influence of strong base; this process is mostly applied for methanol production (Balat and Balat, 2010). As a result, production of biodiesel by alkali-catalyzed transesterification method requires low processing pressure, temperature, and economical option for achieving a 98% conversion in production (Balat and Balat, 2010). In contrast, enzyme-catalyzed method is gaining attention due to its high purity of glycerol, low energy expenditure, and lowering soap formation (Christopher et al., 2014). Although two main difficulties to expand these processes are its small rate of reaction and expensive enzyme cost. The very familiar biodiesel-blended products are B5 (95% petroleum diesel and 5% biodiesel), B2 (98% petroleum diesel and 2% biodiesel), and B20 (80% petroleum diesel and 20% biodiesel).

1.5.2 SUSTAINABILITY ISSUES OF THE FIRST-GENERATION FEEDSTOCKS

The composition and conversion of biomass feedstocks have several environmental and socioeconomic impacts. Even though, market and mature technologies have spread worldwide in the first-generation biofuels. The potential climate change mitigation benefits of first-generation biofuels have been another contentious aspect of the biofuel debate (Gasparatos et al., 2013). Vegetable oil, cereal, and sugar globally they share progressively more importance representing the use of biofuels.

1.6 DEDICATED ENERGY CROPS

Energy crops are grown specifically and developed for biofuel production which consists of short rotation forestry (poplar and willows) and perennial grasses (switchgrass and *Miscanthus*). These crops provide higher energy production because they can grow in degraded soil and steady supply stream.

1.6.1 PERENNIAL GRASSES

Southeast Asia and North America are the origin of switchgrass; it is the best choice for bioenergy production in low input because of their nutrition requirements, cool temperature, ability to grow in low water condition, applying conventional farming practices (Lewandowski et al., 2003).

It generally needs 3 years to arrive at profitable development and generate dry substance production (Heaton et al., 2004). Correspondingly, *Miscanthus* takes 2–3 years to achieve complete production and needed rhizome cuttings and excessive expenditure related to propagation.

Perennial grasses crop yields strongly depends on limited situation, that is, land quality, climate, management system, irrigation, and fertilization. Other prospective of herbaceous crops include alfalfa, giant reed, and reed canary grass modified to temperate zone, Johnson grass, bana grass, and Napier grass, in subtropical and tropical regions (Ra et al., 2012). The perennial grasses are beneficial in maintaining wildlife environment and reduce soil erosion because they are efficient in soil stabilization and carbon sequestration (Zhang et al., 2011).

1.6.2 SHORT ROTATION WOOD CROPS

In biofuels production, a number of rapid flourishing trees have also been useful because of their broad geographical division, high yield, low labor consuming, and low costs compared to annual crops (Hauk et al., 2014), for example, *Eucalyptus*, poplar, and willow species. Over annual agricultural crops advantages of perennial grasses and short rotation forestry are acquitted, these enthusiastic energy crops are land-based, and consequently not completely evasion the food versus fuel discuss. Simply fiber and food crops are not reasonable that would dormant energy crops to be advantageous.

1.6.2.1 JATROPHA

In biodiesel production, *Jatropha* (*Jatropha curcas*) has been seen as a cheap ideal crop. It is natively found in tropical America; their seed has a high oil content and is drought impenetrable that grows well on insignificant land. As a result, it benefits remote areas of developing countries. Oil of *Jatropha* plant can be used in diesel generators, cooking gas, and as a fuel in vehicles, transesterification is required in this case (Koh and Mohd, 2011). Mahua, linseed, *Pongamia*, and castor are other species that are potentially used for biodiesel production.

1.6.3 FORESTRY/AGRICULTURAL RESIDUES

Forestry and agricultural residues are characterized as an incredible source of easily obtainable biomass for biofuel production which do not require land cultivation. Forestry residues comprise extracts from fuel, wood-processing mill residues, and logging residues, while agricultural residues comprise bagasse, wheat straw, and corn stove. Hemicelluloses (a mix of hexose and pentose sugars, 20–35%), lignin, and polysaccharides cellulose (hexose sugars, 35–50%) are mainly consisting of biomass residues (Singh et al., 2010). These factors are more impenetrable to be hydrolyzed than sugar, oils, and starch in the predictable food crops, making the methods more expensive and complicated.

1.6.4 MUNICIPAL AND INDUSTRIAL WASTES

Municipal solid waste (MSW) composed of chiefly cardboards, papers, plastics, and putrescibles has been formed in 2012 (IEA Bioenergy, 2013); whereas, the composition of MSW is highly changeable, its main portion is biodegradable with a suitable energy recovery operation and significant calorific (heat) value. Additionally, the paper and food industries also produced a large number of by-products and residues that can be used as biomass for bioenergy production. Industrial solid wastes comprise but are not inadequate to scraps and peelings from vegetables and fruit, poultry, and meat waste, coffee grounds, fiber and pulp from starch and sugar extraction, etc., and utilized as an energy source. It suggests abundant bioenergy applications such as changing fossil fuels with the probable environmental advantage, for example, decline in greenhouse gas emission and landfill space savings.

1.6.5 THIRD-GENERATION FEEDSTOCKS

The most accepted source for biofuels production is the algal biomass. Algae are aquatic photosynthetic microorganisms that develop very fast on coastal seawater, municipal wastewater, and saline water, and also on land inappropriate for farming and agriculture (Pittman et al., 2011). They have the capability to change sunlight and carbon dioxide by cellular actions to manufacture a diversity of chemicals including pigments, carbohydrates, proteins, lipids, and vitamins having abundant applications in various industries such as pharmaceutical and chemical industries, health food, feed supplements, and cosmetics (Costa and de Morais, 2011). Lipids are mostly accumulated by microalgal species (e.g., TAG). The classes which provide a high lipid substrate (50–80%) for the biodiesel production are *Botryococcus* and *Chlorella*. In comparison to other feedstocks which are cultivated once or other year, algae can twice their biomass in 2–5 days (Costa and de Morais, 2011). The algal biomass is five times higher than that attained from oil palm and highest producing oil crop plant (Day et al., 2012). Moreover, the reduction in cost is due to the increased hydrolysis efficiency, high production of fermentation because of no lignin, and low hemicelluloses amount in algae (Li et al., 2014). Other than biodiesel and bioethanol, the diverse kinds of renewable biofuels are formed by the algae. Another popular product used in fuel cells is biohydrogen; although biomethane formed as a component of incorporated methods can be used for electricity generation, transportation, and heating purposes (Costa and de Morais, 2011).

There are still various objections related with algal-biofuel production which required in following type of methods: photobioreactor design, algal cultivation downstream treatment processes, and mode of production (Chen et al., 2011). Microalgae culture believes as a main confinement to marketable development. Usually, production can be done also by open bonds demanding minimum costs but having closed bioreactors or hybrid systems or low biomass yield, with high production and high capital costs (Chen et al., 2011). For that reason, there is an exchange between algal biomass cultivation and investment cost. Besides, strains and algal species differ significantly in terms of photosynthetic efficiency, growth rate, nutrient requirements, productivity, and capacity to adjust in unfavorable circumstances (John et al., 2011).

1.7 PRODUCTION OF BIODIESEL

Direct use and blending, microemulsion process, thermal cracking process, and transesterification processes are the different methods which can be used for the production of biodiesel. This is because these methods are comparatively easy, best in conversion efficiency, and conceded at normal conditions, in addition to superiority of the converted fuel (Shahid et al., 2012).

1.7.1 DIRECT USE AND BLENDING

In diesel engine, direct use of vegetable oil is not good and is difficult because it has numerous innate failings. Although biodiesel fuel and vegetable oil have similar properties, it requires a few chemical alterations prior to its use into the engine. In recent years, only researched has decades but approximately from 100 years it has experimented. Even though pure vegetable oil is used to run various diesel engines, for example, turbocharged direct injection engine, consumption of energy is same as biodiesel fuels as well as vegetable oil. Ratio of 1:10 to 2:10 oil to diesel has been found successful for short-term use (Arifin, 2009).

1.7.2 MICROEMULSION PROCESS

The problem associated with high viscosity of vegetable oils was cured by microemulsion with solvents such as ethanol, 1-butanol, and methanol (Ghaly et al., 2010). Microemulsion is a colloidal mixture of optically isotropic fluid which has an extent in range of 1–150. Biodiesel microemulsion includes appropriate mixture of ionic or nonionic (Prasada, 2014). Vegetable oil, alcohol, cetane (improvers used in appropriate ratio), surfactant, and diesel fuel are included in the part of a biodiesel microemulsion. Ethanol and methanol, for example, alcohols (as viscosity lowering additives), alkyl nitrates (as cetane improvers), and higher alcohols (as surfactants) are used in biodiesel microemulsion. Spray properties of microemulsions can improve by unstable vaporization of low boiling elements in the micelles. Excellent spray characters and cetane number is the result of decrease in viscosity of microemulsion. On the other hand, incomplete combustion, injector needle sticking, and carbon deposit formation are some problems occurring in microemulsified diesel engine when it is continuously used (Parawira, 2010).

1.7.3 PYROLYSIS

The alteration of one material into other through heat or heating with the support of a catalyst is known as pyrolysis. The breakdown of chemical bonds takes place in the absence of oxygen to produce small molecules by heating. Various studies of pyrolysis have been done for the production of biofuels from vegetable and they start to produce alkadienes, alkanes, aromatics, carboxylic acids, and alkenes in different ratio. In developing countries, for modest biodiesel production the most expensive apparatus of pyrolysis is used. Moreover, the oxygen removed during these thermal processes too removes several environmental profit of using an oxygenated fuel (Parawira. 2010).

1.7.4 TRANSESTERIFICATION

Transesterification method is the well-known way of biodiesel production. It refers to the catalytic chemical reactions connecting alcohol and vegetable oil to give glycerol and biodiesel (esters of fatty acid alkyl). Potassium hydroxide, sulfuric acid, sodium, and sodium methylate are generally used as catalyst (strong base) in this process (Saribiyik et al., 2012). In this experiment, acid catalysts are too slow to reform triglycerides to biodiesel; on the other hand, acid catalysts are fairly valuable at reforming free fatty acids (FFAs) to biodiesel. As a result, FFAs are converted into esters in pretreatment step of acid-catalysis pursued by alkali-catalysis process to change the triglycerides (Rodriguez et al., 2012). Viscosity of oil can be reduced very easily by transesterification process. Yield and reaction rate can be frequently improved by the use of catalyst. From the time when the reaction is reversible, surfeit alcohol is used to balance the result side. Due to its low cost and physicochemical advantages methanol is used as alcohol. In vegetable oil, NaOH dissolves easily and methanol quickly reacts (Antony et al., 2011). Large amount of alcohol is used to enhance the production of the alkyl esters and to permit its phase division from the glycerol formed (Schuchardt et al., 1998).

1.8 USE OF BIODIESEL IN OBTAINABLE AND PROMISING TECHNOLOGIES

1.8.1 CONVENTIONAL AGRICULTURAL PRODUCTS

In the world, 60% constituents of sugar-rich crops are used in ethanol production. Globally, beet root and sugarcane are used as the main sources of ethanol production. The production of ethanol from sugarcane is a deep-rooted method (Godbole and Dabhadkar, 2016). The used sweet sorghum in this process is very cost-effective. In biofuels production, starch-rich crop without difficulty is changed to smaller sugars. In crops like cassava, wheat, sweet potato, and maize production of fuels is successfully observed. Worldwide, maize is the biggest contributor of ethanol. Sunflower oil (Autolin, 2002), rapeseed oil (Peterson and Charles, 1996), and palm oil (Widyan et al., 2002), are some oil seeds which are used in industries for the production of biodiesel. The fundamental principle for variety of seeds is the quantity of FFA group present. With the help of FFA, transesterification process can be done very easily. Lot of goods are produced from agricultural residues which create competition and they make enormous profit in biofuel production.

1.8.2 LIGNOCELLULOSIC PRODUCTS

The feedstocks of the cellulosic fuel do not compete with the food products, which is the main advantage of the lignocelluloses. Energy plantations, barks, branches, and agro-based residues are the main sources. Wood gives 60% of primary demand of total energy in the developing countries. Pesticides and fertilizers are required in very low input in terms of cultivation of energy plantation. It decreases soil nutrient that is why it is not well accepted as a feedstock for fuels production. Feedstocks such as cane trash, bagasse, and rice husk are also used. For this reason, India is connecting to the world economy and is the main producer of agricultural residues in large extent (Godbole and Dabhadkar, 2016).

1.8.3 INEDIBLE FEEDSTOCK FOR BIODIESEL

The price of biodiesel is one to three times greater than the conventional petroleum consequent fuels. The way to solve this difficulty is to utilize the feedstock that does not contend with food order.

In the production of biodiesels, inedible vegetable oils are used. After extraction, *Jatropha curcas* L. oil 40–60% is used and contains 75% FFA (unsaturated) (Qiul et al., 2011). Mahua oil is a good example of saturated fats and it produces high amount of FFA.

1.8.4 USE OF BROWN GREASE

Triglycerides, diglycerides, monoglycerides, and FFA (35%) are composed of brown grease. However , FFA increases the speed of transesterification, due to the formation of foam its existence in harmful base catalyzed trans-esterification. As a result, Canacki et al. recommended a two-step method. The amount of biodiesel increases when oil is accelerated with sulfuric acid to esterify the FFA to biodiesel. To check the effectiveness of this phenomenon a pilot plant arrangement was used (Qiul et al., 2011). This method was very efficient; however, large quantity of base was needed for catalysis. As a result cost of manufacturing is increased.

1.8.5 MICROALGAE FOR PRODUCTION OF BIODIESEL

Production of biodiesel holds a very good prospective for large-scale invention of biodiesel, because the production is very high and non-challenging in nature with food security. Algae has very high lipid content and growth rate. *Chrysophyceae, Bacillariophyceae,* and *Chlorophyceae* are some common algae used for fuel production (Qiul et al., 2011). KOH, NaOH, and CH_3OH are commonly used as strong base. It has been experimentally proved that base-catalyzed reactions are 4000 times quicker than acid-catalyzed reactions. Enzymatic catalysis allows no side reactions, enzyme can be reused for longer time, simple product separation, and increases the purity of glycerol that is why they are most promising.

1.9 APPLICATION OF BIOTECHNOLOGY IN BIOFUELS PRODUCTION

Biofuels production is the most eye-catching feature which increases the production with no increase in the energy production. In recent year, with the help of molecular biology, large enhancement has been made to make improvements in enzyme and microbial activity (Tseten and Murthy, 2014). In the case of lignocellulosic biomass, genetic modification is found very competent and quick method to progress biofuels conversion (Gressel, 2008). Composition of lignocellulosic biomass in plant cell and structure of cell wall can be modified with the help of biotechnology to increase ethanol yield (Carpita, 1996). Biotechnology can manipulate yield density by varying their architecture, plant physiology, ability to lessen agronomic inputs for instance pesticides and herbicide, and their photosynthetic effectiveness. Quick improvement has been made on characters which consume nutrients more efficiently and enable to grow crop with smaller quantity of fertilizer. In marginal acres biomass crop production is done such as poor soil characteristics or highly dry land, without any manipulation on food production acres it can increase the rate biofuels production. Plant lives on a broad range of soil conditions such as hot, cold, drought, and salty so the aim of biotechnology is to focus on their improvement. Cellulose and hemicellulose production of plant biomass feedstock in higher level give better fermentation production and therefore, ethanol production is in gallons per ton. This results in added net energy per acre and more revenue. Various studies has been done on genes cloning that code for polygalacturonase and cellulases enzymes to develop cost-effective biorefinery plan to attain improved gas chromatography–mass spectrophotometry methods and highest biomass conversion (Ibrahim et al., 2013). In biofuels production process they have faced problems such as reduction in fermentation effectiveness and microbial digestion by improved biotechnological processes.

1.10 CONCLUSIONS

According to a recent state of knowledge, in next 30–40 years greenhouse gases can be reduced and the demand for low carbon dioxide emission fuel in all forms of automobiles will be enormous. The demand of biofuels production has increased due to various undesirable effects on the environment; the price of crude oil is growing very fast. Biodiesel and bioethanol are liquid biofuels and gaseous biofuels (biohydrogen and biomethane) have manufactured as a valuable alternative of fossil fuel. Biofuels are found resourceful option to decrease the addiction on fossil fuel. Fossil fuel production is the reason of greenhouse effect and to replace fossil fuel, biofuels are very useful. Worldwide agriculture, durable pressure in the first-generation biofuel has been observed. Second-generation biofuel (thermochemical and chemical biofuels) uses two different fundamental approaches and non-crop biomass is used. Cellulose for the reason of its low price, easy availability, degradability by cellulolytic bacteria, has stunning biomass for the manufacturing of second-generation biofuels.

A broad range of industrial residues and agricultural, domestic waste, and lignocellulosic forestry can be used as precursors of biofuels with the help of microbial enzymes. Biofuels production heterogeneous catalysts plant design, protein engineering of lipases, enzyme immobilization techniques, alcohol hydrolases or dehydrogenases to raise their activity, and the production and distillation of the biofuels. Biotechnology plays a fundamental role in biofuel manufacture. It helps in optimizing process characteristics, decreasing agronomic inputs, and in engineering the plants to generate high yield. With enhancement in the biotechnological field, improvements in production of biofuels can be expected.

KEYWORDS

- **biofuels**
- **biodiesel**
- **lipase**
- **biotechnological application**
- **bioethanol**

REFERENCES

Ahmad, A. L.; Mat Yasin, N. H.; Derek, C. J. C.; Lim, J. K. Microalgae as a Sustainable Energy Source for Biodiesel Production: a Review. *Renew. Sustain. Energy Rev*. **2011,** *15*, 1.

Alemayehu, G.; Tewodros, G.; Abile, T. A Review on Biodiesel Production as Alternative Fuel. *J. Forest Prod. Ind.* **2015,** *80*, 85.

Al-Widyan, Mohamad I.; Ali O. Al-Shyoukh. Experimental Evaluation of the Transesterification of Waste Palm Oil into Biodiesel. *Biores. Technol.* **2002,** *253*, 256.

Antolin, G.; et al. Optimisation of Biodiesel Production by Sunflower Oil Transesterification. *Biores. Technol.* **2002,** *111*, 114.

Balat, M.; Balat, H. Progress in Biodiesel Processing. *Appl. Energy* **2010,** *87*, 1815–1835.

Carpita, N. C. Structure and Biogenesis of the Cell Walls of Grasses. *Ann. Rev. Plant Biol.* **1996,** 445–476.

Chen, C. -Y.; Yeh, K. -L.; Aisyah, R.; Lee, D. -J.; Chang, J. -S. Cultivation, Photobioreactor Design and Harvesting of Microalgae for Biodiesel Production: a Critical Review. *Bioresour. Technol.* **2011,** *71*, 81.

Cho , S. S.; Park, D. J.; Simkhada, J. R.; Hong, J. H.; Sohng, J. K.; Lee, O. H.; Yoo, J. C. A Neutral Lipase Applicable in Biodiesel Production from a Newly Isolated *Streptomyces* sp. CS326. *Bioprocess Biosyst. Eng.* **2012,** 227–234.

Christopher, L. P.; Hemanathan, K.; Zambare, V. P. Enzymatic Biodiesel: Challenges and Opportunities. *Appl. Energy* **2014**, 497–520.

Costa, J. A. V.; de Morais, M. G. The Role of Biochemical Engineering in the Production of Biofuels from Microalgae. *Bioresour. Technol.* **2011,** *102*, 29.

Day, J. G.; Slocombe, S. P.; Stanley, M. S. Overcoming Biological Constraints to Enable the Exploitation of Microalgae for Biofuels. *Bioresour. Technol.* **2012,** *109*, 245–251.

Demirbas A. Biodiesel: a Realistic Fuel Alternative for Diesel Engine. Springer Publishing Co.: London ; 2008.

Demirbas A. Progress and Recent Trends in Biodiesel Fuels. *Energy Convers. Manag.* **2009,** *14*, 34.

Godbole, E. P.; Dabhadkar, K. C. Review of Production of Biofuels. *J. Biotechnol. Biochem*. **2016,** 2, 6.

Gasparatos, A.; Stromberg, P.; Takeuchi, K. Sustainability Impacts of First-generation Biofuels. *Animal Front*. **2013,** 12–26.

Gressel, J. Transgenics Are Imperative for Biofuel Crops. *Plant Sci.* **2008,** 246–263.

Ibrahim, E.; Jones, K. D.; Hossenya, E. N. Molecular Cloning and Expression of Cellulase and Polygalacturonase Genes in *E. coli* as a Promising Application for Biofuel Production. *J. Pet. Environ. Biotechnol.* **2013,** 33–44.

IEA, *International Energy Agency*, *World Energy Outlook 2013*, International Energy Agency. IEA/OECD, Paris, 2013.

John, R. P.; Anisha, G. S.; Nampoothiri, K. M.; Pandey, A. Micro and Macroalgal Biomass: a Renewable Source for Bioethanol. *Bioresour. Technol.* **2011,** *102*, 186.

Kafuku, G.; Mbarawa, M. Biodiesel Production from Croton Megalocarpus Oil and Its Process Optimization. *Fuel* **2011,** *89*, 60.

Karmakar, A.; Karmakar, S.; Mukherjee, S. Properties of Various Plants and Animals Feedstocks for Biodiesel Production. *Bioresour Technol*. **2010,** *19*, 10.

Kibazohi, O.; Sangwan, R. S. Vegetable Oil Production Potential from *Jatropha curcas*, *Croton megalocarpus, Aleurites moluccana, Moringa oleifera* and *Pachira glabra*: Assessment of Renewable Energy Resources for Bio-energy Production in Africa. *Biomass Bioenergy* **2011,** *35*, 6.

Koh, M. Y.; Mohd. Ghazi, T. I. A Review of Biodiesel Production from *Jatropha curcas* L. Oil. *Renew. Sust. Energ. Rev*. **2011,** *15*, 224.

Lewandowski, I.; Scurlock, J. M. O.; Lindvall, E.; Christou, M. The Development and Current Status of Perennial Rhizomatous Grasses as Energy Crops in the US and Europe. *Biomass Bioenerg*. **2003,** *25*, 335.

Li, K.; Liu, S.; Liu, X. An Overview of Algae Bioethanol Production. *Int. J. Energy Res*. **2014,** *38*, 977.

Lin, L.; Cunshan, Z.; Vittayapadung, S.; Xiangqian, S.; Mingdong, D. Opportunities and Challenges for Biodiesel Fuel. *Appl. Energy* **2011,** *88*, 31.

Masjuki Hj. Hassan; Md. Abul Kalam. An Overview of Biofuel as a Renewable Energy Source: Development and Challenges. *Procedia Eng.* **2013,** *39*, 53.

Mittelbach, M. Lipase Catalyzed Alcoholysis of Sunflower Oil. *J. Am. Oil Chem. Soc.* **1990,** 168–170.

Peterson, Charles L.; et al. Ethyl Ester of Rapeseed Used as a Biodiesel Fuel—a Case Study. Biomass Bioenergy **1996,** *331*, 336.

Pittman, J. K.; Dean, A. P.; Osundeko, O. The Potential of Sustainable Algal Biofuel Production Using Wastewater Resources. *Bioresour. Technol.* **2011,** *102*, 17.

Qiul, J.; Xiaohu, F.; Hongyu, Z. Development of Biodiesel from Inedible Feedstock Through Various Production Processes. Review. *Chem. Technol. Fuels Oils* **2011,** *102*, 111.

Ra, K.; Shiotsu, F.; Abe, J.; Morita, S. Biomass Yield and Nitrogen Use Efficiency of Cellulosic Energy Crops for Ethanol Production. *Biomass Bioenergy* **2012,** *37*, 330.

Ramos, M. J.; Fernández, C. M.; Casas, A.; Rodríguez, L.; Pérez, Á. Influence of Fatty Acid Composition of Raw Materials on Biodiesel Properties. *Bioresour. Technol.* **2009,** 261–268.

REN21. *Renewables 2013—Global Status Report*, REN21 Secretariat, Paris, 2013.

Ribeiro, B. D.; De Castro, A. M.; Coelho, M. A. Z.; Freire, D. M. G. Production and Use of Lipases in Bioenergy: a Review from the Feedstocks to Biodiesel Production. *Enzyme Res.* **2011,** 1–16.

Shahid, E. M.; Jamal, J. Production of Biodiesel: a Technical Review. *Renew. Sustain. Energy Rev.* **2011,** *15*, 9.

Singh, A.; Pant, D.; Korres, N. E.; Nizami, A. -S.; Prasad, S.; Murphy, J. D. Key Issues in Life Cycle Assessment of Ethanol Production from Lignocellulosic Biomass: Challenges and Perspectives. *Bioresour. Technol.* **2010,** *101*, 500.

Singh, S. P.; Singh, D. Biodiesel Production Through the Use of Different Sources and Characterization of Oils and Their Esters as the Substitute of Diesel: a Review. *Renew. Sustain. Energy Rev.* **2010,** *14*, 16.

Srivastava, A.; Prasad, R. Triglycerides-based Diesel Fuels. *Renew. Sustain. Energy Rev.* **2000,** *111*, 133.

Tenzin T.; Murthy, T. P. K. Advances and Biotechnological Applications in Biofuel Production: A Review. *Open J. Renew. Sustain. Energy* **2014,** 29–34.

Thirumarimurugan, M.; Sivakumar, V. M.; Merly Xavier, A.; Prabhakaran, D.; Kannadasan, T. Preparation of Biodiesel from Sunflower Oil by Transesterification. *Int. J. Biosci. Biochem. Bioinform.* **2012,** *441*, 444.

Villeneuve, P.; Muderhwa, J. M.; Graille, J.; Haas, M. J. Customizing Lipases for Biocatalysis: a Survey of Chemical, Physical and Molecular Biological Approaches. *J. Mol. Catal. B: Enzym.* **2000,** 113–148.

Yahya, A. R. M.; Anderson, W. A.; Moo-Young, M. Ester Synthesis in Lipase Catalyzed Reactions. *Enzyme Microb. Technol.* **1998,** 438–450.

Zhang, B.; Weng, Y.; Xu, H.; Mao, Z. Enzyme Immobilization for Biodiesel Production. *Appl. Microbiol. Biotechnol.* **2012,** 61–70.

Zhang, G.; Yang, Z.; Dong, S. Interspecific Competitiveness Affects the Total Biomass Yield in an Alfalfa and Corn Intercropping System. *Field Crops Res.* **2011,** *66*, 73.

Zhou, A.; Thomson, E. The Development of Biofuels in Asia. *Appl. Energy 86, Suppl.* **2009,** *11*, 20.

CHAPTER 2

Biohydrogen Production from Dark Fermentation of Lignocellulosic Biomass

ZUMAR M. A. BUNDHOO*

Department of Chemical and Environmental Engineering, Faculty of Engineering, University of Mauritius, Réduit, Mauritius
**E-mail: zumar.bundhoo@gmail.com*

ABSTRACT

Biohydrogen production from dark fermentation (DF) of lignocellulosic biomass has been gaining much attention over the past years. This chapter provides an insight of the DF process and the different parameters that influence biohydrogen production from lignocellulosic biomass such as pH and temperature, among others. The chapter also reviews some of the factors that have been commonly reported to inhibit the DF process while also outlining some of the strategies that have been studied for enhancing dark fermentative biohydrogen production such as inoculum and substrates pretreatment. This chapter also provides an overview of the potential use of the DF effluent as substrates for further bioenergy production through coupling of the DF process with other techniques such as anaerobic digestion, photofermentation, or bioelectrochemical systems. Finally, the chapter outlines some of the attempts made toward the upscaling of the DF process.

2.1 INTRODUCTION

Fossil fuels continue to dominate global energy supply despite its negative environmental concerns, with 81.4% of the total primary energy requirement of the world being met through fossil fuels in 2015 (IEA, 2017). Due to its highly variable prices and its depleting reserves, full dependence on fossil fuels may cause energy security issues which subsequently severely impacts on commercial and industrial activities and affects the economy of a country (Balat, 2011; Bundhoo, 2018a). As such, many countries are now shifting toward renewable energies such as hydro-power, wind energy, solar energy, ocean-based energy, geothermal energy, and bioenergy. Among these renewable sources of energy, bioenergy (from biomass and waste materials) is the main contributor, accounting for 9.7% of global primary energy supply in 2015 (IEA, 2017). Bioenergy may be defined as energy produced directly from biomass or wastes (for example, through incineration) or from biofuels (solid, liquid, or gases) derived from the biomass or waste materials (IEA, 2017). These biofuels, which may be produced from either thermochemical or biochemical technologies, include, among others, bioethanol, biodiesel, biogas (methane), bio-oil, biochar, and biohydrogen (Balat, 2011).

Long viewed as the "fuel of the future" owing to its clean nature (Das and Veziroğlu, 2001), biohydrogen production has been gaining much attraction over the past years. Biohydrogen can be produced through biological technologies, namely, biophotolysis, dark fermentation (DF), photofermentation, and fuel cells (Levin et al., 2004; Manish and Banerjee, 2008). Among these techniques, this chapter focuses only on DF and will be restricted to biohydrogen production from lignocellulosic materials (biomass and wastes). An insight of the principles of the DF process is provided together with the main operating parameters of the process. The chapter also provides an outline of the factors inhibiting the DF process and strategies employed for enhancing dark fermentative biohydrogen production. Finally, the use of the DF effluent for bioenergy production is reviewed prior to concluding on the commercialization aspect of the technology.

2.2 PRINCIPLES OF DARK FERMENTATION

The principles of DF have been comprehensively described in previous studies (Li and Fang, 2007a; Hallenbeck et al., 2012; Bundhoo and Mohee, 2016; Bundhoo, 2017) and will therefore be briefly overviewed in this chapter. Basically, DF refers to the anaerobic decomposition of organics (particularly carbohydrates) in the absence of light and oxygen to produce a biogas (H_2 and CO_2) and an effluent (rich in volatile fatty acids and alcohols) (Levin et al., 2004; Karthic and Shiny, 2012). Glucose, obtained following hydrolysis of lignocellulosic materials, is converted into pyruvate which is then decomposed to produce biohydrogen through two main metabolic pathways, namely, pyruvate:ferredoxin oxidoreductase (Pfor) or pyruvate:formate lyase (Pfl) (Hallenbeck and Benemann, 2002; Manish and Banerjee, 2008; Hallenbeck, 2009; Hallenbeck et al., 2012). Under either pathway, end products are produced and may include acetic acid, butyric acid, and ethanol, among others (Hallenbeck et al., 2012). Depending on the end products formed, the biohydrogen yield varies accordingly, with a maximum yield of 4 mol H_2/mol glucose when acetic acid is the end product and 2 mol H_2/mol glucose with butyric acid as the end product (Hallenbeck and Benemann, 2002; Levin et al., 2004). Dark fermentative biohydrogen has been produced from a variety of substrates including glucose and sucrose while other more complex substrates have also been investigated such as sludge or industrial wastewater. Nonetheless, this chapter will focus only on biohydrogen production from lignocellulosic materials such as agricultural residues and food wastes.

2.3 PROCESS PARAMETERS

There are several factors that impact on dark fermentative biohydrogen production including the type of inoculum, pH, temperature, hydraulic retention time (HRT), organic loading rate (OLR), the hydrogen partial pressure, and the nutritional requirements (Wang and Wan, 2009; Guo et al., 2010; Mohammadi et al., 2012; De Gioannis et al., 2013; Yasin et al., 2013; Wong et al., 2014; Arimi et al., 2015; Ghimire et al., 2015). Some of the most important parameters are discussed in the following subsections.

2.3.1 MICROFLORA

Inoculum used in the DF process may be either mixed microflora or pure H_2-producing cultures, with mixed microflora being more viable to use on larger scale (Valdez-Vazquez et al., 2005; Li and Fang, 2007a; Wang and Wan, 2011). Nonetheless, mixed cultures also possess certain drawbacks such as they consist of H_2 consumers (besides H_2 producers) which eventually decrease the net biohydrogen yield (Li and Fang, 2007a; Saady, 2013).

As such, pretreatment is often applied to mixed microflora to suppress H_2 consumers while enriching H_2 producers (Li and Fang, 2007a; Wong et al., 2014; Bundhoo et al., 2015). Common examples of mixed microflora studied for dark fermentative biohydrogen production include sludge and animal manure while pure H_2-producing cultures include *Clostridium* sp. and *Enterobacter* sp. (Wang and Wan, 2011).

2.3.2 pH

pH is one of the most crucial operating parameter of the DF process since it may impact on microbial activities as well as fermentation routes taken for biohydrogen production, thereby affecting the net biohydrogen yield (Wang and Wan, 2009; Mohammadi et al., 2012; De Gioannis et al., 2013). A highly acidic pH may inhibit microbial activities including those of H_2-producing bacteria (Hwang et al., 2004) while an alkaline operating pH may promote the activities of hydrogenotrophic methanogens and the formation of methane at the expense of biohydrogen (Fang and Liu, 2002). As such, regulating the pH within an optimal range is of utmost importance for maximum biohydrogen production. Several studies have reported a pH of 5.5 as the optimal value for maximum biohydrogen yield (Fang and Liu, 2002; Chen et al., 2005; Chong et al., 2009a). As opposed to the operating pH, the optimal initial pH of the fermentation medium is often higher and close to neutral (~7) (Nazlina et al., 2009; Wang and Wan, 2011). Besides its impact on biohydrogen production, the operating pH also influences the fermentation pathway taken by the microflora as observed by the concentration of end-product metabolites in the DF effluent. Hwang et al. (2004) found higher butyrate, ethanol, and propionate concentrations in pH ranges of 4–4.5, 4.5–5.0, and 5.0–6.0, respectively, while Fang and Liu (2002) reported higher butyrate concentrations in a pH range of 4.0–6.0. Although the value of 5.5 is often reported as the optimal operating pH for maximum biohydrogen production, this is not necessarily true since ultimately, the biohydrogen yield also depends on other operating factors.

2.3.3 TEMPERATURE

Besides pH, temperature is another fundamental operating parameter of the DF process since it can affect the biohydrogen yield due to its effect on the activities of H_2-producing bacteria (Wang and Wan, 2009; Yasin et al., 2013; Wong et al., 2014). There exist different temperature ranges that may be applied to the DF process and these can be classified into mesophilic (25–40°C), thermophilic (40–65°C), extreme thermophilic (65–80°C), and hyperthermophilic (>80°C) (Levin et al., 2004), although most studies have been conducted under mesophilic and thermophilic conditions. The reactions under these two common temperature zones are thus mediated by mesophiles (for mesophilic condition) and thermophiles (for thermophilic conditions) (Wong et al., 2014). Wang and Wan (2008) varied the operating temperature from 20°C to 55°C and reported that maximum biohydrogen production was obtained at 40°C. Shaterzadeh and Ataei (2017) investigated the effects of temperature ranging from 25°C to 40°C on DF of glucose and observed that maximum H_2 production was achieved at a temperature of 37°C. O-Thong et al. (2008) evaluated the potential of using the thermophile *Thermoanaerobacterium thermosaccharolyticum* for biohydrogen production within a temperature range of 45–70°C and obtained maximum H_2 production at a temperature of 60°C. As observed, the optimum temperatures in the different studies are variable since the biohydrogen yield is not only dependent on temperature but also on other operating parameters as well as substrates and inoculum used.

2.3.4 HYDRAULIC RETENTION TIME

HRT is the time that the influent takes to leave the reactor. HRT may influence biohydrogen production since it can impact on microbial activities and metabolic reactions including substrate hydrolysis (Mohammadi et al., 2012; Ghimire et al., 2015). High HRTs are normally required for substrates with low degradation rates as opposed to low HRTs for easily digestible substrates (Li and Fang, 2007a). Badiei et al. (2011) compared four HRTs (24, 48, 72, and 96 h) for the DF process and reported that using a retention time of 72 h, maximum biohydrogen was produced. Earlier, Karlsson et al. (2008) evaluated the effects of varying HRTs (2, 5, and 8 days) on dark fermentative biohydrogen production and found that the lowest retention time of 2 days generated the highest amount of biohydrogen. As opposed, Shin and Youn (2005) varied the HRT from 2 to 5 days and obtained maximum biohydrogen yield from the highest retention time of 5 days.

2.3.5 ORGANIC LOADING RATE

OLR can be defined as the organic matter concentration of the substrates fed into the reactor over a particular HRT (Mohammadi et al., 2012; Arimi et al., 2015). OLR influences microbial growth and biohydrogen production and is therefore a crucial parameter of the DF process (Mohammadi et al., 2012). Shen et al. (2009) varied the OLR from 4.0 to 30 g chemical oxygen demand (COD)/L/day and reported that an OLR of 22 g COD/L/day resulted in maximum biohydrogen production. Likewise, Hafez et al. (2010) compared different OLRs (6.5–206 g COD/L/day) and reported that an OLR of 103 g COD/L/day gave the maximum biohydrogen yield. Sreethawong et al. (2010) investigated the impacts of different OLRs on biohydrogen production from glucose-containing wastewater by varying the loading rates from 10 to 50 g COD/L/day and reported an OLR of 40 g COD/L/day as the optimum loading rate for maximum biohydrogen production. Mariakakis et al. (2011) varied the OLR from 10 to 30 g sucrose/L/day and obtained maximum biohydrogen at an OLR of 20 g sucrose/L/day. In all these studies, the biohydrogen yield increases with increasing OLR followed by a decrease at higher OLRs. This may be attributed to an overloading of the bioreactor, resulting in volatile fatty acids accumulation which negatively impacts on microbial activities and biohydrogen production (Sreethawong et al., 2010).

2.4 BIOHYDROGEN PRODUCTION FROM LIGNOCELLULOSIC MATERIALS

Dark fermentative biohydrogen has been produced from a wide variety of substrates including model substrates such as glucose and sucrose as well as more complex substrates such as lignocellulosic materials, industrial wastewater, and sludge. Lignocellulosic material, which is the focus of this chapter, can be further classified as organic fraction municipal solid wastes (OFMSW), agricultural residues, as well as algal biomass. Table 2.1 lists some of the lignocellulosic materials that have been investigated for their biohydrogen potential from the DF process.

TABLE 2.1 Biohydrogen Yield of Different Lignocellulosic Materials.

Substrates	Inoculum	Process conditions	H_2 yield	References
Sweet sorghum	Indigenous microflora	Temperature: 35°C HRT: 12 h	10.4 mL/g substrates	Antonopoulou et al. (2008)
Waste ground wheat	Heat-treated anaerobic sludge	Temperature: 37°C Initial pH: 7	223 mL/g substrates	Argun et al. (2009)
Rice straw	Heat-treated sludge	Temperature: 55°C Initial pH: 6.5	24.8 mL/g total solids	Chen et al. (2012)
Hyacinth	Digested sludge	Temperature: 55°C Initial pH: 5.5	122.3 mL/g substrates	Cheng et al. (2006)
Wheat stalk	Anaerobic digested dairy manure	Temperature: 35°C Operating pH: 6.5	37.0 mL/g VS	Chu et al. (2011)
Potato	Heat-treated anaerobic sludge	Temperature: 37°C Initial pH: 5.5	106 mL/g VS	Dong et al. (2009)
Lettuce	Heat-treated anaerobic sludge	Temperature: 37°C Initial pH: 5.5	50 mL/g VS	Dong et al. (2009)
OFMSW	Anaerobic sludge	Temperature: 34°C Initial pH: 6.0 HRT: 5 days	67.6 NmL/g $VS_{removed}$	Gomez et al. (2006)
Wheat straw hydrolysate	Thermophilic mixed culture	Temperature: thermophilic 5% (volume/volume hydrolysate)	318.4 mL/g sugars (batch)	Kongjan et al. (2010)
Household solid waste	Anaerobic sludge	Temperature: 37°C	43 mL/g VS	Liu et al. (2006)
Dried distillers grain	*Clostridium thermocellum* ATCC 27405	Working volume: 26 mL Substrate concentration: 5 g/L	1.27 mmol/ glucose	Magnusson et al. (2008)
Barley hulls	*Clostridium thermocellum* ATCC 27405	Working volume: 26 mL Substrate concentration: 5 g/L	1.24 mmol/ glucose	Magnusson et al. (2008)
Beet pulp	Mixed anaerobic culture	Temperature: 35°C Initial pH: 6	90.1 mL/g COD	Ozkan et al. (2011)
Food waste	Anaerobic sludge	Temperature: 50°C Food: microorganism ratio: 7	57 mL/g VS	Pan et al. (2008)

TABLE 2.1 *(Continued)*

Substrates	Inoculum	Process conditions	H_2 yield	References
Microalgal biomass	Anaerobic digester sludge	Temperature: 35°C Initial pH: 7.4	31.1 mL/g dry cell weight	Yun et al. (2013)
Cassava powder	Cattle dung compost	Temperature: 37°C Initial pH: 6.8	199 mL/g substrates	Zong et al. (2009)
Cornstalk wastes	Cow dung compost	Temperature: 36°C	3.16 mL/g VS	Zhang et al. (2007)

COD, chemical oxygen demand; HRT, hydraulic retention time; OFMSW, organic fraction municipal solid wastes; VS, volatile solids.

2.5 INHIBITION OF THE DF PROCESS

Several factors have been reported to inhibit the DF process and subsequently biohydrogen production and some of these include the presence of H_2 consumers in mixed microflora, metal ions concentrations above a minimum threshold value, inhibitors from substrates pretreatment, and concentration of soluble metabolites in the DF system (Bundhoo and Mohee, 2016; Elbeshbishy et al., 2017).

2.5.1 PRESENCE OF H_2 CONSUMERS IN MIXED MICROFLORA

As previously mentioned, mixed microflora contain H_2-consuming bacteria besides H_2 producers and these ultimately reduce the net biohydrogen production. H_2 consumers may exist as hydrogenotrophic methanogens, homoacetogens, sulfate-reducing bacteria, nitrate-reducing bacteria, and propionate producers, among others (Saady, 2013; Bundhoo and Mohee, 2016). These H_2 consumers use the biohydrogen produced during the DF process as substrates for conversion into methane (hydrogenotrophic methanogenesis), acetate (homoacetogenesis), hydrogen sulfide (sulfate-reducing bacteria), ammonia (nitrate-reducing bacteria), and propionate (propionate producers), among others and this ultimately reduces the net biohydrogen yield. To counter this problem, many studies have pretreated the inoculum or mixed microflora prior to the DF process. The aim of pretreatment is to suppress the H_2 consumers while enriching the H_2-producing bacteria (Li and Fang, 2007a; Wong et al., 2014; Bundhoo et al., 2015; Wang and Yin, 2017). The principle behind inoculum pretreatment is that some H_2-producing bacteria are spore formers and under the severe conditions of pretreatment, they form spores and are thus resistant as opposed to H_2 consumers that are vulnerable and thus, suppressed (Foster and Johnstone, 1990). Under suitable conditions, the endospores then germinate and are able to produce H_2 (Foster and Johnstone, 1990; Yasin et al., 2013; Valdez-Vazquez and Poggi-Varaldo, 2009; Bundhoo and Mohee, 2016; Wang and Yin, 2017). Consequently, with inoculum pretreatment, the consumption of biohydrogen by H_2 consumers is reduced and this results in a higher H_2 yield.

2.5.2 METAL IONS CONCENTRATIONS

Both light and heavy metal ions are required in the DF process for effective biohydrogen production as part of the nutritional requirements of the H_2-producing bacteria. Metal ions enhance metabolic activities, cell growth, activation of coenzymes, and eventually biohydrogen production (Li and Fang, 2007b; Chong et al., 2009b; Sinha and Pandey, 2011; Bundhoo and Mohee, 2016; Elbeshbishy et al., 2017). However, while a minimum amount of these metal ions is required, excess concentrations may result in process inhibition and severely suppress biohydrogen production (Li and Fang, 2007b; Wang and Wan, 2009). The optimum and threshold concentrations of metal ions for effective dark fermentative biohydrogen production have been previously reviewed in past studies (Wang and Wan, 2009; Sinha and Pandey, 2011; Bundhoo and Mohee, 2016; Elbeshbishy et al., 2017). Some of the strategies to counter process inhibition due to high metal ion concentrations are to dilute the reactor contents or acclimatize the inoculum to the metals prior to the DF process (Kim et al., 2009; Bundhoo and Mohee, 2016).

2.5.3 CONCENTRATION OF SOLUBLE METABOLITES

Soluble or end-product metabolites are dependent on the metabolic pathway taken for biohydrogen production and include acetate, butyrate, propionate, or ethanol, among others (Wong et al., 2014; Bundhoo and Mohee, 2016). Depending on the end products formed, two pathways are generally identified, namely, acidogenesis and solventogenesis (Valdez-Vazquez and Poggi-Varaldo, 2009; Wong et al., 2014). High concentrations of these soluble metabolites may negatively impact on microbial activities and significantly suppress dark fermentative biohydrogen production. High concentrations of organic acids may cause cell lysis, disturb intracellular pH, inhibit cell growth, and eventually, impact on microbial activities and suppress biohydrogen production (Bundhoo and Mohee, 2016). Likewise, solvents formed as a result of solventogenesis, which may itself be initiated due to high organic acids concentration (Jones and Woods, 1986; Millat et al., 2013), also hinders microbial activities, suppresses cell growth, and causes cell death and these ultimately impact on biohydrogen production (Tang et al., 2012; Ciranna et al., 2014). Optimum as well as threshold concentrations of soluble metabolites have been previously reviewed by Bundhoo and Mohee (2016), with the authors also reporting that one of the main strategies to counter inhibition due to soluble metabolites concentrations is to regulate the pH in the DF process.

2.6 PRETREATMENT OF LIGNOCELLULOSIC MATERIALS FOR ENHANCED BIOHYDROGEN PRODUCTION

Pretreatment is one of the most commonly used strategies for enhancing dark fermentative biohydrogen production and can be as either inoculum or substrates pretreatment (Bundhoo et al., 2015; Rafieenia et al., 2018). The aim of inoculum pretreatment has been explained previously and will therefore not be considered in this section. Low biohydrogen yield from lignocellulosic materials may be attributed to the unavailability of cellulose molecules for enzymatic hydrolysis due to the lignin protection surrounding these cellulose molecules (Hendriks and Zeeman, 2009; Zheng et al., 2009). As such, the main aim of pretreating lignocellulosic materials is to break this lignin barrier and increase the amenability of cellulose molecules for biological attack (Sun and Cheng, 2002; Mosier et al., 2005). In addition, pretreatment techniques also assist in the disruption of the crystalline structure of cellulose molecules, solubilization of organics, and increased reducing sugar formation which ultimately results in enhanced biohydrogen production (Sun and Cheng, 2002; Bundhoo et al., 2015; Bundhoo, 2018b; Bundhoo and Mohee, 2018). Several technologies have been studied for improving dark fermentative biohydrogen production and these have been classified as physical, physicochemical, chemical, and biological techniques (Bundhoo et al., 2015). Some of these techniques and their impacts on biohydrogen production are summarized in Table 2.2.

TABLE 2.2 Effects of Pretreatment Technologies on Dark Fermentative Biohydrogen Production from Lignocellulosic Materials.

Pretreatment technology	Substrates	Pretreatment conditions	H_2 yield	References
Thermal	*Laminaria japonica* (marine algae)	Temperature: 150–180°C Time: 5–40 min	H_C: 67.6 mL/g COD H_P (170°C, 20 min): 110.2 mL/g COD	Jung et al. (2011)
Ultrasound irradiation	Food waste	Power: 500 W Frequency: 20 kHz Specific energy: 79 kJ/g solids	H_C: 42 mL/g VS H_P: 97 mL/g VS	Elbeshbishy et al. (2011)
Microwave irradiation + alkali	Cornstalk	Power: 2 kW Time: 5–90 min Alkali: NaOH (0.02–0.16 g/g cornstalk)	H_C: 68.2 mL/g cornstalk H_P (90 min, 0.16 g NaOH/g cornstalk): 107.6 mL/g cornstalk	Li et al. (2014)
Ultrasound irradiation + acid	Food waste	Power: 500 W Frequency: 20 kHz Specific energy: 79 kJ/g solids Acid: 1 N HCl	H_C: 42 mL/g VS H_P: 118 mL/g VS	Elbeshbishy et al. (2011)
Ionic liquid	Cellulose	Ionic liquid: [C4mim]Cl Ratio: 5–15% (w/w) Solvent: water, ethanol, methanol	H_C: 1.59 mol/mol glucose H_P: 2.20 mol/mol glucose (ratio: 10%, solvent: ethanol, with N_2 sparging)	Nguyen et al. (2008)
Acid	Organic wastes	Acid: 1 N HCl pH: 2.0 (24 h)	H_C: 6.99 mL/g VS H_P: 86.00 mL/g VS	Ruggeri and Tommasi (2012)
Alkali	Organic wastes	Alkali: 1 N NaOH pH: 12.5 (24 h)	H_C: 6.99 mL/g VS H_P: 95.93 mL/g VS	Ruggeri and Tommasi (2012)
Ozonolysis	Barley straw	Time: 0–90 min	H_C: 70.1 mL H_P: 186.8 mL	Wu et al. (2013)

COD, chemical oxygen demand; H_C, H_2 yield from untreated substrates (control); H_P, H_2 yield from pretreated substrates; VS, volatile solids.

While substrates pretreatment may enhance dark fermentative biohydrogen production as observed from Table 2.2, pretreatments may also result in the production of inhibitors that decrease biohydrogen production (Bundhoo and Mohee, 2016). These inhibitors may include furan derivatives such as furfural and 5-hydroxymethyl furfural (5-HMF), phenolic compounds, and weak acids (Palmqvist and Hahn-Hägerdal, 2000). Presence of furfural, 5-HMF, and phenolic compounds such as vanillin and syringaldehyde has been reported to inhibit biohydrogen production from acid-treated corn stover (Cao et al., 2010). Nonetheless, several detoxification strategies have been investigated in order to reduce the impact of pretreatment inhibitors on biohydrogen production including the use of activated carbon (Cheng et al., 2015; Orozco et al., 2012) and sodium borohydride (Lin et al., 2015). The acclimation of inoculum to the inhibitors prior to the DF process has also been reported as a potential strategy to reduce the suppression of biohydrogen production (Behera et al., 2014; Cheng et al., 2015).

2.7 BIOENERGY RECOVERY FROM DF EFFLUENTS

As aforementioned, the DF process also generates an effluent besides biohydrogen production. The DF effluent is rich in volatile fatty acids which may be used as substrates for other biochemical or bioelectrochemical systems for further bioenergy generation. These biochemical systems (anaerobic digestion or photofermentation) or bioelectrochemical systems (microbial fuel cell or microbial electrolysis cell) are often coupled with the DF process such that the DF effluent is then further processed into bioenergy (Gómez et al., 2011; Guwy et al., 2011; Azwar et al., 2014; Bundhoo, 2017). When coupled with the anaerobic digestion process, acetic acid from the DF effluent is converted into methane via methanogenesis while other volatile fatty acids and alcohols present in the DF effluent are converted by acetogens into H_2, CO_2, and acetic acid which are subsequently converted into methane by methanogens. With the coupled system of DF/photofermentation, photosynthetic bacteria use the volatile fatty acids present in DF effluents as electron donors in the presence of light to produce biohydrogen (Azwar et al., 2014; Bundhoo, 2017). In a microbial fuel cell, the organic acids from the DF effluent are oxidized to CO_2, electrons, and protons at the anode and the movement of the electrons through an outer circuit to the cathode generates bioelectricity (Du et al., 2007). In the microbial electrolysis cell, protons generated as a result of oxidation of organic acids (present in DF effluents) are reduced by the application of an additional electric current at the cathode, resulting in the production of biohydrogen (Kundu et al., 2013; Kadier et al., 2016). Several studies have coupled the DF process with other biochemical or bioelectrochemical systems as summarized in Table 2.3.

Besides two-stage coupling, bioenergy production from the DF process has also been enhanced in a three-stage process. Rózsenberski et al. (2017) coupled the DF process with anaerobic digestion and microbial fuel cell for enhanced energy production from municipal solid wastewater. Similarly, Wang et al. (2011) obtained enhanced bioenergy production from cellulose using a three-stage DF/microbial fuel cell/microbial electrolysis cell process.

2.8 COMMERCIALIZATION OF THE DF PROCESS

The main obstacle to the large-scale implementation of the DF process is the relatively low biohydrogen yield which decreases its economic feasibility (Das et al., 2008; Das, 2009; Hallenbeck and Ghosh, 2009). Other factors reported to hinder the large-scale commissioning of the DF technology include the high processing costs of lignocellulosic biomass and vulnerability of microorganism and enzymes to oxygen (Das et al., 2008). Furthermore, Ren et al. (2011) reported that the design of bioreactor represents another major bottleneck for commercialization of the DF process. Nonetheless, some strategies are already being studied in an attempt to increase the biohydrogen yield as previously reported. Other strategies investigated to increase dark fermentative biohydrogen production include using metabolically engineered organisms or engineering hydrogenases (Das et al., 2008; Hallenbeck et al., 2012). With respect to large-scale implementation of the DF process, some pilot-scale plants have been tested with industrial wastewaters, as reviewed by Das (2009), whereas pilot-scale implementation of the DF process with the anaerobic digestion process is becoming more common, as reviewed by Bundhoo (2017). Yang et al. (2006), for instance, studied biohydrogen production from citric acid wastewater in a 50 m^3 upflow anaerobic sludge blanket. Another project "HyTIME," consisting of coupling the DF with the

TABLE 2.3 Coupling of Dark Fermentation with Anaerobic Digestion, Photofermentation, Microbial Fuel Cell or Microbial Electrolysis Cell.

Coupled technology	Substrates	Inoculum	Findings	References
Anaerobic digestion	Cassava residues	Anaerobic digester sludge (boiled for 30 min for dark fermentation process)	Y_1: 92.3 mL H_2/g VS Y_2: 79.4 mL CH_4/g VS	Cheng et al. (2017)
Anaerobic digestion	Wheat straw hydrolysate	Anaerobic medium (dark fermentation) Methanogenic granules and digested manure (anaerobic digestion)	Y_1: 89 mL H_2/g VS Y_2: 307 mL CH_4/g VS	Kongjan et al. (2011)
Photofermentation	Rice straw (microwave irradiation/NaOH pretreatment)	Anaerobic activated sludge (boiled for 30 min for dark fermentation) *Rhodopseudomonas palustris* (photofermentation)	Y_1: 135 mL H_2/g VS Y_2: 328 mL H_2/g VS	Cheng et al. (2011)
Photofermentation	Water hyacinth (microwave irradiation/acid pretreated)	Anaerobic activated sludge (boiled for 30 min for dark fermentation) *Rhodopseudomonas palustris* (photofermentation)	Y_1: 112.3 mL H_2/g VS Y_2: 639.3 mL H_2/g VS	Cheng et al. (2013)
Microbial fuel cell	Glycerol	Microbial inoculum from sediment samples (heated at 90–95°C for 30 min for dark fermentation process) Anaerobic sludge (heated at 90–95°C for 30 min for microbial fuel cell)	Y_1: 0.55 mol/mol glycerol Y_2: 92 mW/m^2 cathode	Chookaew et al. (2014)
Microbial electrolysis cell	Waste peach pulp	–	Y_1: 461.7 mL H_2/g COD Y_2: 30.1 mL/mL reactor	Argun and Dao (2016)

COD, chemical oxygen demand; VS, volatile solids; Y_1, hydrogen yield from dark fermentation process; Y_2, additional bioenergy yield (biomethane from anaerobic digestion, biohydrogen from photofermentation or microbial electrolysis cell, or bioelectricity from microbial fuel cell).

anaerobic digestion process for biohydrogen and biomethane production from biomass on a pilot scale, was completed in 2015 (Claassen, 2015).

2.9 CONCLUDING REMARKS

Lignocellulosic biomass represents a huge potential for bioenergy production, with biohydrogen being constantly regarded as an advantageous fuel for the future. This chapter provided an overview of biohydrogen production from DF of lignocellulosic biomass including process parameters, inhibitors, as well as strategies that impact on biohydrogen yields. Among the process parameters reviewed, the pH and inoculum are often the most important ones as they have a significant impact on biohydrogen yield. With respect to the inhibitors of the DF process, the presence of H_2 consumers in mixed microflora is often detrimental to the net biohydrogen production. While substrate pretreatment enhances dark fermentative biohydrogen production, the inhibitors generated from some of these pretreatments also reduce the biohydrogen yield and thus, detoxification mechanisms need to be applied prior to the DF process. Although the DF process has been widely studied at lab scale, its implementation on large scale has been stagnant. Nonetheless, few attempts at moving from lab to pilot scales are now being successful while coupling of the DF process with the anaerobic digestion process on pilot scale is also becoming more common. However, there is still much work to be done on bioreactor design, engineering of microorganisms, or reduced cost of the whole DF process prior to this reaching commercialization status.

ACKNOWLEDGMENTS

The anonymous reviewers are acknowledged for their constructive criticism and helpful comments, recommendations, or suggestions.

KEYWORDS

- **dark fermentation**
- **biohydrogen**
- **lignocellulosic biomass**
- **biofuel**
- **pretreatment**
- **inhibition**

REFERENCES

Antonopoulou, G.; Gavala, H. N.; Skiadas, I. V.; Angelopoulos, K.; Lyberatos, G. Biofuels Generation from Sweet Sorghum: Fermentative Hydrogen Production and Anaerobic Digestion of the Remaining Biomass. *Bioresour. Technol.* **2008,** *99*, 110–119.

Argun, H.; Dao, S. Hydrogen Gas Production from Waste Peach Pulp by Dark Fermentation and Electrohydrolysis. *Int. J. Hydrog. Energy* **2016,** *41*, 11568–11576.

Argun, H.; Kargi, F.; Kapdan, I. K. Microbial Culture Selection for Bio-hydrogen Production from Waste Ground Wheat by Dark Fermentation. *Int. J. Hydrog. Energy* **2009,** *34*, 2195–2200.

Arimi, M. M.; Knodel, J.; Kiprop, A.; Namango, S. S.; Zhang, Y.; Geißen, S. U. Strategies for Improvement of Biohydrogen Production from Organic-rich Wastewater: A Review. *Biomass Bioenergy* **2015,** *75*, 101–118.

Azwar, M. Y.; Hussain, M. A.; Abdul-Wahab, A. K. Development of Biohydrogen Production by Photobiological, Fermentation and Electrochemical Systems: A Review. *Renew. Sust. Energy Rev.* **2014,** *31*, 158–173.

Badiei, M.; Jahim, J. M.; Anuar, N.; Abdullah, S. R. S. Effect of Hydraulic Retention Time on Biohydrogen Production from Palm Oil Mill Effluent in Anaerobic Sequencing Batch Reactor. *Int. J. Hydrog. Energy* **2011,** *36*, 5912–5919.

Balat, M. Production of Bioethanol from Lignocellulosic Materials via the Biochemical Pathway: A Review. *Energy Convers. Manag.* **2011,** *52*, 858–875.

Behera, S.; Arora, R.; Nandhagopal, N.; Kumar, S. Importance of Chemical Pretreatment for Bioconversion of Lignocellulosic Biomass. *Renew. Sust. Energy Rev.* **2014,** *36*, 91–106.

Bundhoo, M. A. Z. Effects of Pre-treatment Technologies on Dark Fermentative Bio-hydrogen Production: A Review. *J. Environ. Manag.* **2015,** *157*, 20–48.

Bundhoo, M. A. Z. Coupling Dark Fermentation with Biochemical or Bioelectrochemical Systems for Enhanced Bio-energy Production: A Review. *Int. J. Hydrog. Energy* **2017,** *42*, 26667–26686.

Bundhoo, M. A. Z. Renewable Energy Exploitation in the Small Island Developing State of Mauritius: Current Practice and Future Potential. *Renew. Sust. Energy Rev.* **2018a,** *82*, 2029–2038.

Bundhoo, M. A. Z. Microwave-assisted Conversion of Biomass and Waste Materials to Biofuels. *Renew. Sust. Energy Rev.* **2018b,** *82*, 1149–1177.

Bundhoo, M. A. Z.; Mohee, R. Inhibition of Dark Fermentative Bio-hydrogen Production: A Review. *Int. J. Hydrog. Energy* **2016,** *41*, 6713–6733.

Bundhoo, M. A. Z.; Mohee, R. Ultrasound-assisted Biological Conversion of Biomass and Waste Materials to Biofuels: A Review. *Ultrason. Sonochem.* **2018,** *40*, 298–313.

Cao, G. L.; Ren, N. Q.; Wang, A. J.; Guo, W. Q.; Xu, J. F.; Liu, B. F. Effect of Lignocelluloses Derived Inhibitors on Growth and Hydrogen Production by *Thermoanaerobacterium thermosaccharolyticum* W16. *Int. J. Hydrog. Energy* **2010,** *35*, 13475–13480.

Chen, W. M.; Tseng, Z. J.; Lee, K. S.; Chang, J. S. Fermentative Hydrogen Production with *Clostridium butyricum* CGS5 Isolated from Anaerobic Sewage Sludge. *Int. J. Hydrog. Energy* **2005,** *30*, 1063–1070.

Chen, C. C.; Chuang, Y. S.; Lin, C. Y.; Lay, C. H.; Sen, B. Thermophilic Dark Fermentation of Untreated Rice Straw Using Mixed Cultures for Hydrogen Production. *Int. J. Hydrog. Energy* **2012,** *37*, 15540–15546.

Cheng, J.; Zhou, J.; Qi, F. ; Xie, B.; Cen, K. In *Bio-hydrogen Production from Hyacinth by Anaerobic Fermentation*, WHEC 16, Lyon, France, June 13–16, 2006.

Cheng, J.; Su, H.; Zhou, J.; Song, W.; Cen, K. Microwave-assisted Alkali Pretreatment of Rice Straw to Promote Enzymatic Hydrolysis and Hydrogen Production in Dark- and Photofermentation. *Int. J. Hydrog. Energy* **2011,** *36*, 2093–2101.

Cheng, J.; Xia, A.; Su, H.; Song, W.; Zhou, J.; Cen, K. Promotion of H_2 Production by Microwave-assisted Treatment of Water Hyacinth with Dilute H_2SO_4 Through Combined Dark Fermentation and Photofermentation. *Energy Convers. Manag.* **2013,** *73*, 329–334.

Cheng, J.; Lin, R.; Song, W.; Xia, A.; Zhou, J.; Cen, K. Enhancement of Fermentative Hydrogen Production from Hydrolyzed Water Hyacinth with Activated Carbon Detoxification and Bacteria Domestication. *Int. J. Hydrog. Energy* **2015,** *40*, 2545–2551.

Cheng, J.; Zhang, J.; Lin, R.; Liu, J.; Zhang, L.; Cen, K. Ionic-liquid Pretreatment of Cassava Residues for the Cogeneration of Fermentative Hydrogen and Methane. *Bioresour. Technol.* **2017,** *228*, 348–354.

Chong, M. L.; Abdul Rahim, R.; Shirai, Y.; Hassan, M. A. Biohydrogen Production by *Clostridium butyricum* EB6 from Palm Oil Mill Effluent. *Int. J. Hydrog. Energy* **2009a,** *34*, 764–771.

Chong, M. L.; Sabaratnam, V.; Shirai, Y.; Hassan, M. A. Biohydrogen Production from Biomass and Industrial Wastes by Dark Fermentation. *Int. J. Hydrog. Energy* **2009b,** *34*, 3277–3287.

Chookaew, T.; Prasertsan, P.; Ren, Z. J. Two-stage Conversion of Crude Glycerol to Energy Using Dark Fermentation Linked with Microbial Fuel Cell or Microbial Electrolysis Cell. *New Biotechnol.* **2014,** *31*, 179–184.

Chu, Y.; Wei, Y.; Yuan, X.; Shi, X. Bioconversion of Wheat Stalk to Hydrogen by Dark Fermentation: Effect of Different Mixed Microflora on Hydrogen Yield and Cellulose Solubilisation. *Bioresour. Technol.* **2011,** *102*, 3805–3809.

Ciranna, A.; Ferrari, R.; Santala, V.; Karp, M. Inhibitory Effects of Substrate and Soluble End Products on Biohydrogen Production of the Alkalithermophile *Caloramator celer*: Kinetic, Metabolic and Transcription Analyses. *Int. J. Hydrog. Energy* **2014,** *39*, 6391–6401.

Claassen, P. *Low Temperature Hydrogen Production from 2nd Generation Biomass: HyTIME*; Fuel Cells and Hydrogen Joint Undertaking, 2015. http://www.fch.europa.eu/project/low-temperature-hydrogen-production-2nd-generation-biomass (accessed Apr 1, 2018).
Das, D. Advances in Biohydrogen Production Processes: An Approach Towards Commercialization. *Int. J. Hydrog. Energy* **2009,** *34*, 7349–7357.
Das, D.; Veziroğlu, T. N. Hydrogen Production by Biological Processes: A Survey of Literature. *Int. J. Hydrog. Energy* **2001,** *26*, 13–28.
Das, D.; Khanna, N.; Veziroğlu, N. Recent Developments in Biological Hydrogen Production Processes. *Chem. Ind. Chem. Eng. Q.* **2008,** *14*, 57–67.
De Gioannis, G.; Muntoni, A.; Polettini, A.; Porni, R. A Review on Dark Fermentative Hydrogen Production from Biodegradable Municipal Waste Fractions. *Waste Manag.* **2013,** *33*, 1345–1361.
Dong, L.; Zhenhong, Y.; Yongming, S.; Xiaoying, K.; Yu, Z. Hydrogen Production Characteristics of the Organic Fraction Of Municipal Solid Wastes by Anaerobic Mixed Culture Fermentation. *Int. J. Hydrog. Energy* **2009,** *34*, 812–820.
Du, Z.; Li, H.; Gu, T. A State of the Art Review on Microbial Fuel Cells: A Promising Technology for Wastewater Treatment and Bioenergy. *Biotechnol. Adv.* **2007,** *25*, 464–482.
Elbeshbishy, E.; Hafez, H.; Dhar, B. R.; Nakhla, G. Single and Combined Effects of Various Pretreatment Methods for Biohydrogen Production from Food Waste. *Int. J. Hydrog. Energy* **2011,** *36*, 11379–11387.
Elbeshbishy, E.; Dhar, B. R.; Nakhla, G.; Lee, H. S. A Critical Review on Inhibition of Dark Biohydrogen Fermentation. *Renew. Sust. Energy Rev.* **2017,** *79*, 656–668.
Fang, H. H. P.; Liu, H. Effect of pH on Hydrogen Production from Glucose by a Mixed Culture. *Bioresour. Technol.* **2002,** *82*, 87–93.
Foster, S. J.; Johnstone, K. Pulling the Trigger: The Mechanism of Bacterial Spore Germination. *Mol. Microbiol.* **1990,** *4*, 137–141.
Ghimire, A.; Frunzo, L.; Pirozzi, F.; Trably, E.; Escudie, R.; Lens, P. N. L.; Esposito, G. A Review on Dark Fermentative Biohydrogen Production from Organic Biomass: Process Parameters and Use of By-products. *Appl. Energy* **2015,** *144*, 73–95.
Gómez, X.; Morán, A.; Cuetos, M. J.; Sánchez, M. E. The Production of Hydrogen by Dark Fermentation of Municipal Solid Wastes and Slaughterhouse Waste: A Two-phase Process. *J. Power Sources* **2006,** *157*, 727–732.
Gómez, X.; Fernández, C.; Fierro, J.; Sánchez, M. E.; Escapa, A.; Morán, A. Hydrogen Production: Two Stage Processes for Waste Degradation. *Bioresour. Technol.* **2011,** *102*, 8621–8627.
Guo, X. M.; Trably, E.; Latrille, E.; Carrère, H.; Steyer, J. P. Hydrogen Production from Agricultural Waste by Dark Fermentation: A Review. *Int. J. Hydrog. Energy* **2010,** *35*, 10660–10673.
Guwy, A. J.; Dinsdale, R. M.; Kim, J. R.; Massanet-Nicolau, J.; Premier, G. Fermentative Biohydrogen Production Systems Integration. *Bioresour. Technol.* **2011,** *102*, 8534–8542.
Hafez, H.; Nakhla, G.; Naggar, M. H. E.; Elbeshbishy, E.; Baghchehsaraee, B. Effect of Organic Loading Rate on a Novel Hydrogen Reactor. *Int. J. Hydrog. Energy* **2010,** *35*, 81–92.

Hallenbeck, P. C. Fermentative Hydrogen Production: Principles, Progress, and Prognosis. *Int. J. Hydrog. Energy* **2009,** *34*, 7379–7389.

Hallenbeck, P. C.; Abo-Hashesh, M.; Ghosh, D. Strategies for Improving Biological Hydrogen Production. *Bioresour. Technol.* **2012,** *110*, 1–9.

Hallenbeck, P. C.; Benemann, J. R. Biological Hydrogen Production; Fundamentals and Limiting Processes. *Int. J. Hydrog. Energy* **2002,** *27*, 1185–1193.

Hallenbeck, P. C.; Ghosh, D. Advances in Fermentative Biohydrogen Production: The Way Forward? *Trends Biotechnol.* **2009,** *27*, 287–297.

Hendriks, A. T. W. M.; Zeeman, G. Pretreatments to Enhance the Digestibility of Lignocellulosic Biomass. *Bioresour. Technol.* **2009,** *100*, 10–18.

Hwang, M. H.; Jang, N. J.; Hyun, S. H.; Kim, I. S. Anaerobic Bio-hydrogen Production from Ethanol Fermentation: The Role of pH. *J. Biotechnol.* **2004,** *111*, 297–309.

IEA. *Key World Energy Statistics*; Chirat: France, 2017.

Jones, D. T.; Woods, D. R. Acetone–Butanol Fermentation Revisited. *Microbiol. Rev.* **1986,** *50*, 484–524.

Jung, K. W.; Kim, D. H.; Shin, H. S. Fermentative Hydrogen Production from *Laminaria japonica* and Optimization of Thermal Pretreatment Conditions. *Bioresour. Technol.* **2011,** *102*, 2745–2750.

Kadier, A.; Simayi, Y.; Abdeshahian, P.; Azman, N. F.; Chandrasekhar, K.; Kalil, M. S. A Comprehensive Review of Microbial Electrolysis Cells (MEC) Reactor Designs and Configurations for Sustainable Hydrogen Gas Production. *Alexandria Eng. J.* **2016,** *55*, 427–443.

Karlsson, A.; Vallin, L.; Ejlertsson, J. Effects of Temperature, Hydraulic Retention Time and Hydrogen Extraction Rate on Hydrogen Production from the Fermentation of Food Industry Residues and Manure. *Int. J. Hydrog. Energy* **2008, 33,** 953–962.

Karthic, P.; Shiny, J. Comparison and Limitations of Biohydrogen Production Processes. *Res. J. Biotechnol.* **2012,** *7*, 59–71.

Kim, D. H.; Kim, S. H.; Shin, H. S. Sodium Inhibition of Fermentative Hydrogen Production. *Int. J. Hydrog. Energy* **2009,** *34*, 3295–3304.

Kongjan, P.; O-Thong, S.; Angelidaki, I. Performance and Microbial Community Analysis of Two-stage Process with Extreme Thermophilic Hydrogen and Thermophilic Methane Production from Hydrolysate in UASB Reactors. *Bioresour. Technol.* **2011,** *102*, 4028–4035.

Kongjan, P.; O-Thong, S.; Kotay, M.; Min, B.; Angelidaki, I. Biohydrogen Production from Wheat Straw Hydrolysate by Dark Fermentation Using Extreme Thermophilic Mixed Culture. *Biotechnol. Bioeng.* **2010,** *105*, 899–908.

Kundu, A.; Sahu, J. N.; Redzwan, G.; Hashim, M. A. An Overview of Cathode Material and Catalysts Suitable for Generating Hydrogen in Microbial Electrolysis Cell. *Int. J. Hydrog. Energy* **2013,** *38*, 1745–1757.

Levin, D. B.; Pitt, L.; Love, M. Biohydrogen Production: Prospects and Limitations to Practical Application. *Int. J. Hydrog. Energy* **2004,** *29*, 173–85.

Li, C.; Fang, H. H. P. Fermentative Hydrogen Production from Wastewater and Solid Wastes by Mixed Cultures. *Crit. Rev. Environ. Sci. Technol.* **2007a,** *37*, 1–39.

Li, C.; Fang, H. H. P. Inhibition of Heavy Metals on Fermentative Hydrogen Production by Granular Sludge. *Chemosphere* **2007b,** *67*, 668–673.

Li, Q.; Guo, C.; Liu, C. Z. Dynamic Microwave-assisted Alkali Pretreatment of Cornstalk to Enhance Hydrogen Production via Co-culture Fermentation of *Clostridium thermocellum* and *Clostridium thermosaccharolyticum*. *Biomass Bioenergy* **2014,** *64*, 220–229.

Lin, R.; Cheng, J.; Ding, L.; Song, W.; Zhou, J.; Cen, K. Sodium Borohydride Removes Aldehyde Inhibitors for Enhancing Biohydrogen Fermentation. *Bioresour. Technol.* **2015,** *197*, 323–328.

Liu, D.; Liu, D.; Zeng, R. J.; Angelidaki, I. Hydrogen and Methane Production from Household Solid Waste in the Two-stage Fermentation Process. *Water Res.* **2006,** *40*, 2230–2236.

Magnusson, L.; Islam, R.; Sparling, R.; Levin, D.; Cicek, N. Direct Hydrogen Production from Cellulosic Waste Materials with a Single-step Dark Fermentation Process. *Int. J. Hydrog. Energy* **2008,** *33*, 5398–5403.

Manish, S.; Banerjee, R. Comparison of Biological Production Processes. *Int. J. Hydrog. Energy* **2008,** *33*, 279–286.

Mariakakis, I.; Bischoff, P.; Krampe, J.; Meyer, C.; Steinmetz, H. Effect of Organic Loading Rate and Solids Retention Time on Microbial Population During Bio-hydrogen Production by Dark Fermentation in Large Lab-scale. *Int. J. Hydrog. Energy* **2011,** *36*, 10690–10700.

Millat, T.; Janssen, H.; Bahl, H.; Fischer, R. J.; Wolkenhauer, O. In *The pH Induced Metabolic Shift from Acidogenesis to Solventogenesis in Clostridium acetobutylicum: From Experiments to Models*, 5th Symposium on Experimental Standard Conditions of Enzyme Characterization, Rüdesheim/Rhein, Germany, Sept 12–16, **2011;** Beilstein-Institut, 2013.

Mohammadi, P.; Ibrahim, S.; Annuar, M. S. M.; Ghafari, S.; Vikineswary, S.; Zinatizadeh, A. A. Influences of Environmental and Operational Factors on Dark Fermentative Hydrogen Production: A Review. *Clean: Soil Air Water* **2012,** *40*, 1297–1305.

Mosier, N.; Wyman, C.; Dale, B.; Elander, R.; Lee, Y. Y.; Holtzapple, M.; Ladisch, M. Features of Promising Technologies for Pretreatment of Lignocellulosic Biomass. *Bioresour. Technol.* **2005,** *96*, 673–686.

Nazlina, H. M. Y.; Nor Aini, A. R.; Ismail, F.; Yusof, M. Z. M.; Hassan, M. A. Effect of Different Temperature, Initial pH and Substrate Composition on Biohydrogen Production from Food Waste in Batch Fermentation. *Asian J. Biotechnol.* **2009,** *1*, 42–50.

Nguyen, T. A. D.; Han, S. J.; Kim, J. P.; Kim, M. S.; Oh, Y. K.; Sim, S. J. Hydrogen Production by the Hyperthermophilic Eubacterium, *Thermotoga neapolitana*, Using Cellulose Pretreated by Ionic Liquid. *Int. J. Hydrog. Energy* **2008,** *33*, 5161–5168.

Orozco, R. L.; Redwood, M. D.; Leeke, G. A.; Bahari, A.; Santos, R. C. D.; MacAskie, L. E. Hydrothermal Hydrolysis of Starch with CO_2 and Detoxification of the Hydrolysates with Activated Carbon for Bio-hydrogen Fermentation. *Int. J. Hydrog. Energy* **2012,** *37*, 6545–6553.

O-Thong, S.; Prasertsan, P.; Karakashev, D.; Angelidaki, I. Thermophilic Fermentative Hydrogen Production by the Newly Isolated *Thermoanaerobacterium thermosaccharolyticum* PSU-2. *Int. J. Hydrog. Energy* **2008,** *33*, 1204–1214.

Ozkan, L.; Erguder, T. H.; Demirer, G. N. Effects of Pretreatment Methods on Solubilization of Beet-pulp and Bio-hydrogen Production Yield. *Int. J. Hydrog. Energy* **2011,** *36*, 382–389.

Palmqvist, E.; Hahn-Hägerdal, B. Fermentation of Lignocellulosic Hydrolysates: II: Inhibitors and Mechanisms of Inhibition. *Bioresour. Technol.* **2000,** *74*, 25–33.

Pan, J.; Zhang, R.; El-Mashad, H. M.; Sun, H.; Ying, Y. Effect of Food to Microorganism Ratio on Biohydrogen Production from Food Waste via Anaerobic Fermentation. *Int. J. Hydrog. Energy* **2008,** *33*, 6968–6975.

Rafieenia, R.; Lavagnolo, M. C.; Pivato, A. Pre-treatment Technologies for Dark Fermentative Hydrogen Production: Current Advances and Future Directions. *Waste Manag.* **2018,** *71*, 734–748.

Ren, N.; Guo, W.; Liu, B.; Cao, G.; Ding, J. Biological Hydrogen Production by Dark Fermentation: Challenges and Prospects Towards Scale-up Production. *Curr. Opin. Biotechnol.* **2011,** *22*, 365–370.

Rózsenberszki, T.; Koók, L.; Bakonyi, P.; Nemestóthy, N.; Logroño, W.; Pérez, M.; Urquizo, G.; Recalde, C.; Kurdi, R.; Sarkady, A. Municipal Waste Liquor Treatment via Bioelectrochemical and Fermentation (H_2 + CH_4) Processes: Assessment of Various Technological Sequences. *Chemosphere* **2017,** *171*, 692–701.

Ruggeri, B.; Tommasi, T. Efficiency and Efficacy of Pre-treatment and Bioreaction for Bio-H_2 Energy Production from Organic Waste. *Int. J. Hydrog. Energy* **2012,** *37*, 6491–6502.

Saady, N. M. C. Homoacetogenesis During Hydrogen Production by Mixed Cultures Dark Fermentation: Unresolved Challenge. *Int. J. Hydrog. Energy* **2013,** *38*, 13172–13191.

Shaterzadeh, M. J.; Ataei, S. A. The Effects of Temperature, Initial pH, and Glucose Concentration on Biohydrogen Production from *Clostridium acetobutylicum*. *Energy Source Part A* **2017,** *39*, 1118–1123.

Shen, L.; Bagley, D. M.; Liss, S. N. Effect of Organic Loading Rate on Fermentative Hydrogen Production from Continuous Stirred Tank and Membrane Bioreactors. *Int. J. Hydrog. Energy* **2009,** *34*, 3689–3696.

Shin, H. S.; Youn, J. H. 2005. Conversion of Food Waste into Hydrogen by Thermophilic Acidogenesis. *Biodegradation* **2005,** *16*, 33–44.

Sinha, P.; Pandey, A. An Evaluative Report and Challenges for Fermentative Biohydrogen Production. *Int. J. Hydrog. Energy* **2011,** *36*, 7460–7478.

Sreethawong, T.; Niyamapa, T.; Neramitsuk, H.; Rangsunvigit, P.; Leethochawalit, M.; Chavadej, S. Hydrogen Production from Glucose-containing Wastewater Using an Anaerobic-sequencing Batch Reactor: Effects of COD Loading Rate, Nitrogen Content, and Organic Acid Composition. *Chem. Eng. J.* **2010,** *160*, 322–332.

Sun, Y.; Cheng, Y. Hydrolysis of Lignocellulosic Materials for Ethanol Production: A Review. *Bioresour. Technol.* **2002,** *83*, 1–11.

Tang, J.; Yuan, Y.; Guo, W. Q.; Ren, N. Q. Inhibitory Effects of Acetate and Ethanol on Biohydrogen Production of *Ethanoligenens harbinese* B49. *Int. J. Hydrog. Energy* **2012,** *37*, 741–747.

Valdez-Vazquez, I.; Poggi-Varaldo, H. M. Hydrogen Production by fermentative Consortia. *Renew. Sust. Energy Rev.* **2009,** *13*, 1000–1013.

Valdez-Vazquez, I.; Ríos-Leal, E.; Esparza-García, F.; Cecchi, F.; Poggi-Varaldo, H. M. Semi-continuous Solid Substrate Anaerobic Reactors for H_2 Production from Organic Waste: Mesophilic vs. Thermophilic Regime. *Int. J. Hydrog. Energy* **2005,** *30*, 1383–1391.

Wang, J.; Wan, W. Effect of Temperature on Fermentative Hydrogen Production by Mixed Cultures. *Int. J. Hydrog. Energy* **2008,** *33*, 5392–5397.

Wang, J.; Wan, W. Factors Influencing Fermentative Biohydrogen Production: A Review. *Int. J. Hydrog. Energy* **2009,** *34*, 799–811.

Wang, J.; Wan, W. Combined Effects of Temperature and pH on Biohydrogen Production by Anaerobic Seed Sludge. *Biomass Bioenergy* **2011,** *35*, 3896–3901.

Wang, J.; Yin, Y. Principle and Application of Different Pretreatment Methods for Enriching Hydrogen-producing Bacteria from Mixed Cultures. *Int. J. Hydrog. Energy* **2017,** *42*, 4804–4823.

Wang, A.; Sun, D.; Cao, G.; Wang, H.; Ren, N.; Wu, W. M.; Logan, B. E. Integrated Hydrogen Production Process from Cellulose by Combining Dark Fermentation, Microbial Fuel Cells, and a Microbial Electrolysis Cell. *Bioresour. Technol.* **2011,** *102*, 4137–4143.

Wong, Y. M.; Wu, T. Y.; Juan, J. C. A Review of Sustainable Hydrogen Production Using Seed Sludge via Dark Fermentation. *Renew. Sust. Energy Rev.* **2014,** *34*, 471–482.

Wu, J.; Ein-Mozaffari, F.; Upreti, S. Effect of Ozone Pretreatment on Hydrogen Production from Barley Straw. *Bioresour. Technol.* **2013,** *144*, 344–349.

Yang, H.; Shao, P.; Lu, T.; Shen, J.; Wang, D.; Xu, Z.; Yuan, X. Continuous Bio-hydrogen Production from Citric Acid Wastewater via Facultative Anaerobic Bacteria. *Int. J. Hydrog. Energy* **2006,** *31*, 1306–1313.

Yasin, N. H. M; Mumtaz, T.; Hassan, M. A.; Abd Rahman, N. Food Waste and Food Processing Waste for Biohydrogen Production. *J. Environ. Manag.* **2013,** *130*, 375–385.

Yun, Y. M.; Jung, K. W.; Kim, D. H.; Oh, Y. K.; Cho, S. K.; Shin, H. S. Optimization of Dark Fermentative H_2 Production from Microalgal Biomass by Combined (Acid + Ultrasonic) Pretreatment. *Bioresour. Technol.* **2013,** *141*, 220–226.

Zhang, M. L; Fan, Y. T.; Xing, Y.; Pan, C. M.; Zhang, G. S.; Lay, J. J. Enhanced Biohydrogen Production from Cornstalk Wastes with Acidification Pretreatment by Mixed Anaerobic Cultures. *Biomass Bioenergy* **2007,** *31*, 250–254.

Zheng, Y.; Pan, Z.; Zhang, R. Overview of Biomass Pretreatment for Cellulosic Ethanol Production. *Int. J. Agric. Biol. Eng.* **2009,** *2*, 51–68.

Zong, W.; Yu, R.; Zhang, P.; Fan, M.; Zhou, Z. Efficient Hydrogen Gas Production from Cassava and Food Waste by a Two-step Process of Dark Fermentation and Photo-fermentation. *Biomass Bioenergy* **2009,** *33*, 1458–1463.

CHAPTER 3

Tailoring Triacylglycerol Biosynthetic Pathway in Plants for Biofuel Production

KSHITIJA SINHA[1], RANJEET KAUR[2], and RUPAM KUMAR BHUNIA[1,*]

[1]*National Agri-Food Biotechnology Institute (NABI), Plant Tissue Culture and Genetic Engineering, Mohali 140306, Punjab, India*

[2]*Department of Biotechnology, Mangalmay Institute of Management and Technology (MIMT), Greater Noida 201306, Uttar Pradesh, India*

**Corresponding author.*
E-mail: rupamb2005@gmail.com; rupamb@nabi.res.in

ABSTRACT

Rising global demands for high energy needs together with a sharp decline in the fossil fuel reserves, has propelled the research for an alternative fuel source that is sustainable yet eco-friendly. Triacylglycerols (TAGs) extracted from the fruit and seeds of oil crop plants, are the major sources of plant based vegetable oils used for food, feed and feedstocks for biodiesel and industrial chemicals. Although, plant vegetative tissues contain minor amount of neutral lipid or TAG, they holding ability for TAG biosynthesis, storage, and mobilization. Thus, TAG accumulation in plant vegetative tissues has become the new revolution in the metabolic engineering for expand production of TAG as a renewable and sustainable energy source. Evaluation of the flux of acyl-changes through different key biosynthetic genes and transcription factors (TFs) has opened up new avenues for advancing our research towards generating alternative fuel resource by targeted and improvised molecular biology interventions. In this chapter, we have summarized the recent advances achieved in TAG biosynthesis, mobilization and its regulation via various genes and TFs. Such knowledge could be judiciously used to generate high throughput transgenic plants for TAG production either by use of single-gene overexpression or by multi-gene stacking.

3.1 INTRODUCTION

Use of fossil fuels has significantly increased the level of CO_2 globally which is ultimately contributing to the increase in the adverse effects of global warming. Being exploited so extensively by the growing population, the energy demand has been estimated to grow up to 50% by 2025. To overcome these problems, researchers are trying to find sustainable alternatives which not only should provide us energy but also should not contribute in any further worsening of the environmental conditions. Biofuels are one of such alternatives and sustainable as well as renewable sources of energy derived from organic matter in the form of biomass (Naik et al., 2010). Since the oil crises occurred in 1973 and 1979, biofuels have been considered as a good replacement of fossil fuels. One of the finest examples of the initiation of the biofuel usage is the National Ethanol Program, which was developed to produce ethanol from the abundant sugarcane stock. Similarly, an oil crisis in the United States led to a corn-based bioethanol in the 1970s (Timilsina, 2014). At present day, such initiations have been reported from all over the globe and implementation of the application of biofuels extensively at a larger scale will not only help in mitigating the global warming issue but also open doors to new income and employment schemes for the rural areas.

Current solutions to the development of renewable biofuels are focused on using bioethanol, derived either from the fermentation of corn starch (an established industry), or the fermentation of lignocellulosic biomass (an emerging concept, with technology hurdles that require additional solutions). Although these potential biofuels offer near-term solutions, each of them suffers from different degrees of inadequacy. Specifically, bioethanol is not a good substitute for the current petroleum fuel due to the fact that it is partially oxygenated. Thus, its inclusion in gasoline thermodynamically dilutes the energy content of the fuel. In addition, ethanol's high affinity for water has implications for the infrastructure that is required to support its transportation, the blending of ethanol–gasoline mixtures, and the dispersion to the consumers (e.g., E85 pumps and E85-compatible cars). Although biodiesel does not suffer from these issues, its utility is abridged by the fact that availability of vegetable oil is limited.

Plants contain stored lipids in the form of glycerol esters of fatty acids, also known as triacylglycerol (TAG). TAGs are regarded as one of the most widely available and energy-rich forms of reduced carbon available in nature (Durrett al., 2008; Thelen and Ohlrogge, 2002). Plant vegetative tissues contain only minor amounts of neutral lipid or TAG. The lipids of plant vegetative tissues primarily consist of polar lipids (glycolipids and phospholipid) and store lesser energy in comparison to neutral lipid. Oil seed crops have formed an inevitable source of neutral lipids due to the presence of high amount of TAG in their seeds and fruits. TAGs are highly concentrated source of metabolic energy because they are reduced and anhydrous, containing double amount of energy of either carbohydrates or proteins that can also be used to supply energy to the body. Chemically, biodiesel is a mixture of fatty acid alkyl esters and the chemical similarities of TAG represent logical substitute for conventional diesel. Thus, a wise approach would be to broaden our knowledge on the intricacies of the processes involved in biosynthesis and accumulation of TAG in seeds of oilseed crops along with its regulation will help us to evolve newer and advanced strategies for enhancing its accumulation in plant vegetative tissues of high biomass crops, thereby leading to development of designer crops for meeting the energy demands of future.

3.2 TAG METABOLISM PATHWAY IN PLANTS: BIOSYNTHESIS AND MOBILIZATION

TAG is composed of three fatty acyl groups esterified to a glycerol backbone at *sn*-1, *sn*-2 and, *sn*-3 positions. The site of TAG biosynthesis is endoplasmic reticulum (ER) and it is also the site of acyl editing of fatty acyl chains. As the TAG synthesis occurs, the droplets of TAG accumulate in the outer leaflet of the ER and ultimately get released from there in the form of small oil bodies (OB) of 0.5–1 μm in diameter (Murphy, 2001). These OBs are surrounded by a phospholipid monolayer and is protected by amphiphilic structural proteins which are embedded in the TAG and phospholipid layer. The mobilization of TAG requires three main steps. It is first hydrolyzed by lipases followed by its breakdown in the peroxisome to fatty acids (by β-oxidation), which is finally converted into acetyl-CoA . This acetyl-CoA is further converted into citrate, which is either used up in respiration or is converted into soluble sugars by the process of glyoxylate cycle and gluconeogenesis which are then consumed in the growth and development of the seed.

3.2.1 BIOSYNTHESIS

The four most important oilseed crops are soybean, oil palm, rapeseed, and sunflower which contain four fatty acids, that is, linoleic acid (18:2 cis-9,12), palmitic acid (16:0), lauric acid (12:0), and oleic acid (18:1), abundantly in their seeds. However, there are five major fatty acids in plants which constitute 90% of the acyl chains of the glycerolipids of almost all membranes found in plants. These fatty acids have 16 or 18 carbons in the chains (i.e., 18:1, 18:2, 18:3, 16:0, and in some species, 16:3), and they contain one to three cis double bonds (Thelen and Ohlrogge, 2001). The de novo biosynthesis of fatty acid is catalyzed by acetyl-CoA carboxylase and the fatty acid synthase known as Kennedy Pathway (Bates, et al., 2013). The unsaturated fatty acids of 4–18 carbons are synthesized in plastid and the fatty acyl chain then is attached to the acyl carrier protein of the fatty acid synthase complex. The fatty acyl chain is released from the carrier protein by thioesterase. Once exported from the plastid, acyl-Coenzyme A (acyl-CoA) synthetase combines fatty acid chains to CoA-forming acyl-CoA on the outer membrane of plastids.

The initiation of synthesis of all glycerolipids includes acylation of *sn*-glycerol-3-phosphate (G-3-P) to lysophosphatidic acid which is catalyzed by *sn*-1 glycerol-3-phosphate acyltransferase (GPAT). The frequent participation of acyl-CoA in comparison to any other substrate in this step leads to an asymmetric distribution of saturated and unsaturated fatty acids in different positions of phospholipids and TAGs. Further elongation of the acyl chains occurs in ER (Fig. 3.1). The role of GPAT in glycerolipid synthesis is still a topic of discussion as it is seen that the in yeast, one of the isoforms, SCT1, when overexpressed, enhances TAG accumulation while the other isoform, GPT1 decreases the accumulation. The homologs of GPAT in plants such as *Arabidopsis* have also been studied. It is seen that mutation in these genes does not affect the oil metabolism. However, it has been known that GPAT4 and GPAT6 participate in other functions such as generation of monoacylglycerol which ultimately leads to the cutin biosynthesis (Zheng et al., 2003). The next step is catalyzed by an

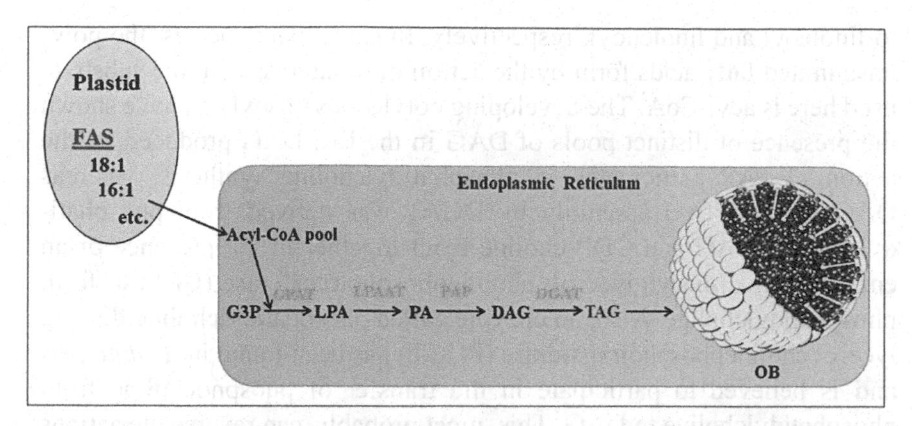

FIGURE 3.1 (See color insert.) Schematic diagram of triacylglycerol biosynthesis in plants. Green compartment represents plastid and orange compartment represents ER. DAG, tiacylglycerol; DGAT, diacylglycerol acyltransferase; FAS, fatty acid synthase; G-3-P, glycerol-3-phosphate; GPAT, glycerol-3-phosphate acyltransferase; LPA, lysophosphatidic acid; LPAAT, lysophosphatidic acid acyltransferase enzyme; PA, phosphatidic acid; TAG, triacylglycerol.

enzyme called lysophosphatidic acid acyltransferase enzyme (LPAAT) which converts lysophosphatidic acid to phosphatidic acid. It is believed that LPAAT favors different types of fatty acyl-CoAs in different type of plants. LPAAT enzyme in Tropaeolaceae and Limnanthaceae family prefer very long fatty acyl-CoAs; similarly, medium-length fatty acyl-CoAs are preferred by this enzyme in plants such as coconut and palm. The third step in this pathway is the formation of diacylglycerol (DAG) from phosphatidic acid (Maisonneuve et al., 2010). The enzyme involved here is known as phosphatidate phosphatase1 (PAP1), which dephosphorylates phosphatidic acid. The final step in the biosynthesis of TAG is catalyzed by an enzyme called diacylglycerol acyltransferase (DGAT) which converts DAG to TAG (Weselake et al., AOCS lipid library).

3.2.1.1 ACYL EDITING IN THE ER AND HOW IT AFFECTS THE FATTY ACID COMPOSITION OF TAG

Polyunsaturated fatty acids are formed from phosphatidylcholine by the actions of membrane-bound fatty acid desaturase 2 and 3 (FAD2 and FAD3) (Bhunia et al., 2016). These enzymes convert the oleoyl groups to linoleoyl and linolenoyl, respectively. In some plant species, the polyunsaturated fatty acids form by the action of desaturases but the substrate used here is acyl-CoA. The developing cotyledons of soybean have shown the presence of distinct pools of DAG in the ER. DAG produced by the action of PAP participates in phosphatidylcholine synthesis, whereas DAG used in TAG assembly by DGAT was derived from phosphatidylcholine. DAG and CDP-choline react together in the presence of an enzyme called diacylglycerol: cholinephosphotransferase (CPT) to form phosphatidylcholine. Also, an enzyme called phosphatidylcholine:diacylg lycerol cholinephosphotransferase (PDCT) has been found in *Arabidopsis* and is believed to participate in the transfer of phosphocholine from phosphatidylcholine to DAG. This, most probably, can reverse the actions by moving DAG-containing oleoyl moieties into phosphatidylcholine and removing DAG-containing polyunsaturated fatty acids (PUFA) from phosphatidylcholine, which would ultimately make DAG available for its conversion into TAG via the actions of DGAT (Weselake et al., AOCS lipid library).

TAG can also be synthesized in an acyl-CoA-independent manner. This reaction requires a acyl donor and an acyl acceptor, that is, phosphatidylcholine and *sn*-1,2-diacylglycerol, respectively, and is catalyzed by an enzyme called phospholipid:diacylglycerol acyltransferase (Ghosal et al., 2007). This reaction could be a useful measure of reducing the unusual fatty acid moieties by transferring them from phosphatidylcholine into TAG. The same purpose is also served by phopholipase A_2 (PLA_2) and acyl-CoA: lysophosphatidylcholine acyltransferase (LPCAT) (Weselake et al., AOCS lipid library).

3.2.2 MOBILIZATION

Germinating seeds are considered to be the best understood system in context of mobilization of storage lipid (TAG). After germination, the seeds enter into a photosynthetically active stage where the rate of growth and development is very high. The storage of reserve compounds in plants is established during this very period. The biomolecules that are required for its growth are supplied from these stored reserves (Bewley et al., 1994). However, variations can be seen in the form of stored reserves depending upon the tissue type but oils, carbohydrates, and proteins are usually predominant. However, the breakdown of these reserves and their efficient utilization play an important role in the development of seedlings from seeds (Graham, 2008).

TAG contains three fatty acids esterified to a glycerol backbone because of which it is considered to be an efficient source of energy. Its oxidation yields more than twice the energy of protein or carbohydrate. It is considered to be an efficient storage reserve in most of the seed oil crops. Being located in the cotyledon or endosperm tissues of the seed, it makes up to 60% of the total weight of the seed. It is found as lipid droplets of diameter ranging from ~0.5 to 1 μm are surrounded by a phospholipid monolayer and structural proteins called oleosin which prevent these droplets from desiccation (Murphy, 2001).

The first step of mobilization is hydrolysis of TAG to free fatty acids and glycerol, which is triggered by a lipase called sugar-dependent-1 (SDP-1) in plants. It belongs to the same gene family which consists of the lipases that initiate the hydrolysis of TAG in other organisms such as mammals, insects, and yeasts. The glycerol that is released here is converted into dihydroxyacetone phosphate (DHAP) via the action of glycerol kinase (GLI-1) located in the cytosol and flavin adenine dinucleotide-linked G-3-P dehydrogenase (SDP-6) located in the inner mitochondrial membrane. This DHAP can now be converted into sugars by gluconeogenesis (Quettier et al. 2009).

Peroxisome or glyoxysomes in seeds are the sites where the fatty acids are activated to form acyl-CoA and are then further broken down by β-oxidation. An ATP-binding cassette transporter called peroxisomal ABC transporter-1 (PXA1), which is located at the peroxisomal membrane, is believed to import fatty acyl group inside the peroxisome where two long-chain fatty acyl CoA synthetases, called long-chain acyl-coenzyme A (CoA) synthetases (LACS6 and LACS7), activate them; for this mechanism to occur fatty acid must first attach to the cytosol. However, it has been suggested by researchers that consumption of ATP by PXA1 is dependent on acyl-CoAs not on free fatty acids (Graham, 2008).

β-oxidation of fatty acid is the process of cleavage of fatty acid carbon units into acetyl-CoA moieties, and it occurs in glyoxysome (Bewley, 2001). The glyoxysome contains the enzymes involved in this β-oxidation pathway, that is, acyl-CoA oxidase (ACX), multifunctional protein (MFP), and 3-ketoacyl-CoA thiolase (KAT). These enzymes catalyze oxidation, hydration and dehydrogenation, and thiolytic cleavage reactions, respectively (Graham, 2008). Other reactions are also involved which are required for the breakdown of unsaturated fatty acids with double or triple bonds in the cis configuration at even- and odd-numbered carbons in the chain. The enzymatic activity of ACX leads to the production of hydrogen peroxide which is a kind of reactive oxygen species and is necessary to be detoxified. This detoxification happens with the help of a catalase enzyme located inside the peroxisome or by ascorbate peroxidase (APX) present on the membrane of peroxisomes (Pracharoenwattana, 2010). After the breakdown of the fatty acids via β-oxidation, the acetyl-coA undergoes a sequence of reactions, altogether called the glyoxylate cycle, in which succinate and oxaloacetate are formed. The enzymes involved in this cycle such as citrate synthase, aconitase, and malate dehydrogenase also participate in the citric acid cycle. However, two enzymes known as isocitrate lyase and malate synthase work only for the glyoxylate cycle in bypassing the steps of decarboxylation in the citric acid cycle (Beevers, 1980). These two enzymes generate malate and succinate, both of which are the citric acid cycle intermediates which can be converted to oxaloacetate. Out of the two molecules of oxaloacetate formed here, one can rejoin the glyoxylate cycle while the other participates in gluconeogenesis (Beevers, 1980).

The movement of organic acids between peroxisome and cytosol affect both glyoxylate cycle and β-oxidation equally but the mechanism of transport is still not known. However, it has been seen that oxaloactetate is incapable of crossing the peroxisomal membrane, and so it is first converted into aspartate by the action of an enzyme called aspartate aminotransferase and then is taken up by the mitochondria in exchange of glutamate and alpha ketoglutarate. These organic acids formed in the glyoxylate pathway further participate in gluconeogenesis. One of the major events of this process is oxaloactetate getting converted into phosphoenolpyruvate (PEP) by PEP carboxykinase. There is a loss of CO_2 and ATP is also consumed in this reaction but then glycolytic enzymes convert PEP to sugar phosphates, allowing their participation in the synthesis of sucrose. The sugars formed by this process then participate in growth and development of the seedlings (Graham, 2008).

The phytohormone gibberellic acid (GA) is released from scutellum and diffuses to the aleurone layers of seed coat and promotes the activation of multiple hydrolytic enzymes. The hydrolytic enzymes further hydrolyze the storage reserve in seeds (Bewley, 2001). The genes in the oil metabolism process are regulated by metabolic status and high levels of sugar repress mobilization. The positive regulation of seed oil mobilization is yet to be studied. However, it is known that in *Arabidopsis*, the domain transcription factor called abscisic acid insensitive 4 (ABI4) participates in as well as sugar repression of oil mobilization in the embryo (Graham, 2008).

The metabolic pathways that lead to the synthesis of TAG have very extensively been studied. But looking up at the increasing demand of plant-derived oils and feedstock, researchers are reconsidering to study these pathways with the point of view of increasing the TAG synthesis in the plants with the help of several techniques falling in the spectrum of plant biology and biotechnology ranging from basic plant biochemistry and genetics to interdisciplinary biophysics and imaging techniques. Although the efforts made by the scientists have led us to understand the metabolic pathways involved in TAG synthesis, some pieces are still left to be put together to complete this vast jigsaw puzzle.

These unsolved loops might be the reason behind the lack of efficiency of the transgenic lines developed to modulate the enhanced accumulation of TAGs. The concept of scattered regulatory steps could be responsible for our misses in the attempt of tracking down the main players of the pathway. Studies on metabolic control analysis as well as top-down control analysis tests have proven two important facts that the flux control is distributed throughout the pathway and is actually exerted by two pathways namely, TAG biosynthetic pathway in chloroplast and TAG assembly pathway in ER (Baud and Lepiniec, 2010). Finally, with the concept of the central metabolism that provides the precursor of TAG synthesis and TAG storage processes, added up with these two angles, we would be capable of getting an overview of the whole process of oil accumulation in the seeds.

3.3 TRANSCRIPTION FACTORS

Seed tissue is the most explored system for the study related to the transcription factors (TFs) that regulate TAG biosynthesis. TFs are mainly believed to regulate the embryogenesis and seed maturation in most of the plants. These TFs are leafy cotyledon-1 (LEC1), LEC1-LIKE (L1L), leafy cotyledon-2 (LEC2), FUSCA-3 (FUS3), and absiscic acid insensitive 3 (ABI3) and are considered as the master regulators of TAG biosynthesis

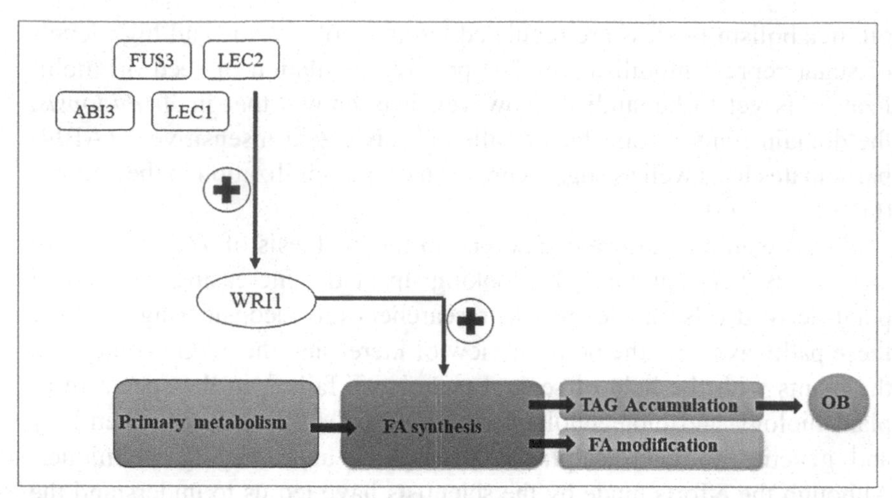

FIGURE 3.2 A schematic diagram defining the role of the main regulatory transcription factors in TAG accumulation. LEC 1, leafy cotyledon 1; LEC 2, leafy cotyledon 2; ABI3, abscisic acid insensitive 3; FUS3, FUSCA 3; TAG, triacylglycerols; WRI1, wrinkled 1.

and seed maturation (Fig. 3.2). Santos-Mendoza et al. (2008) have shown that mutation in the genes of these master regulators affects not only the abundance of oil in the seeds but it also exhibits characteristics like cotyledons appearing such as leaves, low desiccation tolerance, and accumulation of abnormal levels of pigments. However, they do not directly act upon the different pathways of seed development; they regulate each other's activity and effect the seed development through downstream activities of other transcription factors (Kagaya et al., 2005; Mu et al., 2008). It is seen that the production of TAG in seeds is regulated in a similar way by a transcription factor called wrinkled1 (WRI1) (Dabbs, 2015). Although these master regulators have been studied to a great extent and are used to manipulate the pathways in several plants, there is still a lot to uncover.

3.3.1 LEAFY COTYLEDON 1

LEC1 and LEC1-LIKE TFs play a major role in embryo development and are mainly expressed during seed development. LEC1 in *Arabidopsis* regulates the expression of genes that are primarily involved in glycolysis and condensation, chain elongation and desaturation of glycerolipid biosynthesis (Mu et al., 2008). The decrease in the expression levels of these TFs has resulted in the lesser abundance of oil seeds and also exhibited negative effects on seed maturation in different plants (Manan et al., 2017). The regulation of these TFs is partially dependent on AB1, FUS3, and WR1. Their combined activity is responsible for the alteration of sucrose metabolism, reduction in glycolytic activity, and increase in carbon flux toward fatty acid synthesis which eventually leads to increase in the TAG production. Similar kind of regulatory mechanism has been seen in *Zea mays* as well as *Brassica napus* (Manan et al., 2017).

3.3.2 LEAFY COTYLEDON 2

This TF regulates the genes that are involved in cotyledon developmental network and the structural genes participating in synthesis of storage proteins and TAGs in seeds. It has been observed that the LEC2 of *Arabidopsis*, when expressed in seedlings, very effectively regulates the synthesis of seed storage proteins along with OB in the vegetative tissues. This normally activates during the maturation phase before increase in the expression of LEC1 and FUS3 (Manan et al., 2017).

3.3.3 ABI3 AND FUSCA3

These two TFs regulate the ABA/GA signaling pathways and therefore, participate positively in seed maturation and dormancy in plants, which ultimately affects other physiological mechanisms such as the stomatal movement, seed desiccation, drought/cold tolerance, and seedling growth inhibition. ABI3 and viviparrous1 (VP1) have shown active participation in regulation of seed desiccation and oil storage in several plants. On the other hand, FUS3 mainly interacts with the genes responsible for embryogenesis (Manan et al., 2017).

3.3.4 WRINKLED 1

WRI1, the most important TFs participating in oil biosynthesis in *Arabidopsis* seeds, is the member of APETALA2 family. It works along with LEC1 in this pathway. It controls the oil biosynthesis by directly regulating genes such as *BCCP* (a subunit of ACCase), Enoyl-acyl carrier protein reductase (ENR), Beta-ketoacyl carrier protein reductase (BKACPR), FAD2, plastidial pyruvate kinase, sucrose synthase, pyruvate dehydrogenase, and acyl-carrier protein which are involved in main metabolic processes, seed germination, and seedling establishment (Cernac et al., 2006; Manan et al., 2017). To et al. (2012) have identified four *WRI1* genes belonging the APETALA2/ethylene-responsive element binding protein (AP2/EREBP) family in *Arabidopsis*, that is, WRI1, WRI2, WRI3, and WRI4 (Dabbs, 2015). The exact functions of these TFs were not identified. But from their studies, they have observed that WRI3 and WRI4 complement the WRI1 mutant but it showed no effect on the levels of oil content in the seeds. Since these TFs have been observed to express more in other tissues of the plants such as roots, stems, and flowers, it is believed that they have tissue-specific functions. The functions of WRI2 are yet to be known (Dabbs, 2015). Hence, it can be said that only WRI1 is known to directly regulate the biosynthesis of TAG in seeds.

3.3.5 MANIPULATION OF TRANSCRIPTION FACTORS FOR ENHANCED TAG ACCUMULATION

Ample amount of research has been done on manipulation of these TFs in order to study their effect on the accumulation of TAG in seed tissues as well as in nonseed tissues. Mutation in maize *LEC1* has shown a detrimental effect on the seed maturation which includes low abundance of seed oil. The overexpression of seed-specific *LEC1* gene resulted in 48% increase in oil accumulation without hampering the agronomic traits of the plants or reducing their nutritional quality (Shen et al., 2010; Tan et al., 2011; Tang et al., 2018). LEC 2 has been known as the master regulator of the developmental mechanism. It has been reported that *Arabidopsis thaliana* LEAFY COTYLEDON2 (LEC2), when expressed in *Arabidopsis* seedlings, regulates the expression of various proteins and OB in the vegetative organs that normally get activated during the maturation before the transcripts of LEC1 and FUS3 increase (Stone et al., 2009). Hence, overexpression of this TF can be a potential approach to increase oil content. Expression of LEC2 in mature tobacco plants has led to accumulation of up to 6.8% per dry weight of total extracted fatty acid (Andrianov et al., 2010). Also, it has been reported that expression of castor bean LEC2 in *Arabidopsis* increases expression of fatty acid elongase 1 (FAE1) and induces accumulation of TAG in vegetative tissues (Kim et al., 2013). Although not much work has been done on FUS3, still it has been reported that FUS3 is a potential candidate which can be used to increase TAG accumulation. FUS3 and ABI3 expression have shown to be positively regulated by LEC1 and LEC2 (To et al., 2006) and overexpression of FUS3 in *Arabidopsis* seedlings have shown 6% increase in the TAG contents (Zhang et al., 2016). WRI1 is a downstream regulating TF which, on overexpression, has provided evidences of increasing oil content in seed as well as in nonseed tissues (Dabbs, 2015). However, it has been seen that WRI1 is important for the expression of genes responsible for TAG accumulation and OB synthesis but overexpression of WRI1 does not upregulate the expression of TAG biosynthesis pathway genes (Maeo et al., 2009). Overexpression of Zea mays WRI1 leads to an increase in the oil content similar to overexpression of LEC1 without affecting germination, seedling growth, or grain yield (Shen, 2010). Due to its ability to enhance oil accumulation, it is also being expressed in combination with genes such as DGAT which would further enhance its activity (Vanhercke, 2013). Recent studies have shown that another TF called MYB96 participates in TAG assembly pathway in seeds and is independent of WRI1 (Lee et al., 2018).

3.4 METABOLIC ENGINEERING OF TAG BIOSYNTHETIC PATHWAY FOR ENERGY DENSIFICATION IN VEGETATIVE TISSUES

Vegetable oil-producing crops have been one of the most dynamic parts of agriculture. Worldwide, over the next 30 years, demand for vegetable oil is expected to double with increasing requirements for food, feed, and fuel (FAO, 2016). Therefore, in order to meet the energy demands of the ever-increasing population, we need to develop additional strategies to produce plant oils in vegetative tissues. One strategy to increase the supply of plant oil is to engineer the oil production in vegetative tissues. Scientists worldwide have identified the key genes required for oil production and accumulation in plant leaves and seeds. Overexpression of these genes resulted in increased oil content in leaves, the most abundant sources of plant biomass. Thus, identification of these genes involved in increasing the energy content of plant based foods is urgently required to achieve the bigger goal of biodiesel production.

In plants, diacylglycerol acyltransferase (DGAT), which catalyzes the final step of oil biosynthesis (the esterification of *sn*-1,2-diacylglycerol with a long-chain fatty acyl-CoA), are located in distinct but separate regions of the ER (Shockey, 2006). In the ER, TAGs are synthesized by the stepwise acylation of G-3-P, known as Kennedy pathway (Shen et al., 2010). In the beginning, fatty acyl moieties are esterified to the G-3-P backbone at the *sn*-1 and *sn*-2 positions. This esterification reaction is catalyzed by G-3-P acyltransferase and lysophosphatidic acid acyltransferase, respectively, to form phosphatidic acid. Phosphatidate phosphatase further hydrolyzed phosphatidic acid to yield DAG. DGAT is the last key rate-limiting enzyme that synthesizes TAG by esterifying third acyl chain to DAG (Voelker and Kinney, 2001).

Genetic manipulation of DGAT has been one of the main approaches to increase the oil accumulation in plants. There are several DGAT enzymes in the plants which have similar functions but have shown variations in their pattern of work. DGAT1 and DGAT2 are related to DGATs involved in TAG synthesis in animals and fungi. When genetically modified in *Arabidopsis*, it exhibited to be one of the major enzymes involved in TAG biosynthesis. DGAT2 on the other hand still has an unidentified purpose in the *Arabidopsis*. However, in other plants such as *Ricinus communis,* it shows a greater extent of expression than DGAT1 during seed maturation (Shockey, et al., 2016). A soluble DGAT, DGAT3, has also been identified which might be involved in TAG recycling in *Arabidopsis* seedlings (Rani et al., 2010).

In plants, after TAG biosynthesis, oleosin is a protein that coats and stabilizes TAGs as OBs. Each oleosin protein has a highly conserved central hydrophobic domain of 72 amino acid residues. The central hydrophobic domain is proposed to be made up of β-strand structure and to interact with the lipids. Protein model shows oleosin has a hairpin-like hydrophobic structure that is inserted inside the TAG, while the amphipathic arms are left outside the OBs. Genes involved in oleosin biosynthesis are not expressed in leaves. Also, vegetative tissues of plants are limited by SDP1 lipase function when plants are genetically engineered to synthesize more oil (Kelly et al., 2013).

The increased accumulation of TAG from vegetative tissues can be enhanced by the metabolic engineering approaches and altering

TAG biosynthetic pathways. Metabolic engineering attempts focused on single-gene manipulations have only achieved moderate increases in levels of storage lipid accumulation and protection in crop plants (Andrianov et al., 2010; Sanjaya et al., 2013). Similarly, when TAG expression is introduced in leaves (e.g., *Arabidopsis thaliana* leaves), because they are not protected with oleosin-coated vesicles, its levels remain limited about 1% of total lipids (Yang and Ohlrogge, 2009). Hence, any attempt to increase the TAG levels in vegetative tissues by metabolic engineering is hindered by this lack of protection of the product. Recent successful application of multi-gene expression approaches have been demonstrated with DGAT1 and oleosin in *A. thaliana* (Winichayakul et al., 2013); WRI1 and DGAT1 in *Nicotiana benthamiana* (Vanhercke et al., 2013); and WRI1, DGAT1, and oleosin in *Nicotiana tabacum* (Vanhercke et al., 2014), which significantly increased leaf TAG accumulation.

3.5 BIODIESEL FROM TAG VIA TRANSESTERIFICATION

Vegetable oils are thought to become a potential replacement to fossil fuels but they are very high in viscosity and reduce its application as a biodiesel. Therefore, various ways have been found to decrease the viscosity of these oils. There are four major techniques being used for this purpose: dilution, microemulsification, pyrolysis, and transesterification

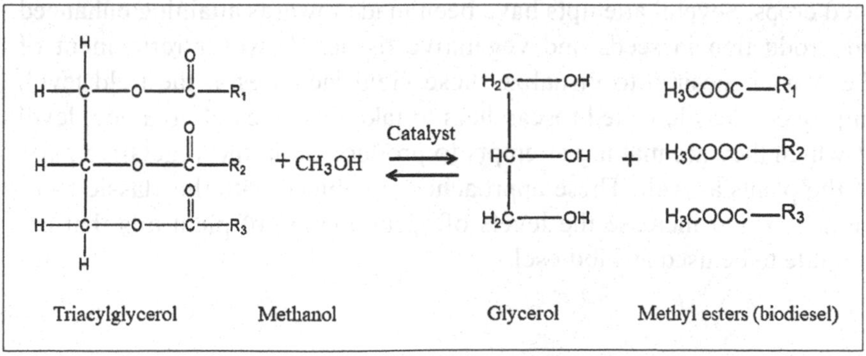

$$\text{Triacylglycerol} + CH_3OH \underset{}{\overset{\text{Catalyst}}{\rightleftharpoons}} \text{Glycerol} + 3\ H_3COOC\text{—}R_{1,2,3}$$

Triacylglycerol Methanol Glycerol Methyl esters (biodiesel)

FIGURE 3.3 Conversion of TAG to biodiesel by the process of transesterification.

(Demirbas, 2008). Out of these, the most common method is transesterification which is the process of conversion of oil into its corresponding fatty acids (Bala, 2005). This reaction of an oil triglyceride with an alcohol, in the presence of a catalyst, ultimately forms esters and glycerol, and therefore is also known as alcoholizes (Fig. 3.3). In this reaction of biodiesel synthesis, the catalyst used is sodium hydroxide which is used to split the oil molecules and an alcohol is used to separate the ethanol or methanol produced. The by-product of this reaction is glycerin (Demirbas, 2008).

3.6 CONCLUSION

The increase in population which has subsequently led to the increase in the quantity of the fossil fuels consumed has ultimately developed a need for renewable sources which could provide a long-term availability. A sustainable source of energy, called biodiesel, has come up as a potential alternative of fossil fuels. Biodiesel is produced from renewable and domestic sources, which not only play a significant role in providing energy for transportation but also contribute in reducing the CO_2 production. But limited supply of vegetable oil-producing feedstock has created a huge hindrance in achieving these goals. Therefore, in order to clear this hindrance, production of plant oil should be increased without worsening the global warming scenario or decreasing the availability of food crops. This can be achieved by several genetic engineering measures. A thorough understanding of the processes involved in TAG biosynthesis has advanced our attempts in producing improved biodiesel from oil seed crops. Several attempts have been made towards attaining enhanced oil production in seeds and vegetative tissues, however refinement of the work is desired to visualize these yield increases at the field level. Improved ideas have led researchers to take this research to a new level at which they are making attempts to produce oil in the vegetative parts of the plants as well. These approaches, combined with the classic techniques, would increase the levels of accumulation of plant oils that are suitable to be used as biodiesel.

KEYWORDS

- **triacylglycerol**
- **oleosin**
- **diacylglycerol acyltransferase**
- **genetic engineering**
- **biodiesel**
- **biofuel**

REFERENCES

Andrianov, V.; Borisjuk, N.; Pogrebnyak, N.; Brinker, A.; Dixon, J.; Spitsin, S.; Flynn, J.; Matyszczuk, P.; Andryszak, K.; Laurelli, M.; Golovkin, M. Tobacco as a Production Platform for Biofuel: Overexpression of *Arabidopsis* DGAT and LEC2 Genes Increases Accumulation and Shifts the Composition of Lipids in Green Biomass. *Plant Biotechnol. J.* **2010,** *8*, 277–287

Demirbas, A. Biodiesel from TAG via Transesterification. In *Biodiesel*. Springer: London, **2008;** pp 121–140

Bala, B. K. Studies on Biodiesels from Transformation of Vegetable Oils for Diesel Engines. *Energy Edu Sci Technol.* **2005,** *15*, 1–43.

Bates, P. D.; Stymne, S.; Ohlrogge, J. Biochemical Pathways in Seed Oil Synthesis. *Curr. Opin. Plant Biol.* **2013,** *16*, 358–364.

Baud, S.; Lepiniec, L. Physiological and Developmental Regulation of Seed Oil Production. *Prog. Lipid Res.* **2010,** *49*, 235–249.

Bewley, J. D.; Black, M. Seeds: Germination, Structure and Composition. *Seeds Physiol. Develop. Germination* **1994,** 1–33.

Bewley, J. D . Seed Germination and Reserve Movilization. *Encyclopedia of Life Sciences;* Nature Publishing Group: London, United Kingdom, 2001.

Beevers, H. The Role of the Glyoxylate Cycle. *Biochem. Plants* **1980,** *4*, 117–130

Bhunia, R. K.; Chakraborty, A.; Kaur, R.; Maiti, M. K.; Sen, S. K. Enhancement of α-linolenic Acid Content in Transgenic Tobacco Seeds by Targeting a Plastidial ω-3 Fatty Acid Desaturase (*fad7*) Gene of *Sesamum indicum* to ER. *Plant Cell Rep.* **2016,** *35*, 213–226.

Cernac, A.; Andre, C.; Hoffmann-Benning, S.; Benning, C. WRI1 Is Required for Seed Germination and Seedling Establishment. *Plant Physiol.* **2006,** *141*, 745–757.

Dabbs, P. Identification of Plant Transcription Factors That Play a Role in Triacylglycerol Biosynthesis. *Electronic Theses and Dissertations,* **2015.** Paper 2496.

Dombkowski, A. A.; Sultana, K. Z.; Craig, D. B. Protein Disulfide Engineering. *FEBS Lett.* **2014,** *588*, 206–212.

Durrett, T. P.; Benning, C.; Ohlrogge, J. Plant TAGs as Feed Stocks for the Production of Biofuels. *Plant J.* **2008,** *54*, 593–607.

Ghosal, A.; Banas, A.; Stahl, U.; Dahlqvist, A.; Lindqvist, Y.; Stymne, S. *Saccharomyces cerevisiae* Phospholipid:Diacylglycerol Acyl Transferase (PDAT) Devoid of Its Membrane Anchor Region is a Soluble and Active Enzyme Retaining Its Substrate Specificities. *Biochimica et Biophysica Acta: Mol. Cell Biol. Lipids* **2007,** *1771*, 1457–1463

Graham, I. A. Seed Storage Oil Mobilization. *Annu. Rev. Plant Biol.* **2008,** *59*, 115–142.

Kagaya, Y.; Toyoshima, R.; Okuda, R.; Usui, H.; Yamamoto, A.; Hattori, T. Leafy Cotyledon1 Controls Seed Storage Protein Genes Through Its Regulation of FUSCA3 and Abscisic Acid Insensitive3. *Plant Cell Physiol.* **2005,** *46*, 399–406.

Kelly, A. A.; van, Erp H.; Quettier, A.-L.; Shaw, E.; Menard, G.; Kurup, S.; Eastmond, P. J. The Sugar-Dependent1 Lipase Limits TAG Accumulation in Vegetative Tissues of *Arabidopsis*. *Plant Physiol.* **2013,** *162*, 1282–1289.

Kim, H. U.; Jung, S. J.; Lee, K. R.; Kim, E. H.; Lee, S. M.; Roh, K. H.; Kim, J. B. Ectopic Overexpression of Castor Bean LEAFY COTYLEDON2 (LEC2) in *Arabidopsis* Triggers the Expression of Genes That Encode Regulators of Seed Maturation and Oil Body Proteins in Vegetative Tissues. *FEBS Open Bio.* **2014,** *4*, 25–32.

Lee, G. H.; Kim, H.; Suh, M. C.; Kim, H. U.; Seo, P. J. The MYB96 Transcription Factor Regulates Triacylglycerol Accumulation by Activating DGAT1 and PDAT1 Expression in *Arabidopsis* Seeds. *Plant Cell Physiol.* **2018,** *59*, 1432–1442.

Maeo, K.; Tokuda, T.; Ayame, A.; Mitsui, N.; Kawai, T.; Tsukagoshi, H.; Ishiguro, S.; Nakamura, K. An AP2-type Transcription Factor, WRINKLED1, of *Arabidopsis thaliana* Binds to the AW-box Sequence Conserved Among Proximal Upstream Regions of Genes Involved in Fatty Acid Synthesis. *Plant J.* **2009,** *60*, 476–487.

Manan, S.; Chen, B.; She, G.; Wan, X.; Zhao, J. Transport and Transcriptional Regulation of Oil Production in Plants. *Crit. Rev. Biotechnol.* **2017,** *3*, 641–655.

Maisonneuve, S.; Bessoule, J.-J.; Lessire, R.; Delseny, M.; Roscoe, T. J.. Expression of Rapeseed Microsomal Lysophosphatidic Acid Acyltransferase Isozymes Enhances Seed Oil Content in *Arabidopsis*. *Plant Physiol.* **2010,** *152*, 670–684.

Mu, J.; Tan, H.; Zheng, Q.; Fu, F.; Liang, Y.; Zhang, J.; Yang X.; Wang T.; Chong K.; Wang X-J. Leafy Cotyledon1 Is a Key Regulator of Fatty Acid Biosynthesis in *Arabidopsis*. *Plant Physiol.* **2008,** *148*, 1042–1054.

Murphy, D. J. The Biogenesis and Functions of Lipid Bodies in Animals, Plants and Microorganisms. *Prog. Lipid Res.* **2001,** *40*, 325–438.

Naik, S. N.; Goud, V. V.; Rout, P. K.; Dalai, A. K. Production of First and Second Generation Biofuels: A Comprehensive Review. *Renew. Sustain. Energy Rev.* **2010,** *14*, 578–597.

Pracharoenwattana, I.; Zhou, W.; Smith, S. M. Fatty Acid Beta-oxidation in Germinating *Arabidopsis* Seeds Is Supported by Peroxisomal Hydroxypyruvate Reductase When Malate Dehydrogenase Is Absent. *Plant Mol. Biol.* **2010,** *72*, 101–109.

Quettier, A. L.; Eastmond, P. J. Storage Oil Hydrolysis During Early Seedling Growth. *Plant Physiol. Biochem.* **2009,** *47*, 485–490.

Rani, S. H.; Krishna, T. H. A.; Saha, S.; Negi, A. S.; Rajasekharan, R. Defective in Cuticular Ridges (DCR) of *Arabidopsis thaliana*, a Gene Associated with Surface Cutin

Formation, Encodes a Soluble Diacylglycerol Acyltransferase. *J. Biol. Chem.* **2010,** *285*, 38337–38347.

Sanjaya, M. R.; Durrett, T. P.; Kosma, D. K.; Lydic, T. A.; Muthan, B.; Koo, A. J.; Bukhman, Y. V.; Reid, G. E.; Howe, G. A.; Ohlrogge, J.; Benning, C. Altered Lipid Composition and Enhanced Nutritional Value of *Arabidopsis* Leaves Following Introduction of an Algal Diacylglycerol Acyltransferase 2. *Plant Cell* **2013,** *25*, 677–693.

Santos-Mendoza, M.; Dubreucq, B.; Baud, S.; Parcy, F.; Caboche, M.; Lepiniec, L. Deciphering Gene Regulatory Networks That Control Seed Development and Maturation in *Arabidopsis*. *Plant J.* **2008,** *54*, 608–620.

Shen, W.; Li, J. Q.; Dauk, M.; Huang, Y.; Periappuram, C.; Wei, Y.; Zou, J. Metabolic and Transcriptional Responses of Glycerolipid Pathways to a Perturbation of Glycerol 3-phosphate Metabolism in *Arabidopsis*. *J. Biol. Chem.* **2010,** *285*, 22957–22965.

Shockey, J. M. Tung Tree DGAT1 and DGAT2 Have Nonredundant Functions in TAG Biosynthesis and Are Localized to Different Subdomains of the Endoplasmic Reticulum. *Plant Cell Online* **2006,** *18*, 2294–313.

Stone, S. L.; Braybrook, S. A.; Paula, S. L.; Kwong, L. W.; Meuser, J.; Pelletier, J.; Hsieh, T. F.; Fischer, R. L.; Goldberg, R. B.; Harada, J. J. *Arabidopsis* Leafy Cotyledon 2 Induces Maturation Traits and Auxin Activity: Implications for Somatic Embryogenesis. *Proc. Natl. Acad. Sci.* **2008,** *105*, 3151–3156.

Tan, H.; Yang , X.; Zhang, F.; Zheng, X.; Qu, C.; Mu, J.; Fu, F.; Li, J.; Guan, R.; Zhang, H.; Wang, G. Enhanced Seed Oil Production in Canola by Conditional Expression of *Brassica napus* Leafy Cotyledon1 and Lec1-like in Developing Seeds. *Plant Physio.* **2011,** *153*, 1577–1588.

Tang, G.; Xu, P.; Ma, W.; Wang, F.; Liu, Z.; Wan, S.; Shan, L. Seed-Specific Expression of AtLEC1 Increased Oil Content and Altered Fatty Acid Composition in Seeds of Peanut (*Arachis hypogaea* L.). *Front. Plant. Sci.* **2018,** *9*, 260.

Thelen, J. J.; Ohlrogge , J. B. Metabolic Engineering of Fatty Acid Biosynthesis in Plants. *Metab Eng.* **2002,** *4*, 12–21

Timilsina, G. R. Biofuels in the Long-run Global Energy Supply Mix for Transportation. *Philos. Trans. Royal Soc. A Math. Phys. Eng. Sci. A* **2014,** *372, 2012032.*

To, A.; Valon, C.; Savino, G.; Guilleminot, J.; Devic, M.; Giraudat, J.; Parcy, F. A Network of Local and Redundant Gene Regulation Governs *Arabidopsis* Seed Maturation. *Plant Cell* **2006,** *18*, 1642–1651.

To, A.; Joubès , J.; Barthole, G.; Lécureuil, A.; Scagnelli, A.; Jasinski, S.; Lepiniec, L.; Baud, S. WRINKLED Transcription Factors Orchestrate Tissue-specific Regulation of Fatty Acid Biosynthesis in *Arabidopsis*. *Plant Cell* **2012,** *24*, 5007-5023.

Vanhercke, T.; El Tahchy, A.; Shrestha, P.; Zhou, X. R.; Singh, S. P.; Petrie, J. R. Synergistic Effect of WRI1 and DGAT1 Coexpression on Triacylglycerol Biosynthesis in Plants. *FEBS Lett.* **2013,** *587*, 364–369.

Vanhercke, T.; El Tahchy, A.; Liu, Q.; Zhou, X. R.; Shrestha, P.; Divi, U. K.; Ral, J. P.; Mansour, M. P.; Nichols, P. D.; James, C. N.; Horn, P. J. Metabolic Engineering of Biomass for High Energy Density: Oilseed-like Triacylglycerol Yields from Plant Leaves. *Plant Biotechnol. J.* **2014,** *12*, 231–239.

Voelker, T.; Kinney, A. J. Variations in the Biosynthesis of Seed-storage Lipids. *Ann. Rev. Plant Biol.* **2001,** *52*, 335–361.

Weselake R.; Zou J.; Taylor D. Plant TAG Synthesis. http://lipidlibrary. aocs.org/Biochemistry/content.cfm?ItemNumber=40314

Winichayakul, S.; Scott, R. W.; Roldan, M.; Hatier, J. H.; Livingston, S.; Cookson, R.; Curran, A. C.; Roberts, N. J.; In Vivo Packaging of Triacylglycerols Enhances *Arabidopsis* Leaf Biomass and Energy Density. *Plant Physiol.* **2013,** *162*, 626–639.

Yang, Z.; Ohlrogge, J. B. Turnover of Fatty Acids During Natural Senescence of *Arabidopsis*, *Brachypodium*, and Switchgrass and in *Arabidopsis* β-oxidation Mutants. *Plant Physiol.* **2009,** *150*, 1981–1989.

Zhang, M.; Cao, X.; Jia, Q.; Ohlrogge, J. FUSCA3 Activates Triacylglycerol Accumulation in *Arabidopsis* Seedlings and Tobacco BY2 cells. *Plant J.* **2016,** *88*, 95–107.

Zheng, Z.; Xia, Q.; Dauk, M.; Shen, W.; Selvaraj, G.; Zou, J. *Arabidopsis* AtGPAT1, a Member of the Membrane-bound Glycerol-3-phosphate Acyltransferase Gene Family, Is Essential for Tapetum Differentiation and Male Fertility. *Plant Cell* **2003,** *5*, 1872–1887.

CHAPTER 4

Industrial Technologies for Bioethanol Production from Lignocellulosic Biomass

AMRITA SAHA*, SOUMYAK PALEI, MINHAJUL ABEDIN, and BHASWATI UZIR

Department of Environmental Science, Amity Institute of Environmental Science, Amity University, Kolkata, West Bengal, India

**Corresponding author. E-mail: asmicrobio@yahoo.in*

ABSTRACT

Research and development in recent years have led to the sustainable production of biofuels and have aimed toward reduction in greenhouse gas production. Lignocellulosic waste has been the most abundant agricultural residue for the production of second-generation bioethanol. Fermentable sugars residing in lignocellulosic wastes and energy crops have been one of the important sources for producing gallons of fuel-grade ethanol per year across the world. It comprises different physical, chemical, and biological processes for pretreatment of various lignocellulosic biomasses aiming at removal of lignin and conversion of cellulose and hemicellulose into reducing sugars for the production of bioethanol or other value-added products. Bioconversion technologies exploit the use of microbial enzymes in large scale ensuring eco-friendly and cost-effective production of biofuel. A great deal of research has been directed in this field in order to improve the overall process economics of pretreatment and fermentation technologies for conversion of lignocellulosic biomass in bioethanol. This would address one of the greatest problems of this century by meeting the growing energy demand without posing threat to the environment. Recent trend has been converting waste to wealth, hence, production of bioethanol from waste residues would also add to the mission of "Swachh Bharat."

4.1 INTRODUCTION

The demand for petroleum-dependent chemicals and materials has been increasing despite the reduction or dwindling of their fossil resources. As the dead end of petroleum-based industry has started to appear, today's modern society has to implement alternative energy and search for cost-effective biofuel products. Biofuels emerged as an ideal option to meet the requirement in sustainable manner. This chapter is based on the ligno-cellulosic biomass, a most abundant and bio-renewable biomass, as an alternative platform to fossil resources.

Since no economically viable technologies are available for their conversion, most is burned by farmers in the field, not only polluting the environment but also causing other problems such as the transportation of the polluted air. It has been acknowledged worldwide that agricultural residues are one of the best choices to replace grains for fuel ethanol production without endangering food security.

Lignocellulosic biomass is a promising feedstock for biofuel production, due to its low cost, vast abundance, and sustainability. Bioethanol is expensive but the lignocellulosic starch-based feedstocks such as corn, wheat, and cassava used for biofuel are easily available and the rate is low. Currently, there are five fuel ethanol producers across the country, producing 1.52 million tons. Several countries have initiated new alternatives for gasoline from renewable feedstocks. In the North American hemisphere, bioethanol has been extracted from starch sources such as corn while in the South American hemisphere, biofuel has been largely produced from sugars including sugarcane and sugar beets. In the United States, biofuel derived from corn has emerged as one of the primary raw materials for bioethanol production. According to the Renewable Fuels Association statistics, the production of bioethanol was 10.9 billion gallons (41.26 billion liters) in the year 2009 representing 55% of the worldwide production and 12.82 billion gallons (48.52 billion liters) in the year 2010. European countries are putting extensive efforts to increase their 5% worldwide bioethanol production. Although, most of the remaining countries in the world collectively account for only 5% of the global bioethanol production, China, Thailand, as well as India are continuing to invest substantially in agricultural biotechnology and emerge as potential biofuel producers.

According to the research results of the International Energy Agency (IEA), biofuels could provide 27% of total transport fuel by 2050, and contribute in particular to the replacement of diesel, kerosene, and jet fuel. The projected use of biofuels could avoid around 2.1 gigatons of CO_2 emission per year if produced sustainably. There are some disadvantages of chemical conversion where biological conversion is dominating such as it can be harmful to the environment—before the sources are converted into useful energy, combustion is required and this can produce harmful by-products which can contribute to pollution, it comes with a high cost—this is another practical disadvantage of using chemical energy where the production of nuclear energy needs a lot of money to invest and set up nuclear power stations, it can produce radioactive waste—this is particularly the case with regards to nuclear power production where the waste produced by reactors needs to be disposed of properly in a secure area, since it is extremely hazardous and can leak radiations if not stored properly, it is nonrenewable—most sources of chemical energy cannot be replenished, it is not good for human health—considering the fact that chemical energy is highly combustible, it is expected to contribute to pollution, which causes various illnesses and is never good for human health. In order to prove biological conversion better, some advantages are as follows: industrial chemicals (organic acids, acetic acids, gibberellic acids, and biopolymers), food additives (amino acids, nucleosides, vitamins, fats, and oils), health care products (antibiotics, steroid, vaccines, and monoclonal antibodies), and industrial enzymes (amylases, proteases, and diastases).

In 2011, roughly 78% of energy consumed in the world was from fossil fuel, 3% from nuclear energy, and the remaining 19% from renewable energy. About 13% of the renewable energy is harnessed from carbon-rich bio-based materials available on earth. Currently, the majority of biofuels are produced using sugars extracted from agricultural feedstock or by converting starch into sugars primarily from edible grains.

In order to attain sustainability, lignocellulosic biomass is used as biofuel because of cost-effective and fast production of feedstocks. There are contradictory points on disturbance to national food supply but proper stock market arrangement can easily mitigate this problem. Very soon we will be seeing less global temperature and advance transportation with no pollution emitting to atmosphere.

As the world population is increasing rapidly, the demand of energy is also increasing and by 2035 the global energy consumption is likely to increase by 40–42%, the depletion of fossil fuels at a high rate and the major environmental problems caused due to burning of fossil fuel (greenhouse effect, global warming, acid rain, etc.) attracted the interest in nonconventional fuel mainly bioethanol, which is obtained from lignocellulosic biomass. The lignocellulosic biomass refers to the plant biomass that is composed of cellulose, hemicelluloses, and lignin. It is increasingly recognized as alternative to petroleum for the production of fuels. Bio fuels can be obtained from agricultural residues (corn stover, straw, etc.), forestry residues, biowaste, etc. A recent survey suggests that the world's annual biomass yield contains inherent energy to contribute to 20–100% of world annual energy consumption. For the past few years, biomass based on bioethanol has caught the attraction of global industries. Because, ethanol burns more cleanly, the use of ethanol-blended fuels can reduce the net emissions of greenhouse gases, it is considered renewable energy source and it is also biodegradable and less toxic than fossil fuels. Ethanol-blended fuel is widely used in Brazil, United States, and European countries. Brazil has replaced almost 42% of its gasoline needs with ethanol. So, bioethanol is emerging as a global biofuel.

4.2 GLOBAL SCENARIO

Lignocellulosic bacteria are harsh substrates which partly ferment due to the lignin degradation and presence of acids. Lignocellulose has diverse group of substrates and the amount of inhibitors are highly dependent on the origin of the material as well as the pretreatment method. In 1980, the projected cost of producing ethanol from lignocellulosic bacteria was $0.95/liter. Many investigations in bacterial contamination during industrial fermentation have been conducted, but a few have dealt with lignocellulosic material as the fermentation substrate which hamper ethanol production and biomass formation in the fermenting organism.

Bacterial contaminations decrease the ethanol yield in commercial lignocellulose-based fermentation plants. One organism was focused named as *Saccharomyces cerevisiae*, more commonly known as "baker's yeast." It has been shown that *S. cerevisiae* have a greater potential to produce ethanol if adapted to the fermentation media prior to fermentation.

Biofuels are energy sources which are made from recently grown biomass (plant or animal matter) but petroleum and coal have been used primarily as energy sources because of their high-energy potential and cheap prices. Fossil fuels such as coal and petroleum also come from biomass, but the difference is that they take millions of years to produce. Biofuels are making transformation to increasing oil prices and the desire to have a renewable source of energy to mitigate the effects of climate change.

There are three types of biofuels: first-, second-, and third-generation biofuels. They are characterized by their sources of biomass, their limitations as a renewable source of energy, and their technological progress. The main drawback of first-generation biofuels is that they come from biomass that is also a food source. This presents a problem when there is not enough food to feed everyone. Second-generation biofuels come from non-food biomass, but still compete with food production for land use. Finally, third-generation biofuels present the best possibility for alternative fuel because they do not compete with food. However, there are still some challenges in making them economically feasible.

4.3 FIRST-GENERATION BIOFUELS

They are also known as conventional biofuels made up of sugar, starch, or vegetable oil. First-generation biofuels are produced through good technologies and processes, such as fermentation, distillation, and transesterification. These processes have been used for hundreds of years. Sugars and starches are fermented to produce small quantities of ethanol, butanol, and propanol. Ethanol has one-third of the energy density of gasoline but is currently used in many countries as an additive to gasoline. The benefit of ethanol is that it burns cleaner than gasoline and therefore, causes less pollution. In addition, first-generation biofuel known as biodiesel is produced when plant oil or animal fat goes through a process called transesterification. This process involves exposing oils with an alcohol such as methanol in the presence of a catalyst and the distillation process involves separating the main product from the by-product. Biodiesel can be used in place of petroleum diesel in many diesel engines or in a mixture of the two. But this generation biofuels have several disadvantages which possess threat to food prices since the biomass used are food crops such as corn and sugar beet. First-generation biofuel production has contributed to increase in world food price. They also show negative impact on biodiversity and competition for water in some regions and this generation biofuels are more expensive than gasoline, making them economically unfavorable. Because of this second generation was initiated.

4.4 SECOND-GENERATION BIOFUELS

The biomass sources in this generation include wood, organic waste, food waste, and specific biomass crops. The fast-growing trees need to undergo a pretreatment step, which is a series of chemical reactions that break down the lignin, in order to make fuel. This pretreatment involves thermochemical or biochemical reactions which unlock the sugar embedded in fibers. Additionally, the forest residues can go through a thermochemical step which produces syngas (a mixture of carbon monoxide, hydrogen, and other hydrocarbons). Hydrogen can be used as a fuel and the other hydrocarbons can be used as additives to gas oil. But it addresses many issues associated which do not compete between fuels and food crops since they come from distinct biomass. Second-generation biofuels also generate higher energy yields per acre than first-generation fuels. They were given very poor quality land where food crops may not be able to grow. The technology was not good enough, but still had potential to reduce the cost and increased production efficiency. However, some biomasses for second-generation biofuels still compete with land use since some of the biomass grow in the same climate as food crops. This leaves farmers and policy makers with the hard decision of which crop to grow. However, use of corn would take away many nutrients from the soil and would need to be replenished through fertilizer. In addition, the second-generation fuel production process is more elaborate than first-generation biofuels because it requires pretreating the biomass to release the trapped sugars. This requires more energy and materials.

4.5 THIRD-GENERATION BIOFUELS

This generation biofuels use specially engineered crops such as algae as the energy source and are more energy dense than first- and second-generation biofuels. They are cultured as low-cost, high-energy, and complete source of renewable energy. Algae can grow in areas which are unsuitable for first- and second-generation crops, which would relieve stress on water and using arable land. It can be grown using sewage, wastewater, and saltwater, such as oceans or salt lakes. One of the favorable properties of algae is the diversity of ways in which algae can be produced or cultivated, few are as follows: open ponds where algae grow in an open air in normal pond which have low capital costs but are less efficient than other methods. Next is closed loop system where pond is not exposed to the atmosphere and sterile source of carbon is used. Algae can produce fuels such as biodiesel, butanol, gasoline, methane, ethanol, vegetable oil, and jet fuel. Because of this, there is no requirement of any water that is used for human consumption. These algae are grown and harvested to extract oil within them to convert into biodiesel through a similar process as first generation biofuels, or it can be refined into other fuels as replacements to petroleum-based fuels. However, further research is to be done to know the extraction process in order to make it financially competitive in this petroleum-based fuel market.

4.6 FOURTH-GENERATION BIOFUELS

These are not only aimed in producing sustainable energy but also to find a way for capturing and storing CO_2. Biomass materials, which have absorbed CO_2, are converted into fuel using the same processes as second-generation biofuels. This process differs from second- and third-generation production because at all stages of production the carbon dioxide is captured using processes such as oxy-fuel combustion. The carbon dioxide then can be captured by storing it in old oil and gas fields or in saline aquifers. This carbon capture makes fourth-generation biofuel production carbon negative rather than simply carbon neutral, as it controls in producing carbon. This system not only captures and stores carbon dioxide from the atmosphere but it also reduces CO_2 emissions by replacing fossil fuels.

4.7 TECHNOLOGIES AND PROCESSES

There are many technologies and processes from where ethanol is being produced such as molecular adaptation mechanism by ethanologenic bacteria, hydrothermal pretreatment of switchgrass or corn Stover, bacterial culture media, microbial synthesis of cellulases, enzymatic hydrolysis, saccharification, fermentation of ethyl alcohol, SLAMSYSTEM simulation network has been developed to represent forest biomass units flow from felling to storage at a mill yard, etc.

Some of the important and large conversion processes are as follows:

Combustion: It is the most common biomass conversion technology which is applied for household- and industrial-level conversion since ancient time. Over the last decade, modern biomass combustion technologies have emerged with automated pellet boilers, co-firing, heat and power production, etc.

Gasification: It offers the advantage to produce homogeneous fuel from an inhomogeneous solid fuel for secondary conversion. Biomass gasification is an endothermic thermal conversion technology where a solid fuel is converted into a combustible gas. A limited supply of steam, oxygen, air or a combination serves as the oxidizing agent. The product, gas, consists of hydrogen, methane, carbon monoxide, carbon dioxide, trace amounts of water, nitrogen, higher hydrocarbon, and various contaminants, such as small char particles, ash, tars, higher alkalis, acids, ammonia, and hydrocarbons.

Fast pyrolysis: It transforms difficult-to-handle biomass of different nature into a clean and uniform liquid called pyrolysis oil. A robust fast pyrolysis process was developed which results in a large fraction of pyrolysis oil, which can be used for heat, chemical production, power, and transport fuel.

Carbonization and torrefaction: Charcoal is produced from wood using the most common carbonization technology, but the cotton stalks can also be carbonized and further upgraded to household fuels. Torrefaction is a partial carbonization process at 200–400°C which makes the biomass crispy. The torrefied biomass is suitable for co-firing in coal-fired power plants. Compared to carbonized biomass, a higher percentage of initial energy content of biomass remains in the product.

Anaerobic digestion: It is the production of a methane-rich biogas from wet biomass sources such as manure, kitchen and garden waste, wastewater, etc. the biogas can be used for heat and power generation using gas engines or upgraded for use in the natural gas grid.

4.8 ECONOMIC PERSPECTIVE

To find an energy source that is sufficient to provide our energy needs and at the same time, this energy source must be dependable, renewable, and non-contributing to climate change is very much difficult. First-generation biofuels initiated a step toward cleaner and renewable energy resource, but they lag behind gasoline because of energy density and economic factors. They also present a report in regards to use food crops, as there are millions of people starving around the world and countries with large populations where corn is grown such as in China, Brazil, and Mexico will face lots of problems. Then, second-generation biofuels provided some benefits, but the biomass requires pretreating procedures and competes with food crops over arable land in various parts of the world which leads to its failure. Then, third-generation biofuels show the most hope in reaching to conclusion, but plenty of research still needs to be done to reduce the cost of the production and make the fuel production commercially viable.

The most important drivers of the biofuel market are:

- security of energy supply,
- reducing the emission of greenhouse gases, and
- instability of the fossil fuel prices.

The recent innovations in the first-generation and second-generation biofuels herald a long-term emphasis on the energy sustainability and efficiency. As of 2014–2015, biofuels have enjoyed a regular and assured growth in the United States, Brazil, and Europe.

By 2014, the biodiesel production in the United States was 135 million gallons with a capacity of 2.1 billion gallons per year. In 2005, ethanol net production cost was $0.40 per energy equivalent liter (EEL) of gasoline, while wholesale gasoline prices averaged $0.44/liter. Estimated soybean biodiesel production cost was $0.51 per diesel EEL, whereas diesel wholesale prices averaged $0.46/liter. Further increases in petroleum prices above 2005 average prices improve the cost competitiveness for biofuels. Ethanol and biodiesel producers also benefit from federal crop subsidies that lower corn prices (which are approximately half of ethanol production's operating costs) and soybean prices. Some of the rates are as follows: petroleum $3.30 per gallon, ethanol $2.086 per gallon, biodiesel $2.84 per gallon, etc.

Biofuels became one of the most promising sources of fuel supply due to their high-quality engine performance, environment friendly in nature, economic development, and agricultural development (i.e., plantation for biomass). The production of biofuels can play a major role in the economic development of a country. According to biotechnology Industry

Organization, by 2022 the biofuel industry is expected to create 1,900,000 direct jobs and 6,100,000 indirect jobs.

4.9 CONCLUSION

Ethanol is an excellent fuel for transportation purpose which can be blended with gasoline either directly or indirectly to reduce greenhouse gases, ozone-forming compounds, and carbon monoxide, or used as a neat and clean fuel to reduce the release of ozone-forming compounds. In both instances, ethanol can be used to improve urban air quality. Because enough waste materials and energy crops can be made available to produce sufficient ethanol to replace all gasoline used in the United States, the potential ethanol supply from biomass is substantial, and it is in large-scale application which would thereby reduce the strategic vulnerability which makes disruptions in oil supply. Technologies got advance significantly for the conversion of lignocellulosic biomass into ethanol, so that the projected price of ethanol from lignocellulosic biomass became competitive with ethanol derived from corn. Furthermore, opportunities have been identified to reduce the cost of ethanol production.

In many developing countries, the framework conditions needed to set up a second-generation biofuel industry are not currently sufficient. The main obstacles that need to be overcome include poor infrastructure, lack of skilled labor, and limited financing possibilities. Agricultural and forestry residues should be the feedstock of choice in the initial stage of the production since they are readily available and do not require additional land cultivation.

The review undertaken in this chapter raises the following issues and findings: the contribution of biomass cost to the overall production cost of lignocellulosic bioethanol proves to be high. So, the standard production cost estimation should be replaced by an approach which covers up value engineering, value resource, and target costing and due to the complexity of the techno-economic evaluation of lignocellulosic ethanol, the perceived risks by private investors will be high.

Strategies to decrease these risks include promoting such projects as integration of second-generation with first-generation bioethanol and thus use existing residues and share equipment.

Sugarcane bagasse is particularly concerned with such a strategy.

Lignocellulosic biorefineries that aim at decreasing the production cost of bioethanol will be attractive only if the perceived risks by the investors are affordable.

KEYWORDS

- **lignocellulosic biomass**
- **bioethanol production**
- **industrial technology**
- **bioconversion**
- **waste material**

REFERENCES

Álvarez, C.; Reyes-Sosa, F.; Diez, B. Enzymatic Hydrolysis of Biomass from Wood. *Microb. Biotechnol.* **2016,** *9*, 149–156.

Antibiotic Resistant Bacteria in Fuel Ethanol Fermentations. Ethanol Producer Magazine, May 2005 Issue. http://www.ethanolproducer.com/article-print.jsp?article_id=511 (accessed Dec 8, 2009).

Boland, S.; Unnasch, S. *GHG Reductions from the RFS2. Life Cycle Associates Report LCA.*6075.116.2015; Prepared for Renewable Fuels Association, 2015.

Borregaard. www.borregaard.com (accessed March 2011).

Brownell, H. H.; Yu, E. K. C.; Saddler, N. Steam-explosion Pretreatment of Wood: Effect of Chip Size, Acid, Moisture Content and Pressure Drop. *Biotechnol. Bioeng.* **1986,** *28*, 792–801.

Domsjö Fabriker. www.domsjoe.com (accessed March 2011).

Hertel , T. W.; Steinbuks, J.; Tyner, W. E. What is the Social Value of Second Generation Biofuels? *Appl. Econ. Perspect. Policy* **2016,** *38*, 599–617.

Lallemand Ethanol Technology. www.lallemandbds.com (accessed June 21, 2013).

Ragauskas, A.; Beckham, G.; Biddy, M.; Chandra, R.; Chen, F.; Davis, M., et al. Lignin Valorization: Improving Lignin Processing in the Biorefinery. *Science* **2014,** *344*, 6185.

Ramos, J. L.; Valdivia, M.; Garcia-Lorente, F.; Segura, A. Benefits and Perspectives on the Use of Biofuels. *Microb. Biotechnol.* **2016,** *9*, 436–440.

Sims, R. E. H.; Mabee, W.; Saddler, J.; Taylor, M. An Overview of Second Generation Biofuel Technologies. *Bioresour. Technol.* **2009,** *101*, 1570–1580.

Stelte, W. *Steam Explosion for Biomass Pre-treatment*; Danish Technological Institute: Denmark, 2013.

Tadesse, H.; Luque, R. Advances on Biomass Pretreatment Using Ionic Liquids: An Overview. *Energy Environ. Sci.* **2011,** *4*, 3913–3929.

Tolonen, A.; Zuroff, T.; Ramya, M.; Boutard, M.; Cerisy, T.; Curtis, W. Physiology, Genomics, and Pathway Engineering of an Ethanol-tolerant Strain of *Clostridium phytofermentans*. *Appl. Environ. Microbiol.* **2015,** *81*, 5440–5448.

UNCTAD, United Nations Conference on Trade and Development, 2016. Second Generation Biofuel Markets: State of Play, Trade and Developing Country Perspectives. UNCTA/DITC/TED/2015/8 United Nations Publication, p. 61. <http://unctad.org/en/PublicationsLibrary/ditcted2015d8_en.pdf> (accessed May 6, 2016).

U.S. Department of Energy. *U.S. Billion-ton Update: Biomass Supply for a Bioenergy and Bioproducts Industry*; Perlack, R. D., Stokes, B. J., Eds.; Oak Ridge National Laboratory: Oak Ridge, TN, 2011; pp 227 (ORNL/TM-2011/224).

Yang, B.; Wyman, C. Pretreatment: The Key to Unlocking Low-cost Cellulosic Ethanol. *Biofuels Bioprod. Bioref.* **2008,** *2*, 26–40.

CHAPTER 5

Production of Bio-Syngas for Biofuels and Chemicals

SHRITOMA SENGUPTA[1,2], DEBMALLYA KONAR[1], DEBALINA BHATTACHARYA[3], and MAINAK MUKHOPADHYAY[1,*]

[1]*Department of Biotechnology, JIS University, Kolkata, West Bengal, India*

[2]*Department of Biochemistry, University of Calcutta, Kolkata, West Bengal, India*

[3]*Department of Biotechnology, University of Calcutta, Kolkata, West Bengal, India*

[*]*Corresponding author. E-mail: m.mukhopadhyay85@gmail.com.*

ABSTRACT

Biofuels are gaining importance increasingly for future sustenance in reduction of Carbon dioxide emissions, production of next generation of fuels, and obtaining security of supply for sustainability. A lot of researches and efforts are being made worldwide to focus on ways for production of these second generation biofuels that are characterized by exceptional environmental activity as well as high flexibility of biomass feedstock. For a sustainable development and environmental management, producing syngas from biomass is widely recognized as it is considered to be a crucial step and a necessary precursor in the production of various second generation biofuels. Syngas is a mixture comprising of carbon monoxide, carbon dioxide, and hydrogen. It is produced by gasification of a carbon containing fuel to a gaseous product with high energy values. It is an important intermediate product in chemical industry, corresponding to almost 2% of the present total worldwide primary energy consumption.

The raw syngas produced from biomass can be purified and conditioned extensively, for catalytic synthesis of second generation biofuels such as methanol, dimethyl ether (DME), mixed alcohols, Fischer–Tropsch fuels, and even pure hydrogen. The main variables studied for syngas production are temperature and weight hourly space velocity in the fixed bed reactor. There are two major approaches for conversion of biomass into syngas: Fluidized bed gasification with subsequent catalytic reforming (operating around 900°C) and entrained flow gasification (at 1300°C) with extensive pre-treatment. The fluidized bed gasification approach is extensively studied and has an advantage that the gasification technology helps in the production of heat and/or electricity from biomass. This study describes the different technical options to produce, clean, and conditioning syngas from biomass and topics related to scaling up production and biomass transports.

5.1 INTRODUCTION

Conventional fuels, generally include oil, coal, and gas, are indispensable resources whose availability over the past few centuries has been an intrinsic part to the rapid progress in technology (Srirangan et al., 2012). Conventional fuels are non-sustainable and are currently having two major issues, that is: (1) the depletion of natural reserves in the years to come and (2) the substantial environmental impacts associated with their uses. There has been a recent shift pattern which expulses conventional fuels with renewable, sustainable and eco-friendly energy sources. Out of these sources, biomass-derived energy appears to be the most attractive and effective (Klass, 1998; Lucian et al., 2007; Srirangan et al., 2012). The production of first generation biofuels, produced primarily from food crops such as sugar crops, cereals, and oil seeds is already well studied and are presently high on the market. It comprises biogas and ethanol from starch and sugar, pure plant oil (PPO), and fatty acid methyl esters (FAME) such as rapeseed methyl ester (RME) (Sims et al., 2010).

Strong dependence on fossil fuels occurs due to consumption of petroleum derivatives and its intensive use which, resulting in diminishing petroleum resources, causing environmental distress. There is clear scientific evidence that greenhouse gases (GHGs) emission, such as carbon dioxide (CO_2), methane (CH_4), and nitrous oxide (N_2O), arises from fossil fuel combustion and due to the result of human activities, which perturb the Earth's climate and conditions. In order to reduce the dependence on oil and mitigate climate change in transport and chemical sectors, and simultaneous alternative production chains are necessary. The economic and sustainable first generation biofuel production was however under close scrutiny and was limited by:

- Competition for land and water used for fiber and food production (Fargione et al., 2008; Searchinger et al., 2008).
- Widely varying assessments of the net GHG reductions once land-use change is taken into account (OECD, 2008, Sims et al, 2010).
- High production and processing costs that often require government undertaking to compete with petroleum products (Doornbusch and Steenblik, 2007).

The growing shortage of fossil fuels and environmental problems makes it necessary to use alternative energy sources leading to the progress of second generation biofuels. Wind and solar power currently have the highest growth rate in continents such as Europe but has a disadvantage of being produced irregularly (Haas et al., 2011). In contrast to these fluctuating sources, biomass can deliver ample power on demand. Furthermore, the production of hydrogen, synthetic natural gas (SNG), liquid biofuels, or basic chemicals such as dimethyl ether is possible using biomass (Lijun et al., 2008; Chang et al., 2012).

Biofuels are becoming increasingly important in the reduction of CO_2 emissions in near future. The cumulative impacts of various concerns for sustainability have stimulated the interest in developing second generation biofuels produced from non-food biomass (Sims et al., 2010). Worldwide researches nowadays focus on ways to produce the second generation biofuels. These second generation biofuels are characterized by excellent environmental performance as well as high biomass feedstock flexibility. It is evident that nowadays second generation biofuels also considered as third generation biofuels largely rely on synthesis gas. Second generation biofuels are generally designated by a higher front-end flexibility than first generation biofuels and a better environmental performance. Second generation biofuels are produced from a variety of non-food crops such as utilization of lignocellulosic materials, agricultural residues from industry, forestry, and specific lignocellulosic crops. Second generation biofuels are Fischer–Tropsch (FT) fuels, DME, methanol, and ethanol from lignocellulosic raw materials such as wood and straw, mixed alcohols produced from synthesis gas, Substitute Natural Gas (SNG) and H_2. Hydrogen is nowadays the most promising energy sources and offers the large potential benefit in reduction of pollutants and GHGs (Ogden, 1999; Steinberg, 1999; Pehr et al., 2001; Du¨lger et al., 2000; Asadullah et al., 2002), derived from burning fossil fuel. Producing H_2 and CO, that is, syngas from biomass is a crucial step in the production of second generation biofuels.

Synthesis gas or syngas is defined as a gas that contains H_2 and CO as main combustible components. The raw form of syngas mostly contains considerable amounts of CO_2 and H_2O. As syngas is generally used for the synthesis of chemicals and fuels at elevated pressures, the concentration of N_2 in the syngas is also minimized. Bio-syngas is chemically similar or identical to syngas, but it is produced from biomass. Syngas from renewable resources, such as biomass, exhibits a promising perspective for sustainable development (Bridgwater and Double, 1991; Dong and Steinberg, 1997; Larson and Jin, 1999; Faaij et al., 2001; Hamelinck and Faaij, 2002; Tijmensen et al., 2002; Chmielniak and Sciazko, 2003). This is because biomass is a CO_2 neutral resource and is distributed extensively in the world (Lv et al., 2006).

Analyzing the literature involved in this syngas production, it can be found that three different routes of bio-syngas production were studied. They were syngas from biomass-derived oil (Panigrahi et al., 2003), syngas from biomass-derived char (Chaudhari ct al., 2001, 2003) and syngas from reforming of biomass gasification gas (Wang et al., 2005). The synthesis gas production from steam gasification of biomass-derived oil was also studied (Panigrahi et al., 2003). Chemically, syngas is different from gases that are generally produced by low-temperature gasification processes such as fluidized bed reactors. The gas produced by the specialized reactors for syngas production is called "product gas." Product gas is basically defined as a combustible gas that contains H_2 and CO. It also contains considerable amounts of hydrocarbons such as methane, ethane, and benzene. Product gas also inevitably contains CO_2 and H_2O and often N_2 as minor by-products (Drift and Boerrigter, 2006). For the various utilization paths for second generation fuel production, the gasification of biomass is the first process step and the production of a nitrogen free product gas is beneficial. To produce such a high quality product gas, allothermal steam gasification in fluidized beds seems to be a promising way (Corella et al., 2007; Murakami et al., 2007).

This study details about the various options to produce bio-syngas or synthesis gas from biomass, which is generally considered as a key intermediate for the second generation biofuels production. It also details about the syngas conditioning required for production and the marketing requirements and techniques of bio-syngas. The schematic outline of syngas production is depicted in Figure 5.1.

5.2 BIO-SYNGAS PRODUCTION

The production of bio-syngas, that is, syngas from biomass is presently considered as an attractive and reliable route to produce chemicals, hydrogen, biofuels, and electricity (Lin and Huber, 2009; Damartzis and Zabaniotou, 2011; Kirkels and Verbong, 2011). Biomass-derived syngas can typically be obtained from gasification of agricultural and forestry residues, along with industrial wastes such as black liquor, which is a major biomass-containing waste produced in pulp and paper manufacturing regions worldwide (Richardson et al., 2012). Biomass basically refers to a wide range of materials, comprising of all kinds of plants, animals, and their wastes and residues, which are especially used for production of energy and chemicals (Zhao et al., 2009). Biomass has been the fourth largest energy resource in the world and the largest and most important renewable energy resource available now. Biomass is presently widely recognized to have a high potential to meet the increased world energy demand (Zhou

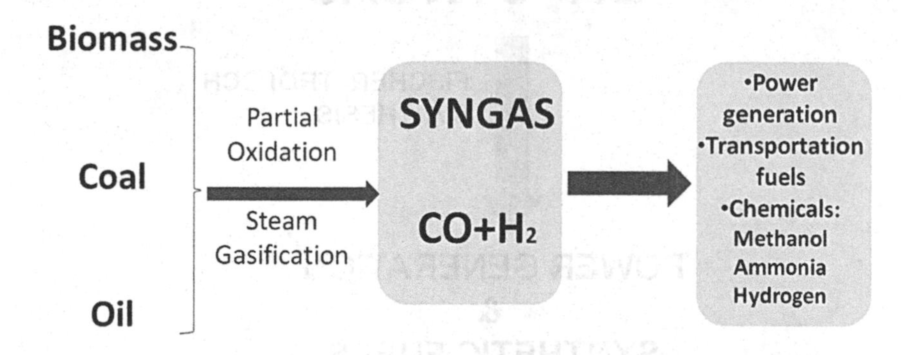

FIGURE 5.1 Production of syngas from various sources by partial oxidation and steam gasification and its downstream application.

et al., 2009). In the present review, the biomass scope is limited to plants including agricultural and forestry wastes. The benefits of biomass utilization are its widespread availability, renewable nature, and potential CO_2 neutrality (Wagner, 1979). The biomass fuels are divided into four primary classes: Wood, straw, grass, and residues. Their properties are as diverse as the sources from which they come. Compared to natural resources such as coal, biomass is usually highly volatile and has high oxygen content, but has low carbon content and heating value. The sulfur content in biomass is also much low, generally around less than 0.5 wt% (Qin, 2012). The major components in the biomass or ash are calcium, potassium, and phosphorous. Sodium, magnesium, iron, and silicon are in lesser amount and some trace elements. The syngas production from biomass, that is, bio-syngas is schematically shown in Figure 5.2.

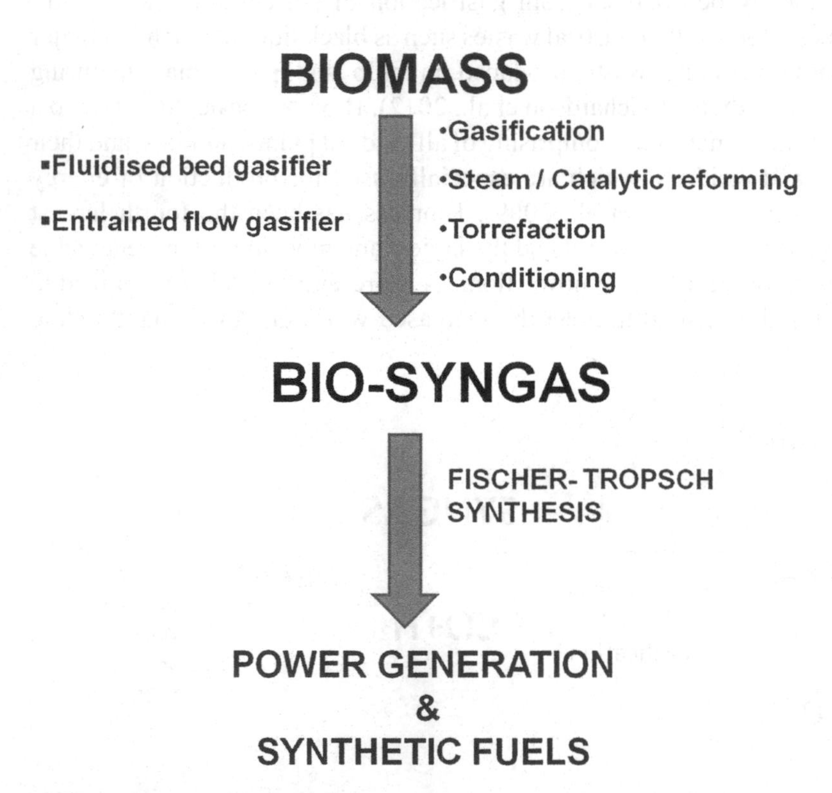

FIGURE 5.2 Process of bio-syngas production from biomass and its downstream application.

5.3 SYNGAS PRODUCTION

5.3.1 CONVENTIONAL PRODUCTION OF SYNGAS

Gasification is an old technology, which was first investigated by Thomas Shirley in 1659 (Higman et al., 2008; Basu, 2010). Gasification is a thermo-chemical process of solid fuel conversion by which a combustible gas is produced, that is, converting carbonaceous material, such as coal and biomass, to a combustible or synthetic gas by partial oxidation at elevated temperatures (Rezaiyan and Cheremisinoff, 2005; Sheth and Babu, 2009). Both quality and quantity of syngas produced through gasification depend upon the properties of feedstock's and the type of gasifier used. In conventional combustion, technology fuel is burned using excess air to ensure complete combustion. In gasification, the amount of oxygen is generally one-fifth to one-third of the amount theoretically required for complete combustion (Collet, 2002). Formerly, steam gasification was used to produce a high-quality synthesis gas. For gasification, R-Gibbs model is used as the main reactor that handles three phase chemical equilibrium. This reactor calculates the syngas composition by minimizing the Gibbs free energy and assumes complete chemical equilibrium. In bio-syngas production, the biomass gasification process is used that can efficiently create syngas with a reaction time nearly 100 times faster than general syngas conversion technologies from biomass. This thermal decomposition of solid carbon and hydrogen containing biomass creates H_2 and CO without requiring any heat inputs (autothermal) and has conversion rates above 99%. The process, known as catalytic partial oxidation, creates synthesis gas from biomass in a total reactor time of tens of milliseconds. The advantage is that it does not accumulate solid carbon (char) on the catalyst as the reaction occurs quickly and is easily scaled and also operable at atmospheric pressure. As very large amounts of heats are produced when the biomass contacts with the catalyst, it makes the process autothermal. These properties of the reactor make this technology especially applicable in small scale applications and in situations where a small portable reactor system containing the catalyst is implemented near renewable biomass sources. There are multiple benefits of the biomass to syngas gasification technique. Some of which are:

- High conversion rates (>99%).
- No solid carbon (char) formation.

- Fast reaction time (< 50 milliseconds).
- Reduced biomass transportation costs.
- Compatible with multiple feed-stocks including solid or liquid biomass; autothermal process.
- Operates at normal atmospheric pressure.

There are various gasification processes, out of which, the moving bed gasifier is the oldest gasifier process. In a moving bed gasifier, fuel particles enter at the top and move slowly downward through several process zones in the reactor vessel while reacting with gases generally from the opposite direction going upward. In moving bed gasifiers, it commonly uses large fuel particles to ensure good bed permeability and efficient heat and mass transfer. It also avoids excess pressure drop during the process. Hence, they require less complex fuel preparation for their advantage. On the other side, they have limited ability to handle fine particles (Olofsson et al., 2005; Higman and Van der Burgt, 2008). It has an advantage that their oxygen consumption is low, but the greatest disadvantage is the formation of large amount of byproducts from pyrolysis, such as tar, which is present in the product gas. It requires more comprehensive gas cleaning (Higman and Van der Burgt, 2008; Ratafia-Brown et al., 2002).

There are two thermo-chemical ways to produce syngas from biomass: Either by using a catalyst at a much lower temperature or by applying high temperatures (Rensfelt, 2005). The first route includes a fluidized bed gasifier and a downstream catalytic reformer, both operating at approximately 900°C. The second route generally requires temperatures as high as 1300°C and generally involves an entrained flow gasifier. So the two major approaches to convert biomass into syngas:

(a) Fluidized bed gasification and a subsequent catalytic reforming, which operates at around 900°C.
(b) Entrained flow gasification, operating at approximately 1300°C with extensive pre-treatment.

5.3.2 FLUIDIZED BED GASIFICATION

The fluidized bed gasification technique is an advantageous process of syngas production. This is because the gasification technology is well-developed and is already used for the bio-syngas production with biomass as the source for production of heat and/or electricity. Presently, researches mainly focus on downstream catalytic reforming (Drift and Boerrigter, 2006).

In fluidized bed reactors, allothermal steam gasification is a favorable way of conversion of biomass into high quality product gas (Mayerhofer et al., 2014). Since it contains a low amount of nitrogen, the gas may be used in a variety of processes. Besides the direct production of heat and power in various process tools, for example, internal combustion engines (ICE), the product gas can also be converted to clean synthesis gas, a second generation biofuels. Presently fluidized bed gasification of biomass is a common way of converting biomass for bio syngas production. Many different technologies are available. The air-blown circulating fluidized bed (CFB) is the most common one. Most fluidized bed applications involve close-coupled combustion with almost negligible intermediate gas cleaning. In this process, electricity and/or heat are the usual end products. The fluidized bed gasifier operating typically at 900°C produces gas such as H_2, CO, CO_2, H_2O, and considerable amounts of hydrocarbons such as methane (CH_4), ethane (C_2H_4), benzene (C_6H_6), and tars. The product gas produced is only suitable for combustion process; but it does not meet up the requirements for production of biofuels or chemicals.

In a fluidized bed gasifier, fuel particles enter at the side of the reactor vessel, and are fluidized by steam and oxidant injected near the bottom with enough velocity. Larger particles are consumed slowly and are recycled internally in the reactor vessel until they are small enough for external recycling. Smaller particles are converted into one pass, or are entrained by the product gas when it leaves the top of the vessel. The product gas requires further treatment in a catalytic reformer. In this process, the hydrocarbons produced in considerable amount are converted into H_2 and CO (and CO_2 and H_2O). As most of the syngas is converted into liquid fuels which requires raw gas and a very little or no amount of inert gas as a by-product, so gasification and reforming process should apply pure oxygen instead of air. In this process, steam is usually added as a moderator. The other way out to avoid N_2 dilution is to use an indirect or allothermal gasifier. In these reactors, gas production and heat generation do not take place in the same reactor. This enables the use of air present in the heat generating reactor, without having the N_2-dilution of the gas coming from the gas generation reactor. In this kind of gasification system, it includes operation on oxygen/steam instead of air, the installation of a high temperature filter, a catalytic reformer, and a shift reactor.

In fluidized bed gasifiers, the biomass is placed over an inert bed of fluidized material (e.g., sand, char, etc.). Such systems are less sensitive to fuel variations but produce larger amounts of tar and dust. They are more compact but also more complex, and usually used at larger scales. Fluidized bed gasifiers other than fixed bed gasifiers, are basically operated at significant high gas flow velocities. The fuel bed and a carrier material (e.g., sand) are fluidized by the gas flow (fumigator and re-circulated product gas). Thus, the gasification reaction takes place in a fluidized bed having a very minimum weight of the bed as fuel. There are three types of fluidized bed gasifiers depending on their degree of fluidization. They are named as bubbling fluidized bed gasifiers (2–3 m/s gas velocity), circulating fluidized bed (CFB) gasifiers (5–10 m/s gas velocity), and transport gasifiers (11–18 m/s gas velocity) (Olofsson et al., 2005; Higman and Van der Burgt, 2008). Fluidized bed gasifiers have higher fuel throughput than other gasification process, but has a disadvantage that their high gas velocity may cause equipment erosion (Olofsson et al., 2005). Due to the moderately high temperatures, a certain amount of tar is present in the product gas. However, the bed material, sand makes it possible to use the in-bed tar in catalytic process. Besides, the product gas is also rich in particulates. When biomass is used as fuel, there is a risk for bed aggregation due to its alkali metals rich ash composition. So to overcome this aggregation problem, three methods are basically employed: decreasing gasification temperature, exchanging bed material with proper intervals, and using some proven mineral binding additives (Olofsson et al., 2005). Another major problem in all downstream applications is the high tar content in the product gas. The tar which is a mixture of mostly aromatic hydrocarbons needs to be removed prior to downstream processes to avoid blocking of equipment by condensed material. Besides the operational parameters, the gas quality depends on the reactor design.

5.3.3 *ENTRAINED FLOW GASIFICATION*

Entrained flow gasification is a commercially available process for large scale production of syngas, mainly from coal and liquid fuels. It also has reached the highest efficiency of syngas production from biomass which is also called bio-syngas, because the non-catalytic production of syngas (H_2 and CO) from biomass generally requires high temperatures, typically 1300°C. The most common reactor operating at this high temperature for bio-syngas production is the entrained flow gasification process (Collet, 2002).

In entrained flow gasification, the gasification reactions take place at a very high reaction rate because of the high operating temperature (1200–1600°C), optimum at 1300°C and pressure (2–8 MPa), and after a few seconds (0.5–4.0 s) the product gas leaves the reactor vessel at the bottom together with the molten slag (Collot, 2006; Higman and Van der Burgt, 2008). Entrained flow gasifiers have the ability to gasify practically any fuels, but fuel with lower moisture and ash content are favored to reduce oxygen consumption (Rezaiyan and Cheremisinoff, 2005; Higman and Van der Burgt, 2008). The product gas formed by the gasifier is generally cooled by two main methods: Quenching the gas with water or using a high temperature radiant cooler, while the molten slag falls to a quench chamber for solidifying and leaves it via a lock hopper (Collot, 2006). In bio-syngas production, ash behavior and feeding have been the major issues in the production process.

There are two types of entrained flow gasifiers that are distinguished as: Slagging and non-slagging. Generally, in the slagging gasifier, the slag mass flow is around or at least 6% of the fuel flow to ensure proper operation. In a non-slagging gasifier, the walls are kept free of slag, which is suitable for fuels with only a little ash. In the high ash melting temperatures of biomass for syngas production, a slagging entrained flow gasifier is preferred over a non-slagging gasifier. The reasons for choosing slagging over non-slagging entrained flow gasifier is as follows:

- Requires fluxing material in order to obtain the proper slag properties at reasonable temperatures.
- Little melt as fouling can never be avoided and
- Slagging entrained flow gasifier is more fuel flexible than non-slagging.

The entrained flow gasifiers have a high oxygen demand and operate in a slagging mode due to the short residence time; high temperatures are required to ensure a good carbon conversion. Since biomass contains mineral matter such as ash, the slagging entrained flow gasifier seems to be the most convenient technology (Drift et al., 2004). In a slagging gasifier, the ash forming components melt within the gasifier, flowing down the walls of the reactor and finally leaving the reactor as a liquid slag. Slagging gasification of biomass requires fluxing material in order to obtain the proper slag properties at reasonable temperatures where silica or clay may be the obvious choice.

5.3.4 TORREFACTION

Torrefaction is a mild thermal treatment of biomass, at 250–300°C that efficiently turns solid biomass such as hemicellulose; one of the most reactive parts of wood which is decomposed into a brittle, easy to reduce material resembling coal (Bergman et al., 2005b; 2005c). The process takes place in oxygen-free conditions and atmospheric pressure. Torrefaction improves the physical and chemical properties of biomass as a fuel (Dudyński et al., 2015). It raises the energy density, by lowering the oxygen to carbon ratio and hydrogen to carbon ratio, and making it lesser hydrophilic (Chew and Doshi, 2011).

The torrefied biomass can also be pelletized very easily and a dense and easy to transport biomass fuel may be obtained (Bergman et al., 2005a). The hydrophobic nature of torrefied material further simplifies logistics. Reduced torrefied biomass can be fed like coal, thus enabling a smooth transition from coal to biomass. It has been studied that torrefaction pretreatment can reduce the electricity consumption during the production process. Although torrefaction is a rather common process but it has never been optimized for efficient production of a brittle "bio-coal." The gases produced during torrefaction may be used to supply the thermal needs of the production process.

5.4 SYNGAS PURIFICATION AND CONDITIONING

Syngas from biomass is obtained after several steps of purification and conditioning stages of the raw gas sample. From various studies, it has been estimated that syngas production from biomass accounts for almost half the cost required for normal biofuel production (Richardson et al., 2012). Firstly, purification and gas conditioning are important stages in the production of electricity, liquid biofuel, or hydrogen from biomass gasification. The raw gas produced mainly consists of CO, H_2, CH_4, CO_2, H_2O, and N_2, when air is used as the oxidizing agent. It also contains secondary products such as tars, solid particles, and inorganic compounds such as alkali metals, nitrogen, sulfur, and chlorine. The tolerance of syngas-using systems due to the formation of secondary products is highly variable. So, purification and conditioning play a major role in bio-syngas post production (Mondal et al., 2011).

Syngas purification includes removal of particles, tars, and inorganic impurities and conditioning involves reforming, water gas shift, followed by H_2 and CO_2 separation (Richardson et al., 2012). Among all the components of syngas, H_2S, COS, and CO_2 are termed as acid gasses. These gasses produce acidic solution after dissolving in water. Hence, they are corrosive under moist conditions when undergoing aerial oxidation. Further, CO_2 is also a major part of GHGs, causing global warming. So the removal of these gasses from syngas is essential to reduce its corrosiveness, as well as, the CO_2 emission from downstream units such as turbines.

Tars are generally or globally explained as a complex mixture of organic compounds, which includes, single ring to multiple ring aromatic compounds. Tars also produce numerous oxygen containing hydrocarbons and complex polycyclic aromatic hydrocarbons (PAH) along with syngas (Milne et al., 1998; Maniatis, 2000; Devi et al., 2005). Serious problems arise when the tar condenses into more complex molecules and forms aerosol, resulting in corrosion, clogging, and fouling of installations in instruments. It also leads to the deactivation of catalysts used in downstream gasification processes (Torres et al., 2007; Han and Kim, 2008; Li and Suzuki, 2009). There are two types of strategies which are typically used to reduce the tar content in syngas (Devi et al., 2003). The first one involves the methods in which during the gasification operation stage tar formation is limited or tars converted to other degradable products in the gasifier or gasification reactor (generally known as "primary treatment"). The second one refers to the post-treatment methods, also called "secondary treatment," which operates downstream of gasification with several purification processes. Secondary methods may involve physical processes such as wet scrubbing, filtration, and electrostatic precipitation. It also involves thermal and/or catalytic processes (Han and Kim, 2008). Physical methods are poorly attractive as they transfer and concentrate tars in a liquid or solid phase, which has the dual drawback of generating toxic wastes and losing the chemical energy contained in tars (Richardson et al., 2012).

To simplify, or even eliminate the costly purification post treatments, primary methods for tar reduction are being increasingly studied and developed. Implementation of this approach is not only optimizing the gasification operating conditions and the reactor design, but also developing efficient catalysts to be used directly in the gasifier, that is, as primary catalysts (Devi et al., 2003). Multi-stage gasification and CFB gasification in the presence of nickel-based catalysts are examples of primary methods which revealed efficient for reducing tar contents in the gasification gas (Pfeifer et al., 2004). New reactor technologies and novel concepts or strategies are being developed in order to considerably reduce syngas or hydrogen production costs by biomass gasification while sidewise favoring the optimization of primary methods for tar removal (Richardson et al., 2012).

5.5 APPLICATION OF SYNGAS

Syngas is used further for producing essential things Such as FT diesel, Methanol/DME, Ammonia, and Hydrogen. It is also being used in chemical industry and for generation of electricity (Boerrigter and Rauch, 2005).

5.5.1 POWER GENERATION

Bio-syngas being a combustible gas may be used for the production of electricity in all prime movers ranging from steam cycles to gas engines, turbines (combined cycle), as well as fuel cells. It used as fuel in ICE that is coupled to a synchronous alternator, which produces electricity. The thermal energy generated in the form of combustion gases, both in the gasifier reactor as well as the ICE by heat recovery is used for generation of thermal energy either in the form of hot water, steam, thermal oil, or organic Rankine cycles (ORCs), etc., ensuring full utilization of the waste power coming into the gasification plant for end use. Generation of thermal energy occurs by direct combustion of syngas in heat generation equipment, such as steam boilers, cement kilns, dryers, etc. This thermal energy can be used in a range of sectors such as the industrial sector, in chemicals, cement, food, etc., as well as in tertiary sectors such as in offices, hotels, residential for district heating, and cooling purposes or agriculture woks for greenhouse condition maintenance. Based on energetic considerations, it may not be an attractive utilization of the syngas. It is more sensible to use product gas instead of bio-syngas for power production. This originates in the lower net energetic efficiency of bio-syngas compared to product gas production due to the higher gasification temperature and electricity consumption for oxygen production (Boerrigter and Rauch, 2005).

5.5.2 TRANSPORTATION FUELS

In the future, bio-syngas will become increasingly important for the production of ultra-clean designer fuels from "gas- to- liquids" (GTL) processes, with the main examples being FT diesel and methanol/DME. These products are used for the production of automotive and transportation fuels, chemicals and plastics, fuel injection into pipelines, etc.

5.5.2.1 FT SYNTHESIS

In the catalytic FT synthesis one mole of CO reacts with two moles of H_2 for mainly formation of paraffin straight-chain hydrocarbons (CxH2x) with minor amounts of primary alcohols along with branched and unsaturated hydrocarbons such as 2-methyl paraffins and α-olefins (Boerrigter and Rauch, 2005). Typical operation conditions for FT synthesis are temperatures of 200–350°C and pressures between 25 and 60 bar (Dry, 1981). Almost 20% of the chemical energy is released as heat from the exothermic FT reaction. Undesirable side reactions may occur such as methanation, coke deposition, oxidation of the catalyst, the Boudouard reaction or carbide formation. FT processes may be used to produce either a light synthetic crude oil (syncrude) and light olefins or heavy waxy hydrocarbons. For direct production of gasoline and light olefins, the FT process is operated at high temperature (330–350°C), for production of waxes and/or diesel fuel, at low temperatures (220–250°C) (Boerrigter and Rauch, 2005). Catalysts for FT synthesis can be damaged with impurities as NH_3, HCN, H_2S, and COS. These impurities poison the catalysts. HCl generally causes corrosion of catalysts whereas alkaline metals are deposited on the catalyst.

5.5.2.2 METHANOL

Methanol may be produced by means of the catalytic reaction of carbon monoxide and some carbon dioxide with hydrogen. The presence of a certain amount of carbon dioxide in the percentage range is necessary to optimize the reaction. Both reactions are exothermic and proceed with volume contraction; where a low temperature and high pressure consequently favors each other. Methanol is currently produced on an industrial scale exclusively by catalytic conversion of synthesis gas (Boerrigter and Rauch, 2005).

5.5.3 CHEMICAL SYNTHESIS

Syngas is subjected to chemical transformation processes, enabling the generation of Bioethanol, Biochemical (ethylene glycol, DME, methanol) Bio-SNG (Synthetic Natural Gas), etc.

5.5.3.1 AMMONIA PRODUCTION FOR FERTILIZER INDUSTRY

The major share of bio-syngas produced which is almost around 85% of the ammonia is used for the production of fertilizers. The rest 15% is used for a broad variety of applications (Boerrigter and Rauch, 2005). The nitrogen used in industrially produced chemicals is directly or indirectly derived from ammonia or its derivative like nitric acid. The ammonia is recovered from the gas as a liquid by cooling and condensation.

5.5.3.2 HYDROGEN PRODUCTION IN REFINERIES

Syngas is one of the main sources for hydrogen used in refineries. The syngas is generally produced by steam reforming of natural gas or in some cases by gasification of heavy oil fractions. In refineries, hydrogen is used for the hydro-treating and hydro-processing operations (Boerrigter and Rauch, 2005).

Other applications of syngas have a relatively small market or the processes are still in development phase. Some of them are: Synthetic natural gas (SNG), mixed alcohols, carbon monoxide, olefins, and aromatics.

5.6 SYNGAS MARKET

Syngas is a versatile building block in chemical industry (Boerrigter et al., 2005). The world market for syngas which is mainly derived from fossil energy sources like coal, natural gas and oil/residues is dominated by the ammonia industry followed by H_2 refineries, methanol industry, electricity, gas-to-liquids and other industries (Drift and Boerrigter, 2005). The H_2 production in oil refineries also represents a significant share in syngas market and a major application, for example, hydrogenation steps, and for the production of methanol. The ability of syngas to be produced from a wide variety of feedstock such as coal, natural gas, pet-coke, and biomass is impacting the market growth positively. The flexibility reduces the dependency on a raw material for producing syngas. The commercially available gasifiers are technologically advanced and possess the ability to process multiple feedstocks for syngas production. Thus, the availability of different types of raw materials across the globe has allowed the syngas producers to leverage the option of choosing feedstock based on the market conditions, eventually contributing significantly to the growth of syngas market during the forecast period. Syngas from renewable biomass can be refined into liquid fuel, chemicals, or fertilizer agents. In the future, syngas will become increasingly important for the production of cleaner fuels to comply with the stringent emission standards. Syngas is the intermediate energy carrier for the production of second generation biofuels like methanol, DME, cellulosic ethanol, and Fischer–Tropsch diesel (Drift and Boerrigter, 2006).

The total global annual use of fossil-derived syngas approximately corresponds to 2% of the total primary energy consumption. The largest part of the syngas is used for the synthesis of ammonia for fertilizer production (~55%), the second largest share is the amount of hydrogen from syngas consumed in oil refining processes (~24%), and smaller amounts are used for methanol production (12%) (Veringa and Boerrigter, 2004; Boerrigter and Rauch, 2005; Boerrigter et al., 2005).

Syngas production from biomass is heading for a great future as renewable energy source. It not only is available in large quantities, but also the only renewable energy source that is suitable for the sustainable production of carbon containing transportation fuels and chemicals. Therefore, the application of biomass as feedstock for the production of fuels and chemicals allows the reduction of fossil fuel consumption and the accompanying CO_2-emmissions.

5.7 CONCLUDING REMARKS

Biomass will be playing an important role in the near future global energy infrastructure not only for its power and heat generation, but also for the production of synthetic fuels and chemicals. Making syngas from biomass is widely recognized as a necessary step in the production of second generation biofuels. Bio-syngas will be the key intermediate in the production of renewable sources such as fuel, chemicals, and electricity, having an envisioned substitution of fossil fuels by biomass. Bio-syngas will be the key-intermediate of renewable transportation fuels, chemicals, and electricity. Hence, it may create large potential markets of fuels. The main application will be the production of Fischer–Tropsch fuels from bio-syngas. The dominant biomass conversion technology will always be the gasification process, as the gases from biomass gasification are intermediates in the high-efficient power production or the synthesis from chemicals and fuels. In biomass gasification process for utilization of gases, it is very much important to understand that the composition of the gasification gas, which is very dependent on the type of gasification process and especially the gasification temperature in which it takes place. Catalytic steam gasification and plasma gasification are the recent developments in this gasification technology which are gaining commercialization. The extent of removal of impurities from syngas is linked to its downstream application. When syngas is used for chemical and fertilizer synthesis, it requires a higher degree of purification compared to its application in power generation. However, clear understanding of the relation between gasification technologies, the generated gas, its typical applications, and the corresponding specifications for bio-syngas production is crucial in today's decision-making processes and future sustainability.

KEYWORDS

- **biomass**
- **renewable energy source**
- **fluidized bed gasification**
- **entrained flow gasification**
- **torrefaction**
- **conditioning**
- **Fischer–Tropsch fuels**

REFERENCES

Asadullah, M.; Ito, S. I.; Kunimori, K.; Yamada, M.; Tomishige, K. Biomass Gasification to Hydrogen and Syngas at Low Temperature: Novel Catalytic System Using Fluidized-Bed Reactor. *J. Cat.* **2002,** *208* (2), 255–259.

Basu, P. *Biomass Gasification and Pyrolysis: Practical Design and Theory*. Academic Press: Oxford, 2010.

a) Bergman, P. C. A. *Combined Torrefaction and Pelletisation, the TOP Process*, ECN: Petten, the Netherlands. 2005, p. 29.

b) Bergman, P. C. A.; Boersma, A. R.; Kiel, J. H. A.; Prins, M. J.; Ptasinski, K. J.; Janssen, F. J. J. G. *Torrefaction for Entrained Flow Gasification of Biomass*, ECN: Petten, The Netherlands, ECN-C--05-067, 2005, p 51.

c) Bergman, P. C. A.; Boersma, A. R.; Zwart, R. W. R.; Kiel, J. H. A. *Torrefaction for Biomass Co-Firing in Existing Coal-Fired Power Stations (BIOCOAL)*, ECN, Petten, the Netherlands, ECN-C-05-013, 2005, p 72 .

Boerrigter, H.; Drift, van der, A.; van Ree, R. Biosyngas; Markets, Production Technologies, and Production Concepts for Biomass-Based Syngas, Energy Research Centre of the Netherlands (ECN), Petten, The Netherlands, report CX--04-013, February 2004, p 37.

Boerrigter, H.; Rauch, R.. *Syngas Production and Utilization*, Chapter-10, Handbook Biomass Gasification. 2005.

Bridgwater, A. V.; Double, J. M. Production Costs of Liquid Fuel From Biomass. *Fuel.* **1991,** *70* (10), 1209–1224.

Chang, J.; Fu, Y.; Luo, Z. Experimental Study for Dimethyl Ether Production From Biomass Gasification and Simulation on Dimethyl Ether Production: Biorefinery. *Biomass Bioenergy*. **2012,** *39* (0), 67–72.

Chaudhari, S. T.; Bej, S. K.; Bakhshi, N. N.; Dalai, A. K. Steam Gasification of Biomass-Derived Char for the Production of Carbon Monoxide-Rich Synthesis Gas. *Energy Fuels*. **2001,** *15* (3), 736–742.

Chaudhari, S. T.; Dalai, A. K.; Bakhshi, N. N. Production of Hydrogen and/or Syngas (H_2 + CO) via Steam Gasification of Biomass-Derived Chars. *Energy Fuels.* **2003,** *17* (4), 1062–1067.

Chew, J. J.; Doshi, V. Recent Advances in Biomass Pretreatment – Torrefaction Fundamentals and Technology, *Renew. Sust. Energ. Rev.* **2011,** 15, 4212–4222.

Chmielniak, T.; Sciazko, M. Co-Gasification of Biomass and Coal for Methanol Synthesis. *Appl. Energy*. **2003,** *74* (3–4), 393–403.

Collet, A. G. *Matching Gasifiers to Coal*. IEA Clean Coal Centre, 2002. p. 63.

Collot, A. G. Matching Gasification Technologies to Coal Properties. *Int. J. Coal Geol.* **2006,** *65*, 191–212.

Corella, J.; Toledo, J. M.; Molina, G. A Review on Dual Fluidized-Bed Biomass Gasifiers. *Ind. Eng. Chem. Res.* **2007,** *46* (21), 6831–6839.

Damartzis, T.; Zabaniotou, A. Thermochemical Conversion of Biomass to Second Generation Biofuels Through Integrated Process Design-A Review. *Renew. Sustain. Energy Rev*. **2011,** *15*, 366–378.

Devi, L.; Ptasinski, K. J.; Janssen, F. J. J. G. A Review of the Primary Measures for Tar Elimination in Biomass Gasification Processes. *Biomass Bioenergy*. **2003,** *24*, 125–140.

Devi, L.; Ptasinski, K. J.; Janssen, F. J. J. G.; van Paasen, S. V. B.; Bergman, P. C. A.; Kiel, J. H. A. Catalytic Decomposition of Biomass Tars: Use of Dolomite and Untreated Olivine. *Renew. Energy* **2005,** *30*, 565–587.

Dong, Y.; Steinberg, M. Hynol—an Economic Process for Methanol Production From Biomass and Natural Gas with Reduced CO_2 Emission. *Int. J. Hydrogen Energy* **1997,** *22* (10–11), 971–977.

Doornbusch, R.; Steenblik R. Biofuels: is the Cure Worse Than the Disease? Paper Prepared for the Round Table on Sustainable Development, Organisation for Economic Co-operation and Development (OECD), Paris, 11–12 September, 2007.

Dry, M. E. The Fischer-Tropsch Synthesis in Anderson. In *Catalysis-Science Technol,* J. R. M., Boudart, M. Eds.; Vol. 1, Springer-Verlag: New York, **1981,** pp. 159–255.

Dülger, Z.; Özcelik, K. R. Fuel Economy Improvement by on Board Electrolytic Hydrogen Production. *Int. J. Hydrogen Energy* **2000,** *25*, 895–897.

Dudyński, M.; van Dyk, J. C.; Kwiatkowski, K.; Sosnowska, M. Biomass Gasification: Influence of Torrefaction on Syngas Production and Tar Formation. *Fuel Process. Technol.* **2015,** *131*, 203–212.

Faaij, A.; Hamelinck, C.; Tijmensen, M. Long Term Perspectives for Production of Fuels From Biomass; Iintegrated Assessment and R&D Priorities – Preliminary Results. In: Proceedings of the First World Conference on Biomass for Energy and Industry; Kyritsis S. et al. Eds.; vol. 1/2, James & James Ltd.: London, UK, **2001,** pp. 687–690.

Fargione, J.; Hill, J.; Tilman, D.; Polasky, S.; Hawthorne, P. Land Clearing and the Biofuel Carbon Debt. *Science.* **2008,** *319* (5867), 1235–1238.

Haas, R.; Panzer, C.; Resch, G.; Ragwitz, M.; Reece, G.; Held, A. A Historical Review of Promotion Strategies for Electricity From Renewable Energy Sources in eu Countries. *Renew. Sustain. Energy Rev.* **2011,** *15* (2), 1003–1034.

Hamelinck, C. N.; Faaij, A. P. C. Future Prospects for Production of Methanol and Hydrogen from Biomass. *J. Power Sources.* **2002,** *111* (1), 1–22.

Han, J.; Kim, H. The Reduction and Control Technology of Tar During Biomass Gasification/ Pyrolysis: An Overview. *Renew. Sustain. Energy Rev.* **2008,** *12* (2), 397–416.

Higman, C.; van der Burgt, M.; *Gasification.* Gulf Professional Publishing: Burlington, 2008.

Kirkels, A. F.; Verbong, G.P.J. Biomass Gasification: Still Promising? A 30-year Global Overview. *Renew Sustain Energy Rev.* **2011,** *15*, 471–481.

Klass, D. L. *Biomass for Renewable Energy, Fuels, and Chemicals.* Academic Press: San Diego, 1998.

Larson, E. D.; Jin, H. Biomass Conversion to Fischer-Tropsch Liquids: Preliminary Energy Balances. In *Proceedings of the Fourth Biomass Conference of the Americas.* Overend R, Chornet E, Eds.; Elsevier Science: Kidlington, UK, **1999,** vol. 1/2, 843–854.

Li, C.; Suzuki, K. Tar Property, Analysis, Reforming Mechanism and Model for Biomass Gasification—An Overview. *Renew. Sustain. Energy Rev.* **2009,** *13*, 594–604.

Lin, Y. C.; Huber, G. W. The Critical Role of Heterogeneous Catalysis in Lignocellulosic Biomass Conversion. *Energy Environ. Sci.* **2009,** *2*, 68–80.

Lucian, L. A.; Argyropoulos, D. S.; Adamopoulos, L.; Gaspar, A. R. Chemicals, Materials, and Energy From Biomass: A Review. In: *ACS Symposium Series 954. Materials, Chemicals, and Energy for Forest Biomass;* Argyropoulos, D. S. Ed.; American Chemical Society: Washington; **2007,** pp. 2–30.

Maniatis, K.; Beenackers, A. A. C. M. Tar Protocols. IEA Bioenergy Gasification Task. *Biomass Bioenergy.* **2000,** *18*, 1–4.
Mayerhofer, M.; Fendt, S.; Spliethoff, H.; Gaderer, M. Fluidized Bed Gasification of Biomass– In Bed Investigation of Gas and Tar Formation. *Fuel.* **2014,** *117*, 1248–1255.
Milne, T. A.; Abatzoglou, N.; Evans, R.J. Biomass Gasifiers "Tars": Their Nature, Formation, and Conversion. Golden, Colorado (US): NREL/TP-570-25357, National Renewable Energy Laboratory; 1998.
Mondal, P.; Dang, G.S.; Garg, M.O. Syngas Production Through Gasification and Cleanup for Downstream Applications—Recent Developments. *Fuel Process Technol.* **2011,** *92*, 1395–1410.
Murakami, T.; Xu, G.; Suda, T.; Matsuzawa, Y.; Tani, H.; Fujimori, T. Some Process Fundamentals of Biomass Gasification in Dual Fluidized Bed. *Fuel.* **2007,** *86* (1–2), 244–255.
OECD, *Economic Assessment of Biofuel Support Policies. Organisation for Economic Co-operation and Development*, OECD: Paris, 2008.
Ogden, J. M. Prospects for Building a Hydrogen Energy Infrastructure. *Annu. Rev. Energy Environ.* **1999,** *24* (1), 227–279.
Olofsson, I.; Nordin, A.; Söderlind, U. Initial Review and Evaluation of Process Technologies and Systems Suitable for Cost-Efficient Medium-Scale Gasification for Biomass to Liquid Fuels. *ETPC Report 05-02*. **2005,** 09–30.
Panigrahi, S.; Dalai, A. K.; Chaudhari, S. T.; Bakhshi, N. N. Synthesis Gas Production From Steam Gasification of Biomass-Derived Oil. *Energy Fuels.* **2003,** *17* (3), 637–642.
Pehr, K.; Sauermann, P.; Traeger, O.; Bracha, M. Liquid Hydrogen for Motor Vehicles—the World's First Public LH_2 Filling Station. *Int. J. Hydrogen Energy* **2001,** *26* (7), 777–782.
Pfeifer, C.; Rauch, R.; Hofbauer, H. In-Bed Catalytic Tar Reduction in a Dual Fluidized Bed Biomass Steam Gasifier. *Ind. Eng. Chem. Res.* **2004,** *43*, 1634–1640.
Qin, K. *Entrained Flow Gasification of Biomass*. Ph.D. Thesis, Department of Chemical and Biochemical Engineering, Technical University of Denmark. **2012,** 1–187.
Ratafia-Brown, J.; Manfredo, L.; Hoffmann, J.; Ramezan, M. Major Environmental Aspects of Gasification-Based Power Generation Technologies. U.S. Department of Energy Final Report. *Science Applications International Corporation.* 2002.
Rensfelt, E. *State of the Art of Biomass Gasification and Pyrolysis Technology 2005*. In: Synbios, the Syngas Route to Automotive Biofuels, Conference Held from 18–20 May 2005, Stockholm, Sweden. 2005.
Rezaiyan, J.; Cheremisinoff, N.P. *Gasification Technologies: A Primer for Engineers and Scientists*. Taylor & Francis: Boca Raton, 2005.
Richardson, Y.; Blin, J.; Julbe, A. A Short Overview on Purification and Conditioning of Syngas Produced by Biomass Gasification: Catalytic Strategies, Process Intensification and New Concepts; *Prog. Energy Combust. Sci.* **2012,** *38* (6), 765–781.
Searchinger, T.; Heimlich, R.; Houghton, R. A.; Dong, F.; Elobeid, A.; Fabiosa, J.; Tokgoz, S.; Hayes, D.; Yu, T. H. Use of US Croplands for Biofuels Increases Greenhouse Gases Through Emissions From Land Use Change. *Science* **2008,** *319* (5867), 1238–1240.
Sheth, P. N.; Babu, B. Experimental Studies on Producer Gas Generation From Wood Waste in a Downdraft Biomass Gasifier. *Bioresour. Technol.* **2009,** *100*, 3127–3133.
Sims, R. E. H.; Mabee, W.; Saddler, J. N.; Taylor, M. An Overview of Second Generation Biofuel Technologies. *Bioresour. Technol.* **2010,** *101*, 1570–1580.

Srirangan , K.; Akawi, L.; Young , M. M.; Chou, C. P. Towards Sustainable Production of Clean Energy Carriers From Biomass Resources. *Appl. Energy*. 2012, *100*, 172–186.

Steinberg, M. Fossil Fuel Decarbonization Technology for Mitigating Global Warming. *Int. J. Hydrogen Energy*. **1999,** *24* (8), 771–777.

Tijmensen, M. J. A.; Faaij, A. P. C.; Hamelinck, C. N.; van Hardeveld, M. R. M. Exploration of the Possibilities for Production of Fischer–Tropsch Liquids and Power via Biomass Gasification. *Biomass Bioenergy.* **2002,** *23* (2), 129–152.

Torres, W.; Pansare, S. S.; Goodwin, J. G. Hot Gas Removal of Tars, Ammonia, and Hydrogen Sulphide From Biomass Gasification Gas. *Catal. Rev. Sci. Eng.* **2007,** *49*, 407–456.

van der Drift, A.; Boerrigter, H.; Coda, B.; Cieplik, M. K.; Hemmes, K. *Entrained Flow Gasification of Biomass; Ash Bbehaviour, Feeding Issues, and System Analyses*, ECN: Petten, The Netherlands, ECN-C-04-039. 2004.

van der Drift, A.; Boerrigter, H. *Synthesis Gas From Biomass for Fuels and Chemicals, IEA Bioenergy Task 33*, SYNBIOS: Stockholm, 2006.

Veringa, H. J.; Boerrigter, H. De syngas-route... van duurzame productie tot toepas-singen, presented at Transitie workshop "Duurzaam Synthesegas", 10 February 2004, Utrecht, the Netherlands. Published in: Energy research Centre of the Netherlands (ECN), Petten, The Netherlands, report RX--04-014.

Wagner, H. Soot formation in combustion. *Proc. Combust. Inst.* 1979, *17*, 3–19.

Wang, L. J.; Weller, C. L.; Jones, D. D.; Hanna, M. A. Contemporary Issues in Thermal Gasification of Biomass and Its Application to Electricity and Fuel Production. *Biomass Bioenergy.* **2008,** *32*(7), 573–581.

Wang, T. J.; Chang, J.; Lv, P. M. Synthesis Gas Production via Biomass Catalytic Gasification with Addition of Biogas; *Energy Fuels.* **2005,** *19* (2), 637–644.

Zhao, Y.; Sun, S.; Tian, H.; Qian, J.; Su, F., Ling, F. Characteristics of Rice Husk Gasification in an Entrained Flow Reactor. *Bioresour. Technol.* **2009,** *100*, 6040–6044.

Zhou, J.; Chen, Q.; Zhao, H.; Cao, X.; Mei, Q.; Luo, Z. Biomass–Oxygen Gasification in a High-Temperature Entrained-Flow Gasifier. *Biotechnol. Adv.* **2009,** *27*, 606–611.

CHAPTER 6

Biofuel Cell from Biomass

RITUPARNA SAHA[1,2], DEBALINA BHATTACHARYA[3], and MAINAK MUKHOPADHYAY[1,*]

[1]*Department of Biotechnology, JIS University, Kolkata, West Bengal, India*

[2]*Department of Biochemistry, University of Calcutta, Kolkata, West Bengal, India*

[3]*Department of Biotechnology, University of Calcutta, Kolkata, West Bengal, India*

[*]*Corresponding author. E-mail: m.mukhopadhyay85@gmail.com.*

ABSTRACT

In recent years, Biofuel cells (BFCs) have become a potent source of alternative and clean energy compared to traditional fuel cells. BFC mainly utilizes microbes and enzymes as electron exchangers and have the ability to convert chemical energy to electrical energy via electrochemical reactions using biocatalysts. Microbial BFCs utilize enzymes that are produced intracellularly or on the cell surface, thus exhibiting advantages over enzymatic BFCs in areas of long-term stability and higher efficiency; whereas controlled environment and accumulation of a load of catalysts in enzymatic BFCs make them have a high turnover rate but a shorter half-life.

Biomass in any form can be utilized for BFCs. Natural biomass either plays the role of a substrate for an enzymatic BFCs or the producer of microbial BFCs. Various BFCs have since been reported which effectively depolymerize various components of biomass to their primitive forms. Also, some definitive biomasses act as electron mediators or co-substrate between the biocatalysts and the electrode. Specific biomass has been known to stimulate the production of certain enzymes, which are further used for either microbial BFCs or enzymatic BFCs. Thus, BFCs from biomass provide an economical approach and can be further used for industrial purposes and also for different environmental issues.

This chapter will provide detailed information about the different types of BFCs along with their production and function related to biomass. The role of biomass as a part of microbial and enzymatic BFC will also be explored. Further, the applications of the BFCs related to industrial and environmental subjects will be discussed in detail.

6.1 INTRODUCTION

Biofuel cells (BFCs), the abbreviated form of biological fuel cells, is the fusion of two technologies—fuel cells and biotechnology (Palmore and Whitesides, 1994). Although it has been there in literature since old times (Davis, 1962; Yahiro et al., 1964), it was not until recently that BFCs have become a potent source for alternative and clean energy compared to traditional fuel cells. Due to the ever-increasing rise in consumption of non-renewable energy resources, the world is facing an energy crisis leading academic and industrial researchers to develop green energy sources for producing electricity, and for which BFCs provide a viable option.

BFCs such as traditional fuel cells consist of an anode where the fuel gets oxidized and produces electrons that are transferred through an external circuit to the cathode, where they reduce ambient oxygen to water by reacting with protons. Unlike in fuel cells, the metal catalyst is replaced with either micro-organisms or purified enzymes in BFCs (Moehlenbrock and Minteer, 2008).

There has been much research in the area of BFCs because of their overlapping characteristics with biosensors, which has led to its mass-market acceptance. However, these two diverge in the area of energy supply and stability. Biosensors are mainly energy consumers with low electronic potential and are made up of low-cost materials which make them easily disposable, thus making long-term stability nonessential, whereas BFCs have been made to be energy producers with high electronic potential (Barton et al., 2004). But BFCs with their biocatalysts make them less stable, which is why new techniques have come up where the fuel cells are encapsulated on a hydrophilic material such as a silica gel or immobilized on an inert support like single-walled carbon nanotubes (SWCNTs), making them highly stable. There have been deep concerns regarding the caging technique or the immobilization support. During current flow, they have been known to act as electron mediators, making them highly susceptible to the surrounding chemical fluids (Mano et al., 2003). This has made the development of BFCs as a biosensor an independent process.

BFCs mainly convert chemical energy into mechanical energy, utilizing microbes and enzymes as electron exchangers. They are mainly classified on the location of the biocatalysts as to whether it is present inside or outside a living cell. When living cells are involved, they are known as microbial BFCs, and when it is not, they are known to be enzymatic BFCs. Microbial fuel cells (MFCs) are more stable and efficient than enzymatic fuel cells (EFCs), but low power densities due to resistance across cell membranes make their applications limited in small-scale electronic devices. Whereas low resistance and high power density enable EFCs to be suitable for use in various microscale devices (Barton et al., 2004).

Biomass forms a natural renewable source of energy. Biofuels derived from biomass has already been extensively used and produced (Xuan et al., 2009). But the idea of using biomass as a source for MFCs or as a substrate for EFCs is extremely new and is getting constantly developed. Unlike non-renewable fuels which are harvested for producing energy and give rise to waste that has various harmful implications on the environment, biomass does not do so. In fact, it is readily available and even the abundant waste biomass that is produced after its primary use can be used as a potential source of BFCs.

One form of readily available biomass is the lignocellulosic biomass which is mainly composed of lignin, cellulose, and hemicellulose, with each component present in varying amounts from one type of source to another (Mcmillan, 1994). Lignocellulosic biomass has been a potent source for various forms of value-added products, which provides the current alternative for sustainable development (Iqbal and Kamal 2012). Each of the components of lignocellulosic biomass plays the role of a valuable source separately and together for the production of valuable chemicals such as C5 and C6 sugars and sugar-containing polymers (Isikgor and Becer, 2015) which are produced with low energy consumption. Though a relatively old process but the degradation of lignocellulosics by chemical hydrolysis, under mild and harsher conditions, produces C5 and C6 sugars but has also been achieved with enzymatic hydrolysis as well, which is a slow process (Demirbas, 2007). One disadvantage of chemical hydrolysis is the release of sludge which has proved to be harmful to the environment. Enzymatic hydrolysis includes microorganisms producing cellulases, hemicellulases, and ligninolytic enzymes, mostly by filamentous fungi, which are capable of converting lignocellulose to monosaccharide sugars, which can be further used for fermentation (Maitan-Alfenas et al., 2015). Different types of conversion and degradation strategy of lignocellulosics have been made to maximize its utilization (Sanchez 2009; Alonso et al., 2017), but hardly converting it for the production of electrical energy. Thus, BFCs utilizing lignocellulosics can be cost-effective and provide a novel market in the near future.

6.2 TYPES OF BFCs

6.2.1 MFCs

MFCs were the first of a kind of natural fuel cell which was thoroughly studied. MFCs include the use of intact microorganisms, which act as an electron source of an electrochemical cell. The electrons get produced as a reduced intermediate during substrate degradation, which then gets transferred to the anode. Allen and Bennetto (1993) chose *Proteus vulgaris* and investigated its current output by immobilizing it on the anode and feeding it with glucose. They observed the glucose getting assimilated in seconds, with an immediate reduction of the mediator compared to fuel cells containing the free bacteria; and concluded that immobilized bacteria produce a constant current output with less resistance and increase in functional lifetimes of the bacteria.

Agricultural, municipal, and industrial wastes form a source of biomass with abundant energy mainly in the form of carbohydrates, which are generally used for the production of bioethanol and ultimately converted to hydrogen. The biomass can be used by MFCs for generating electricity but is incapable because it needs a redox mediator for electron transfer and cannot completely oxidize the sugars. But *Rhodoferax ferrireducens* as a Fe (III)-reducing bacteria is able to fully oxidize glucose to CO_2 along with transferring electrons without the help of a redox mediator (Chaudhuri and Lovley, 2003).

The metabolic pathways used by a microorganism to produce electrons and intermittently protons have long been a mystery. So, to fully understand the mechanism involved the anodic current was reduced which led to the finding that microorganisms actually use the electron transport chain for its source of electrons and protons (Rabaey and Verstraete, 2005). Characteristics of MFCs such as power density, electrode potential, coulombic efficiency, and energy recovery depend on the ionic strength of the solution, electrode spacing and its composition, and temperature (Liu et al., 2005). One such example is the application of polytetrafluoroethylene (PTFE) layers on a carbon/PTFE base layer, to the cathode side exposed to air in an MFC chamber, which has improved coulombic efficiency significantly, with increased power densities and reduction of water loss through the cathode (Cheng et al., 2006). There are various types of bacteria that are used in MFC, and these include α-, β-, γ- , and -Proteobacteria, Firmicutes, and other uncharacterized clones of different bacteria which are termed as "exoelectrogens"—a group of bacteria capable of transferring electrons exocellularly (Logan and Regan 2006).

An important aspect of MFC technology incorporates a large number of substrates which includes various kinds of real and artificial wastewaters and different forms of lignocellulosic biomass (Pant et al., 2010). Though the current and power generated are low, future development to increase current yield may result in MFCs providing a sustainable resource for converting biomass to energy.

MFC containing mixed cultures has been found to produce more power density compared to when a single species of microorganism is used. MFCs are generally used for oxidizing different sources of carbon (Lovley, 2006). Applicatory disadvantages of using MFCs include the failure in simplifying the fuel cell for cost-effective up-scale purposes and success in wastewater treatment only under specific conditions such as temperature, the concentration of organic matter, and the absence of toxic materials (Logan et al., 2006(I)).

Over the years, MFCs have become a potential source for generating electricity using wastewater and biomass. Further research is needed for increasing the power output as well as decreasing the cost of electrodes for industrial bioreactor purposes.

6.2.2 EFCs

EFCs, also known as the primary or direct BFCs, consist of biocatalysts directly involved in the redox reaction for electricity generation. The biocatalysts or the enzymes generally get immobilized on electrodes, promoting its repeated use, and where direct electron transfer (DET) occurs between the two. The cathode contains the oxidizing enzyme which generates water upon reacting with protons and electrons whereas, the fuel at the anode produces electrons and protons upon getting enzymatically oxidized (Kim et al., 2006). Some advantages of using EFCs over MFCs are that they do not require any additional nutrients, produce more power output and can be easily engineered for miniaturized systems (Fischback et al., 2006).

The idea of using a purified enzyme in a fuel cell arose from the desire to have controlled and specific reactions in the fuel cell. Some uses of this type of system include in vivo applications, like a self-powering sensing system using glucose for detecting diabetes or using it as an implantable power supply for a cardiac pacemaker (Bullen et al., 2006). Though several developments have been made in this regard, yet many issues were encountered as well, like slow current generation and low efficiency. To maximize current density, multi-directional pore structures in the form of scaffolds have been used to provide support to the enzyme. This stabilizes the enzyme, increases the reactive surface area and permeability to liquid phase fuel transport (Minteer et al., 2007).

Many different but highly specific enzymes have been employed for the anodic and cathodic reaction in a biofuel cell. Glucose oxidase (EC 1.1.3.4, GOx) and different dehydrogenases are readily available and are mostly used for anodic reactions for glucose oxidation, whereas for oxygen reduction at the cathode, plant, and fungal laccases (EC 1.10.3.2) are generally studied (Ivanov et al., 2010). On the basis of electron transfer between the enzyme and the electrode, EFCs have been classified into direct and mediated electron transfer (DET and MET, respectively). When a mediator is used for enhancing electron transfer between the active part of the enzyme and the electrode, it is usually referred to as MET, which is generally used for most of the enzymes. A DET includes the direct transfer of electrons from the enzyme to the electrode (Osman et al., 2011; Falk et al., 2012).

The complete functioning of the enzyme-based electrode depends upon the physical and chemical properties of the enzyme immobilized layer. Though both physical and chemical methods are available for enzyme immobilization, it's mostly the chemical immobilization, which is used for fabricating the enzyme onto the electrode or the polymer matrix attached to it (Yu and Scott, 2010). The ability to achieve high efficiency of electron transfer has always been challenging. This is why the choice of electrode material and its surface chemistry has been extremely difficult. In this context, the field of nanobiotechnology has been the most promising. The availability of nanostructured conducting materials like the carbon nanotubes (CNTs), both single-walled and multi-walled, has contributed to important breakthroughs in the field of fuel cells (Holzinger et al., 2012). The highly accessible electrochemical surface area of the CNTs along with their high electronic conductivity and useful mechanical properties make these CNTs overtly attractive to use them as electrodes (Baughman et al., 2002).

Though more development has been done in the area of MFCs in case of biomass-derived fuel cells, recent research has focused on fuel processor systems which rely on enzymes (Xuan et al., 2009).

6.3 ROLE OF BIOMASS

6.3.1 AS A PRODUCER OF BFCs

The increasing demand for energy has shifted the focus from nonrenewables to renewables. Recent advances in conversion technology have primed the spotlight on lignocellulosic biomass, due to its ability to play a central role as a renewable carbon source for transportation fuels and chemicals (Binder and Raines 2009). The considerable amount in which lignocellulosics are produced as waste byproducts through agricultural practices especially in different agro-based industries has gained importance and research interests (Anwar et al., 2014). Detailed studies on conversion of lignocellulosic biomass into different products such as biofuels, value-added chemicals, and cheap energy sources for microbial fermentation and enzyme production has already been established, but the area of using the material as a fuel cell source is still in its infancy (Mosier et al., 2005; Lin and Huber 2009; Kang et al., 2004; Dien et al., 2000; Krishna and Chowdary 2000, Wang et al., 2009).

Different forms of lignocellulosic biomass are chemically treated or chemically pretreated followed by biological treatment to extract sugars for bio-energy production (Limayem and Ricke 2012). But over the years, researchers have derived many microbial communities including bacteria, fungi, and yeasts, from biomass produced as waste (Ryckeboer et al., 2003). These microorganisms later found many applications in biomass conversion or in its subsequent application, due to their ability to thrive within the biomass (Mayer and Hillebrandt, 1997). But, difficulty has been faced when they were used as MFCs, as they need a redox mediator and cannot completely oxidize the sugars. But special microorganisms such as the white rot fungi are capable of degrading the lignin in the biomass, exposing the cellulose and hemicellulose underneath, which can then be used by the other microorganisms, such as bacteria and yeasts, for generating electricity (Robinson et al., 2001). Using a microbial consortium can definitely be the way ahead, for using the biomass for direct energy production.

The biomass can be established as a source or producer of microbes which can be used in MFC, due to the suitable number of autochthonous microbes that have been identified (Maki et al., 2009). The microbes which grow on the biomass, using it as a solid substrate have a unique ability to produce enzymes required for the degradation of the biomass, which is important for its own growth. Lignocellulosic biomass is known to induce a variety of enzyme discharge such as cellulase, a string of peroxidases such as manganese peroxidase, lignin peroxidase, versatile peroxidase, and laccase (Guerra et al., 2008, Reddy et al., 2003, Vares et al., 1995). These enzymes because of their oxidative capacity can be used for EFCs; although some have already established its use in and as EFCs (Farneth and D'Amore 2005).

A key aspect of using lignocellulosic biomass is that they are available in plenty both as waste and as agri products. Though the main constituents remain the same between different types of biomass, yet they differ in the amounts in which the important constituents are present (Hendriks and Zeeman 2009; Hamelinck et al., 2005). Just like the microbes, a mix of the biomass can be used for the growth of microbes which will induce a mix of enzyme discharge but in high amounts, and vice versa, a single type of biomass will induce a specific type of enzyme discharge along with the growth of a specific type of microbes. Another advantage is utilizing the typical characteristic of the biomass to act both as a producer for the microbes and its subsequent use as a substrate in the microbial fuel cell. Also, it will be able to provide the nutrients required for the sustenance of the microbes.

Thus, biomass can readily have effective uses in the field of MFCs. Though further developments are needed for specifications of cathodic and anodic discharge, the material and the fuel that needed to be used, still biomass can prove to be highly valuable.

6.3.2 *AS A SUBSTRATE FOR BFCs*

The use of biomass to produce electricity has increased over the past 20 years. Biomass has helped to evade the global primary concern by providing energy security and by mitigating carbon pollution (Evans et al., 2010). But the technologies which are used such as pyrolysis, gasification, or direct combustion have limitations of huge land and water usage along with various social impacts (Ruiz et al., 2013). This is why the field of MFCs has provided new opportunities for sustainable production of bioenergy from biodegradable and reduced compounds. Using MFCs is highly advantageous in terms of the high conversion efficiency of substrate energy, working at ambient or even low temperatures, does not require any form of gas treatment or energy input and have the potential for widespread applications in remote areas lacking electrical infrastructures (Rabaey and Verstraete 2005).

In recent years, MFCs have gained a lot of attention as a mode of converting lignocellulosic biomass and wastewaters into electricity. Though the current generation is low at present, with development in knowledge and technology about these systems, various potential substrates can be used in the future (Pant et al., 2010). Lignocellulosic biomass contains energy stored in the form of carbohydrates which are utilized by MFCs or EFCs to generate electric power (Chaudhuri and Lovley, 2003).

Biomass such as corn stover is produced in huge amount as both agricultural and municipal waste. It mainly contains 15–20% lignin with 70% cellulose and hemicellulose and has been used extensively to produce ethanol and biohydrogen. Zuo et al. (2006) examined the production of electricity generation using MFCs from corn stover waste biomass by converting the hemicellulose to soluble sugars through either neutral or acid stream-exploded hydrolysis. Power densities were examined using an air-cathode MFC, where the electrochemically active bacteria oxidize the various carbohydrates at the anode surface and transfers the electron through an external circuit and the protons directly through the solution, to the cathode. But electricity can also be generated by using mixed microbial consortia which are acclimated for biomass breakdown. Mixed culture generally provides a high saccharification rate when added to the single chamber, air cathode MFCs producing an increase in power yield (Wang et al., 2009).

Wheat straw biomass is another type of renewable source produced in abundance and is composed of 35–45% cellulose and 20–30%

hemicellulose with relatively low lignin content. A mixed consortium was used to generate electricity from the wheat straw hydrolysate in a two-chamber MFC and was found that different types of bacteria played different roles in electricity generation (Zhang et al., 2009). Other cellulose-based lignocellulosic biomass has also been used for MFCs, although mostly hydrolysates were used (Ahmad et al., 2013).

Other sources of biomass-based substrates include furan derivatives and phenolic compounds which are produced upon the pretreatment process of lignocellulosic biomass. The study has shown 5-hydroxymethyl furfural (5-HMF) and two phenolic compounds, trans-cinnamic acid, and 3,5-dimethoxy-4-hydroxy-cinnamic acid proved to be effective in producing electricity from glucose in air-cathode MFCs (Catal et al., 2008). Biomass-based substrates can be indirect too. Like lignocellulosics-based agricultural products undergo industrial treatment and produce a massive amount of sludge or wastewater which contains lignins, cellulose, and hemicellulose mostly, along with other phenolics. These untreated wastewaters have proved to be harmful to the human beings and environment, as they easily leach out into the soil and can mix with the groundwater. Research has been made for effective treatment of these waste and MFCs were found to be useful. Though low power density was recovered, treatment of rice mill wastewaters by MFCs removed about 92–96% in case of chemical oxygen demand (COD), whereas there was succcssful lignin removal of about 84% and phenol removal of 81% (Behera et al., 2010).

Various forms of wastes have also been used for electricity generation. Though the components aren't mainly lignocellulosics, they have provided a range of substrates upon which MFCs can work. One such example is brewery wastewater, which is produced as a result of different treatment processes in the industry. Biological treatment is effective but the energy and cost required have been a huge problem over the years. Typically, MFC is the device capable of directly converting chemical energy in the organic matter into electrical energy. The MFC is designed with an anaerobic anodic zone and an aerobic cathodic zone separated by a proton exchange membrane (PEM). Brewery wastewater provides an excellent opportunity to generate electricity because of the low quantity of ammonium- nitrogen-related compounds, as compared to the high concentration of carbohydrates (Wang et al., 2008). Different types of food wastes contain a range of organic matter and are able to sustain different species of microbes. They have provided an invaluable source for waste-to-energy conversion. MFC has proven to be a unique way to dispose of food wastes by degrading organic matter and generating electricity, with the microbial community showing species diversity and richness (Jia et al., 2013). Carbon-based food wastes such as orange peel waste, an agricultural, and organic waste were utilized by MFCs for bioelectricity production without any chemical pretreatment or addition of extra mediators. A high current generation was achieved through the combined effort of fermentative microorganisms and exoelectrogens along with effective waste degradation (Miran et al., 2016).

The advancements in MFCs have raised the potential of lignocellulosic biomass which is produced as waste or sewage sludge. A variety of substrates can be used for sustainable energy generation using this technology. Although pretreatment of the biomass helps in an increased current density, direct use of biomass with the occasional addition to the system has tended to increase current yield. A single species of microorganism can be used, but mostly a mixed batch has produced more power density due to different species primed to degrade or use a certain type of component in the biomass. Future developments should include increasing the efficiency of biomass saccharification and electricity generation. In addition, we will need to optimize the degradation of lignocellulosics and power generation, if possible, in separate chambers so to improve the overall performance. Furthermore, development in the MFCs related to energy recovery is needed as the current yield from biomass is far lesser as compared to biofuels such as bioethanol.

6.3.3 AS A MEDIATOR OR CO-SUBSTRATE FOR BFCs

Fuel cell technologies have become the most sought after, due to their desirable feature of direct electrochemical conversion avoiding thermodynamic limitations, in addition to being more environmentally friendly (Osman et al., 2011; Leech et al., 2012). But developments are still needed in the efficiency of the conversion process coupling an oxidation reaction at the anode with a reduction reaction at the cathode. Just like it was found in MFC, the DET from the microorganisms to the anode is highly inefficient because of the selective permeability of the cell wall and membrane, as a result of which redox mediators were included to get rapidly reduced by the microorganisms and re-oxidised at the anode, thus increasing the efficiency of MFCs (Delaney et al., 1984). In early studies, various chemical compounds such as phenoxazine, phenothiazine, phenazine, indophenol, and bipyridinium derivatives were tested as redox mediators, and thiamine was found to be highly effective in MFCs containing *Proteus vulgaris* as the biological agent, with glucose as the oxidizable substrate (Delaney et al., 1984).

In case of EFCs, many oxidoreductase enzymes which are used have not been able to promote the transfer of electrons themselves, and many low molecular weight redox active compounds and polymers including organic dyes have been incorporated to mediate the transfer (Moehlenbrock and Minteer, 2008). This process is generally termed as mediated electron transfer (MET) and the mediators are selected on the basis of their stability, selectivity of oxidized and reduced species, and reversibility of their redox chemistry, requiring low over-potential. Mediators are mainly used in biofuel systems, either through polymerization on the electrode surface before enzyme immobilization, co-immobilization with an enzyme, or keeping the mediator free in solution (Moehlenbrock and Minteer, 2008; Yu and Scott, 2010). The vitally important part is to keep maintaining a continuous supply of fuel to the active sites and an efficient electron transfer process from the bacteria/enzyme to the electrode via the mediator (Osman et al., 2011).

Initial studies with mediators or co-substrates have been promising. Co-immobilization of *Trametesversicolor* laccase with osmium-based redox polymer coupled to glucose oxidase-mediated oxidation of glucose in a membrane-less biofuel cell resulted in an increase in current density (Barriere et al., 2004). The same redox polymer has been used for aldose dehydrogenase, providing further results of its stability and improved power output (Jenkins et al., 2012). Similar results were also achieved in MFCs when co-substrate was used with a readily biodegradable organic substrate for decolorization of an azo dye (Sun et al., 2009).

Biomass as a renewable source has already been established as a prominent substrate both for MFCs, and to a certain extent in EFCs. But they have been used far less as a co-substrate or mediator. Research has been made into using biomass such as wheat straw hydrolysate in MFCs as fuel to achieve both energy production and domestic wastewater purification, with a surge in power output as compared when synthetic co-substrates were used (Thygesen et al., 2011). The combined effort of treating pollutants with energy recovery was also utilized to remove sulfide and recover power with corn stover filtrate acting as a co-substrate with the MFCs. It was found that corn stover filtrate concentrations and electrolyte conductivity had significant effects on the performance of the MFCs (Zhang et al., 2013). A variety of lignocellulosics can be used as co-substrates or mediators for MFCs and can be further developed to be used in EFCs as well.

Using lignocellulosic biomass as a co-substrate or mediator has its disadvantages as well. The studies reported on the process to be slow. And the biomass cannot be directly used in the process. Generally, an initial step of pretreatment is required to speed up the process (Thygesen et al., 2011; Zhang et al., 2013). Research is required in the range of stability, time, and energy recovery. Recent developments in cathode and anode materials, immobilization techniques, use of nanomaterials, and mediated electron transfer may help in using biomass with EFCs. As compared with MFCs, EFCs have an advantage of having a specific enzyme, and including the list of oxidoreductases which are used, EFCs may be able to speed up the process of energy recovery, along with generating a comparatively higher power output. Both MFCs and EFCs have the potential of scaling-up the application for industrial purposes.

6.4 BIOMASS AS A STIMULANT FOR PRODUCTION OF BFCs

Biomass, especially lignocellulosic biomass has shown promise in the field of fuel cells. Studies have proved its broad application by both MFCs and EFCs alike. It has established itself by playing various roles along with acting as a source of bioenergy recovery. Fuel cells have emerged as a challenging technology due to its many facets in the role of bioconversion.

MFCs have been at the forefront of this technological evolution. The same yet different species of bacteria which are used in MFCs have attracted a range of substrates. Out of which, their ability to convert organic matter into electricity has got the most attention (Pant et al., 2010). Lignocellulosic biomass also does play the role of a stimulant, like providing nutrients for the stimulation of growth of the bacteria or acting as a co-substrate to induce current output. Whereas in EFCs, it generally helps in stimulating the enzyme secretion, mainly oxidoreductases. The type of fungi secreting these enzymes, which are mainly known as the white-rot, are lignin degraders, which are found in a prominent amount in the lignocellulosic biomass, and have long been established as an inducer for the growth and stimulator for lignin-degrading enzymatic discharge from them, which mainly include the group of oxidoreductases.

Substrates are regarded as one of the most important factors affecting electricity generation, which is why a range of substrates has been used as feed (Pant et al., 2010). In most cases, lignocellulosic biomass has not been used directly, they have been mostly pretreated or sometimes reduced to monosaccharides to increase the availability of sugars to the bacteria in MFCs, so as to generate current output, at an increased rate. Wheat straw hydrolysate has been used as a potential fuel to generate electricity in a two-chamber MFC system. The analysis was also made into the change in phylogenetic diversity that was taking place in terms of the presence of hydrolytic and respiratory anaerobes which couples the hydrolysis oxidation with proton reduction in the anode chamber (Zhang et al., 2009). But when co-treated for domestic wastewater purification, a higher power density was observed with 95% degradation of the xylan and glucan present in the hydrolysate (Thygesen et al., 2011).

Another biomass-based substrate which is most often used without pretreatment is rice straw, where a mixed culture of cellulose-degrading bacteria is used in a two-chambered MFC. This led to an increase in power output, almost equal to that generated when wheat straw hydrolysate was used, whereas the maximum power was increased by 3-fold in parallelly stacked MFCs. Thus, the results demonstrated that cellulose-degrading bacteria has the potential to be used as a biocatalyst with electricity produced from rice straw (Hassan et al., 2014). Pretreated rice straw has also been used as a fuel, for simultaneously treating wastewater and generating electric power. Where the fuel cells showed maximum power density, it also showed a reduction of the microbial diversities of the anodic biofilm and increase in diversity of the cathodic biofilm. Though most prevalent was *Proteobacteria*, the abundance of *Desulfobolus* was also noted. The result demonstrated how the rice straw hydrolysate along with acting as a fuel also enriches the microbial community structure (Wang et al., 2014).

Lignocellulosic biomass may not directly stimulate EFCs. But they do so indirectly. The enzymes mostly used for EFCs are oxidoreductases or oxidases, which are secreted by white rot basidiomycetes fungi. They have a unique function of degrading the lignin component of the biomass, releasing what is commonly known as lignin-degrading enzymes (Pollegioni et al., 2015). Over the years, research has established the effect that lignocellulosic biomass has on the discharge of these enzymes including sometimes on the fungus (Kumar and Sharma, 2017). White rot grows much better on biomass under solid substrate fermentation as compared to submerged and defined media. This is due to the stringent conditions of the substrate, which actually mimics the deficient conditions under which the fungus growing in the wild. But these are the environmental conditions under which the fungus survives the best (Zhou et al., 2015). Lignocellulosics, as mentioned before, contains a high percentage of lignin followed by cellulose and hemicellulose. The lignin-degrading enzymes are mostly released at first to degrade the lignin first, to expose the cellulosic and hemicellulosic mass underneath, for which a different set of enzymes are then released so as to get to the sugar after the breakdown (Montoya et al., 2015; Su et al., 2018). For a long period, white rots have been used for the utilization of lignocellulosics to generate biofuels. Until recently, it has found application in fuel cells, and is still getting studied to remove its limitations and improve and develop it for a broad range of application in fuel cells (Alonso et al., 2017).

6.5 CONVERSION OF BIOMASS TO USEFUL PRODUCTS BY BFCs

It has already been established that different forms of biomass play various roles when it comes to BFCs, starting from being a producer, substrate, and, a co-substrate or mediator. In turn, BFCs also play an important role with the biomass, directly or indirectly. BFCs modify the biomass either through pretreatment or via degrading certain recalcitrant polymer and producing useful products. For a long time, biomass was getting physically or chemically treated for its conversion or extraction of applicable products. But due to the harmful after-effects of the procedures, in terms of the harm, it was causing to the environment, there was a search for biologically natural methods (Akin et al., 1993). One easy find was the white rot fungi which paved the way for numerous applications based on biomass, especially with lignocellulosic biomass (Sindhu et al., 2015). Another principal application came into light with the introduction of BFCs. it was noted that MFCs and EFCs along with generating current, simultaneously convert the biomass as well. Though not much has been reported in this regard till now, it can find a wide applicability in the future.

As mentioned before, white rot fungi have been used for pretreatments of biomass and have successfully converted them into value-added products like biofuels, biomass-derived sugars, and other important potential products (Sanchez, 2009). Fungi have also been used as MFCs in many cases for bioelectricity production. Another way can be used to these fungi to grow on the biomass and will act both as an MFC and concurrently convert the biomass by degrading lignin (Schamphelaire et al., 2008). Whereas in the case of EFCs, the fungi and some bacteria are known to discharge a class of peroxidase enzymes, which are, if not all, most commonly used in EFCs (Logan, 2009). These enzymes have also been used for valuable biomass conversion as well (Dashtban et al., 2009).Adding these two functions together, the biomass can be used as a substrate with the enzymes getting used in EFCs and producing current. But these mechanistic procedures have their disadvantages as well. Some of which includes the slow generation of current, lesser electricity yield, and small-scale use of the biomass.

The various kinds of applicability of BFCs can be used to build a closed circuit for the development of sustainable electricity and as a source of renewable products. For example, white rot fungi used as MFCs on biomass can help for successful conversion of the biomass along with producing electricity (Pant et al., 2009). The fungi will, in turn, exude the enzymes which will be used in EFCs and where another batch of biomass could be used as a substrate for simultaneous electricity generation and conversion into key products for successful application in the industry. But a long time would be required for this kind of closed cycle device development. At first, an upscale of the process will be required with subsequent development of the fungal enzyme production either through biophysical or molecular modification (Logan, 2010). Following this, an evolution of BFCs will be important to counteract with the high level of enzyme and to generate current. A detailed study of the types of biomass will also be necessary for its successive use in the system, as to whether an earlier pretreatment will be necessary or a mixture of fungus will prove better for its use in the circuit, or addition of a substrate will increase the speed of the process and ultimately have an improved current yield. Thus, different modifications can render a valuable system for the production of a network of results.

6.6 INDUSTRIAL APPLICATIONS OF BFCs RELATED TO DIFFERENT TYPES OF BIOMASS

The development of BFCs and its application with biomass have opened a new domain for further study and ample improvement. BFC has shown a variety of applications with different kinds of biomass. Though not much progress has been made in all the cases, few have been developed, whereas other applications related to BFCs and biomass, are still evolving.

Lignocellulosic materials have been used in novel ways with MFCs, such as nutrient recovery from wastewaters. Agricultural residues, such as giant canes and maize stalks were used as porous separators, which helped in the mobility of electro-osmotic ions. This in combination with the power generated helped in depositing valuable elements such as Na, Ca, Mg, K, etc., and drove partial biodegradation of the lignocellulosic biomass as well (Marzorati et al., 2018).MFCs were also used for electricity generation from wastewater, although at first artificial wastewater was used with the continuous feeding of a sucrose solution as the electron donor, giving rise to insufficient substrate diffusion resulting in transport limitation, further proved by cyclic voltammetry (CV) (He et al., 2005). MFCs have become a fruitful bioreactor capable of breaking down organic matter present in a high concentration in wastewater (Du et al., 2007). Different types of MFCs have been developed to cope with the various types of wastewater. Like, a flat plate MFC was designed to work as a plug flow reactor by using a combined electrode/PEM and was continuously fed a solution containing domestic wastewater or wastewater enriched with a specific substrate (Min and Logan, 2004). Types of MFCs developed include a single chamber MFC (Liu et al., 2004), some include functioning at both ambient and mesophilic temperatures (Ahn and Logan 2010), liter-scale MFCs (Zhang et al., 2013), etc. The kind of wastewater used includes swine wastewater (Min et al., 2005), brewery wastewater (Feng et al., 2008), and others containing mostly lignin-related products like in rice mill wastewater (Behera et al., 2010).

Although the same has not been achieved yet in case of EFCs, progress has been made. Efforts were made to generate electricity from specific types of biomass, such as carboxymethyl cellulose (CMC) biomass. In this case, cellulase was used as the biocatalyst in order to catalyze the hydrolysis of CMC into glucose, which was then further used as the biomass to produce electricity through BFCs (Cheng et al., 2012). The same has been attained with other types of biomass such as the corn stover biomass, wheat straw biomass as well (Wang et al., 2009; Thygesen et al., 2011). The range of applications proves the versatility of power generation of BFCs and its uniqueness in the field of substrate utilization (Pandey et al., 2016).

There have been many well-studied applications when they come to different types of biomass and BFCs separately. In contrast, very few have come up when they are combined together. Though progress is slow, efforts are getting continuously made to develop and exploit this combination further. In situ conversion of stored chemical energy into electrical energy by BFCs should be thoroughly studied, for BFCs to perform and exhibit a comparable or a higher than other types of energy conversion technologies with biomass. Enhancement of BFC system is necessary for its efficient use with complex biomass in comparison to simple substrates. This combined technology can provide a breakthrough in the development of renewable technologies.

6.7 CONCLUDING REMARKS

With the advancement of the BFC technology in the past decade, it has opened a new frontier in terms of renewables. The energy crisis has led to the research and development into viable technologies for adoption in the future. This is why the focus has increased on BFCs and its counterpart—the MFCs and EFCs, for the development of sustainable technologies. The ability of BFCs to convert any form of chemical energy to electrical energy has rendered it with the ability for its use with biomass, especially lignocellulosic biomass. This form of biomass forms a potent source of renewable energy and gets generally produced in large amounts, mainly in the form of agriproducts, and has been the cause of environmental pollution due to over-accumulation and subsequent leaching; which led to its use for the harvesting of electrical energy by BFCs.

Any form of the glucose-based substrate has been used in BFCs for suitable transfer of electrons and has proved to be the best mediator in the system. This provided the background for the study with biomass, whose main components were mostly found to be glucose polymers. Most breakthroughs came from the sector where these kinds of polymers-based substrates or biomass were used. Hence, the fusion of biomass with BFCs has found a sure footing in various applications and can be evolved further. However, this fusion has certain disadvantages as well, like that mentioned earlier in this chapter. Future studies should be based on the improvement of this newfound applicability of this technology for the development of a simple, clean, and effective, along with being an environmentally benign sustainable energy conversion technology.

KEYWORDS

- biofuel cells
- microbial fuel cells
- enzymatic fuel cells
- lignocellulosic biomass
- power generation
- bioelectrochemical systems

REFERENCES

Ahmad, F.; Atiyeh, M. N.; Pereira, B.; Stephanopoulos, G. N. A Review of Cellulosic Microbial Fuel Cells: Performance and Challenges. *Biomass Bioenergy* **2013,** *56*, 179–188.

Ahn, Y.; Logan, B. E. Effectiveness of Domestic Wastewater Treatment Using Microbial Fuel Cells at Ambient and Mesophilic Temperatures. *Bioresour. Technol.* **2010,** *101*, 469–475.

Akin, D. E.; Sethuraman, A.; Morrison III, W. H.; Martin, S. A.; Eriksson, K. E. L. Microbial Delignification with White Rot Fungi Improves Forage Digestibility. *Appl. Environ. Microbiol.* **1993,** *59*, 4274–4282.

Allen, R. M.; Benetto, H. P. Microbial Fuel-Cells: Electricity Production from Carbohydrates. *Appl. Biochem. Biotechnol.* **1993,** *39/40*, 27–40.

Alonso, D. M.; Hakim, S. H.; Zhou, S.; Wangyun, W.; Hosseinaei, O.; Tao, J.; Garcia-Negron, V.; Motagamwala, A. H.; Mellmer, M. A.; Huang, K.; Houtman, C. J.; Labbe, N.; Harper, D. P.; Maravelias, C.; Runge, T.; Dumesic, J. A. Increasing the Revenue From Lignocellulosic Biomass: Maximizing Feedstock Utilization. *Sci. Adv.* **2017,** *3*: e1603301

Anwar, Z.; Gulfraz, M.; Irshad, M. Agro-Industrial Lignocellulosic Biomass a Key to Unlock the Future Bio-Energy: A Brief Review. *J. Radiat. Res. Appl. Sci.* 2014, http://dx.doi.org/10.1016/j.jras.2014.02.003

Barriere, F.; Ferry, Y.; Rochefort, D.; Leech, D. Targetting Redox Polymers as Mediators for Laccase Oxygen Reduction in a Membrane-Less Biofuel Cell. *Electrochem. Commun.* **2004,** *6*, 237–241.

Barton, S. C.; Gallaway, J.; Atanassov, P. Enzymatic Biofuel Cells for Implantable and Microscale Devices. *Chem. Rev.* **2004,** *104*, 4867–4886.

Baughman, R. H.; Zakhidov, A. A.; de Heer, W. A. Carbon Nanotubes – the Route Toward Applications. *Science* **2002,** *297*, 787–792.

Behera, M.; Jana, P. S.; More, T. T.; Ghangrakar, M. M. Rice Mill Wastewater Treatment in Microbial Fuel Cells Fabricated Using Proton Exchange Membrane and Earthen Pot at Different pH. *Bioelectrochemistry* **2010,** *79*, 228–233.

Binder, J. B.; Raines, R. T. Simple Chemical Transformation of Lignocellulosic Biomass into Furans for Fuels and Chemicals. *J. Am. Chem. Soc.* **2009,** *131*, 1979–1985.

Bullen, R. A.; Arnot, T. C.; Lakeman, J. B.; Walsh, F. C. Biofuel Cells and Their Development. *Biosens. Bioelectron.* **2006,** *21*, 2015–2045.

Catal, T.; Fan, Y.; Li, K.; Bermek, H.; Liu, H. Effects of Furan Derivatives and Phenolic Compounds on Electricity Generation in Microbial Fuel Cells. *J. Power Sources* **2008,** *180*, 162–166.

Chaudhuri, S. K.; Lovley, D. R. Electricity Generation by Direct Oxidation of Glucose in Mediatorless Microbial Fuel Cells. *Nat. Biotechnol.* **2003,** *21*, 1229–1232.

Cheng, H.; Qian, Q.; Wang, X.; Yu, P.; Mao, L. Electricity Generation From Carboxymethyl Cellulose Biomass: A New Application of Enzymatic Biofuel Cells. *Electrochimica Acta* **2012,** *82*, 203–207.

Cheng, S.; Liu, H.; Logan, B. E. Increased Performance of Single-Chamber Microbial Fuel Cells Using an Improved Cathode Structure. *Electrochem. Commun.* **2006,** *8*, 489–494.

Dashtban, M.; Schraft, H.; Qin, W. Fungal Bioconversion of Lignocellulosic Residues; Oppurtunities & Perspectives. *Int. J. Biol. Sci.* **2009,** *5*, 578–595.

Davis, J. B.; Yarborough, H. F. Preliminary Experiments on a Microbial Fuel Cell. *Science* **1962,** *137*, 615–616.

Delaney, G. M.; Benetto, H. P.; Mason, J. R.; Roller, S. D.; Stirling, J. L.; Thurston, C. F. Electron-Transfer Coupling in Microbial Fuel Cells. 2. Performance of Fuel Cells Containing Selected Microorganism-Mediator-Substrate Combinations. *J. Chem. Tech. Biotechnol.* **1984,** *34B*, 13–27.

Demirbas, A. Products From Lignocellulosic Materials via Degradation Processes. *EnergySources, Part A* **2008,** *30*, 27–37.

Dien, B. S.; Nichols, N. N.; O'Bryan, P. J.; Bothast, R. J. Development of New Ethanologenic *Escherichia coli* Strains for Fermentation of Lignocellulosic Biomass. *Appl. Biochem. Biotechnol.* **2000,** *84–86*, 181–196.

Du, Z.; Li, H.; Gu, T. A State of the Art Review on Microbial Fuel Cells: A Promising Technology for Wastewater Treatment and Bioenergy. *Biotechnol. Adv.* **2007,** *25*, 464–482.

Evans, A.; Strezov, V.; Evans, T. J. Sustainability Considerations for Hlectricity Generation From Biomass. *Renew. Sustain. Energy Rev.* **2010,** *14*, 1419–1427.

Falk, M.; Blum, Z.; Shleev, S. Direct Electron Transfer Based Enzymatic Fuel Cells. *Electrochimica Acta* **2012,** *82*, 191–202.

Farneth, W. E.; D'Amore, M. B. Encapsulated Laccase Electrodes for Fuel Cell Cathodes. *J. Electroanal. Chem.* **2005,** *581*, 197–205.

Feng, Y.; Wang, X.; Logan, B. E.; Lee, H. Brewery Wastewater Treatment Using Air-Cathode Microbial Fuel Cells. *Appl. Microbiol. Biotechnol.* 2008, *78*, https:doi.org/10.1007/s00253-008-1360-2

Fischback, M. B.; Youn, J. K.; Zhao, X.; Wang, P.; Park, H. G.; Chang, H. N.; Kim, J.; Ha, S. Miniature Biofuel Cells with Improved Stability Under Continuous Operation. *Electroanalysis* **2006,** *18*, 2016–2022.

Guerra, G.; Domininguez, O.; Ramos-Leal, M.; Manzano, A. M.; Sanchez, M. I.; Hernandez, I.; Palacios, J.; Arguelles, J. Production of Laccase and Manganese Peroxidase by White-Rot Fungi From Sugarcane Bagasse in Solid Bed: Use for Dyes Decolorization. *Sugar Tech.* **2008,** *10*(3), 260–264.

Hamelinck, C. N.; van Hooijdonk, G.; Faaij, A. P. C. Ethanol From Lignocellulosic Biomass: Techno-Economic Performance in Short-, Middle- and Long Term. *Biomass Bioenergy.* **2005,** *28*, 384–410.

Hassan, S. H. A.; El Rab, S. M. F. G; Rahimnejad, M.; Ghasemi, M.; Joo, J.; Sik-Ok, Y.; Kim, I. S.; Oh, S. Electricity Generation From Rice Straw Using a Microbial Fuel Cell. *Int. J. Hydrogen Energy* **2014,** *3*, 1–7.

He, Z.; Minteer, S. D.; Angement, L. T. Electricity Generation From Artificial Wastewater Using an Upflow Microbial Fuel Cell. *Environ. Sci. Technol.* **2005,** *39*, 5262–5267.

Hendriks, A. T. W. M.; Zeeman, G. Pretreatments to Enhance the Digestibility of Lignocellulosic Biomass. *Bioresour. Technol.* **2009,** *100*, 10–18.

Holzinger, M.; Le Goff, A.; Cosnier, S. Carbon Nanotube/enzyme Biofuel Cells. *Electrochimica Acta* **2012,** *82*, 179–190.

Iqbal, H. M. N.; Kamal, S. Economical Bioconversion of Lignocellulosic Materials to Value-Added Products. *J. Biotechnol. Biomater.* 2012, *2*: e112. Doi: 10.4172/2155-952X.1000e112

Isikgor, F. H.; Becer, C. R. Lignocellulosic Biomass: A Sustainable Platform for Production of Bio-Based Chemicals and Polymers. 2015

Ivanov, I.; Vidakovic-Koch, T.; Sundmacher, K. Recent Advances in Enzymatic Fuel Cells: Experiments and Modeling. *Energies* **2010,** *3*, 803–846.

Jenkins, P.; Tuurala, S.; Vaari, A.; Valkiainen, M.; Smolander, M.; Leech, D. A Comparison of Glucose Oxidase and Aldose Dehydrogenase as Mediated Anodes in Printed Glucose/Oxygen Enzymatic Fuel Cells Using ABTS/Laccase Cathodes. *Bioelectrochemistry* **2012,** *87*, 172–177.

Jia, J.; Tang, Y.; Liu, B.; Wu, D.; Ren, N.; Xing, D. Electricity Generation From Food Wastes and Microbial Community Structure in Microbial Fuel Cells. *Bioresour. Technol.* **2013,** *144*, 94–99.

Kang, S. W.; Park, Y. S.; Lee, J. S.; Hong, S. I.; Kim, S. W. Production of Cellulases and Hemicellulases by *Aspergillus niger* KK2 from Lignocellulosic Biomass. *Bioresour. Technol.* **2004,** *91*, 153–156.

Kim, J.; Jia, H.; Wang, P. Challenges in Biocatalysis for Enzyme-Based Biofuel Cells. *Biotechnol. Adv.* **2006,** *24*, 296–308.

Krishna, S. H.; Chowdhary, G. V. Optimization of Simultaneous Saccharification and Fermentation for the Production of Ethanol From Lignocellulosic Biomass. *J. Agric. Food Chem.* **2000,** *48*, 1971–1976.

Kumar, A. K.; Sharma, S. Recent Updates on Different Methods of Pretreatment of Lignocellulosic Feedstocks: A Review. *Bioresour. Bioprocess.* **2017,** *4*, 7.

Leech, D.; Kavanagh, P.; Schuhmann, W. Enzymatic Fuel Cells: Recent Progress. *Electrochimica Acta* **2012,** *84*, 223–234.

Limayem, A.; Ricke, S. C. Lignocellulosic Biomass for Bioethanol Production: Current Perspectives, Potential Issues and Future Prospects. *Prog. Energy Combust. Sci.* **2012,** *38*, 449–467.

Lin, Y.; Huber, G. W. The Critical Role of Heterogeneous Catalysis in Lignocellulosic Biomass Conversion. *Energy Environ. Sci.* **2009,** *2*, 68–80.

Liu, H.; Cheng, S.; Logan, B. E. Power Generation in Fed-Batch Microbial Fuel Cells as a Function of Ionic Strength, Temperature and Reactor Configuration. *Environ. Sci. Technol.* **2005,** *39*, 5488–5493.

Liu, H.; Ramnarayanan, R.; Logan, B. E. Production of Electricity During Wastewater Treatment Using a Single Chamber Microbial Fuel Cell. *Environ. Sci. Technol.* **2004,** *38,* 2281–2285.

Logan, B. E. Exoelectrogenic Bacteria that Power Microbial Fuel Cells. *Nat. Rev. Microbiol.* **2009,** *7,* 375–381.

Logan, B. E. Scaling Up Microbial Fuel Cells and Other Bioelectrochemical Systems. *Appl. Microbiol. Biotechnol.* 2010, *85*. https://doi.org/10.1007/s00253-009-2378-9

Logan, B. E.; Hamelers, B.; Rozendal, R.; Schroeder. U.; Keller, J.; Fregula, S.; Aelterman, P.; Verstraete, W.; Rabaey, K. Microbial Fuel Cells: Methodology and Technology. *Environ. Sci. Technol.* **2006,** *40* (1), 5181–5192.

Logan, B. E.; Regan, J. M. Electricity-Producing Bacterial Communities in Microbial Fuel Cells. *TRENDS Microbiol.* **2006,** *14,* 512–518.

Lovley, D. R. Microbial Fuel Cells: Novel Microbial Physiologies and Engineering Approaches. *Curr. Opin. Biotechnol.* **2006,** *17,* 327–332.

Maitan-Alfenas, G. P.; Visser, E. M.; Guimaraes, V. M. Enzymatic Hydrolysis of Lignocellulosic Biomass: Converting Food Waste in Valuable Products. *Curr. Opin. Food Sci.* **2015,** *1,* 44–49.

Maki, M.; Leung, K. T.; Qin, W. The Prospects of Cellulase-Producing Bacteria for the Bioconversion of Lignocellulosic Biomass. *Int. J. Biol. Sci.* **2009,** *5,* 500–516.

Mano, N.; Mao, F.; Heller, A. Characteristics of a Miniature Compartment-Less Glucose-O_2 Biofuel Cell and Its Operation in a Living Plant. *J. Am. Chem. Soc.* **2003,** *125,* 6588–6594.

Marzorati, S.; Schievano, A.; Colombo, A.; Lucchini, G.; Cristiani, P. Lignocellulosic Materials as Air-Water Separators in Low-Tech Microbial Fuel Cells for Nutrients Recovery. *J. Clean. Prod.* **2018,** *170,* 1167–1176.

Mayer, F.; Hillebrandt, J.-O. Potato Pulp: Microbiological Characterization, Physical Modification, and Application of this Agricultural Waste Product. *Appl. Microbiol. Biotechnol.* **1997,** *48,* 435–440.

Mcmillan, J. D. In Enzymatic Conversion of Biomass for Fuels Production. *ACS Symposium Series.* American Chemical Society: Washington DC, 1994.

Min, B.; Logan, B. E. Continuous Electricity Generation From Domestic Wastewater and Organic Substrates in a Flat Plate Microbial Fuel Cell. *Environ. Sci. Technol.* **2005,** *38,* 5809–5814.

Min, B.; Kim, J.; Oh, S.; Regan, J. M.; Logan, B. E. Electricity Generation From Swine Wastewater Using Microbial Fuel Cells. *Water Res.* **2005,** *39,* 4961–4968.

Minter, S. D.; Liaw, B. Y.; Cooney, M. J. Enzyme-Based Biofuel Cells. *Curr. Opin. Biotechnol.* **2007,** *18,* 228–234.

Miran, W.; Nawaz, M.; Jang, J.; Lee, D. S. Conversion of Orange Peel Waste Biomass to Bioelectricity Using a Mediator-Less Microbial Fuel Cell. *Sci. Total Environ.* **2016,** *547,* 197–205.

Moehlenbrock, M. J.; Minteer, S. D. Extended Lifetime Biofuel Cells. *Chem. Soc. Rev.* **2008,** *37,* 1188–1196.

Montoya, S.; Sanchez, O. J.; Levin, L. Production of Lignocellulolytic Enzymes From Three White-Rot Fungi by Solid-State Fermentation and Mathematical Modeling. *Afr. J. Biotechnol.* **2015,** *14,* 1304–1317.

Osman, M. H.; Shah, A. A.; Walsh, F. C. Recent Progress and Continuing Challenges in Bio-Fuel Cells. Part I: Enzymatic Cells. *Biosens. Bioelectron.* **2011,** *26,* 3087–3102.

Palmore, G. T. R.; Whitesides, G. M. In Enzymatic Conversion of Biomass for Fuels Production. *ACS Symposium Series; American Chemical Society; Washington DC,* 1994.

Pandey, P.; Shinde, V. N.; Deopurkar, R. L.; Kale, S. P.; Patil, S. A.; Pant, D. Recent Advances in the Use of Different Substrates in Microbial Fuel Cells Toward Wastewater Treatment and Simultaneous Energy Recovery. *Appl. Energy* **2016,** *168*, 706–723.

Pant, D.; Bogaert, G. V.; Diels, L.; Vambroekhovan, K. A Review of the Substrates Used in Microbial Fuel Cells (MFCs) for Sustainable Energy Production. *Bioresour. Technol.* **2010,** *101*, 1533–1543.

Pollegioni, L.; Tonin, F.; Rosini, E. Lignin-Degrading Enzymes. *FEBS J.* **2015,** *282*, 1190–1213.

Rabaey, K.; Verstraete, W. Microbial Fuel Cells: Novel Biotechnology for Energy Generation. *Trends Microbiol.* **2005,** *23*, 291–298.

Reddy, G. V.; Babu, P. R.; Komaraiah, P.; Roy, K. R. R. M.; Kothari, I. L. Utilization of Banana Waste for the Production of Ligninolytic and Cellulolytic Enzymes by Solid Substrate Fermentation Using Two *Pleurotus* Species (*P. ostreatus* and *P. sajor-caju*). *Process Biochem.* **2003,** *38*, 1457–1462.

Robinson, T.; McMullan, G.; Marchant, R.; Nigam, P. Remediation of Dyes in Textile Effluent: A Critical Review on Current Treatment Technologies With a Proposed Alternative. *Bioresour. Technol.* **2001,** *77*, 247–255.

Ruiz, J. A.; Juarez, M. C.; Morales, M. P.; Munoz, P.; Mendivil, M. A. Biomass Gasification for Electricity Generation: Review of Current Technology Barriers. *Renew. Sustain. Energy Rev.* **2013,** *18*, 174–183.

Ryckeboer, J.; Mergaert, J.; Coosemans, J.; Deprins, K.; Swings, J. Microbiological Aspects of Biowaste During Composting in a Monitored Compost Bin. *J. Appl. Microbiol.* **2003,** *94*, 127–137.

Sanchez, C. Lignocellulosic Residues: Biodegradation and Bioconversion by Fungi. *Biotechnol. Adv.* **2009,** *27*, 185–194.

Schamphelaire, L. D.; Bossche, L. V. D.; Dang, H. S.; Hofte, M.; Boon, N.; Rabaey, K.; Verstraete, W. Microbial Fuel Cells Generating Electricity From Phizodeposits of Rice Plants. *Environ. Sci. Technol.* **2008,** *42*, 3053–3058.

Sindhu, R.; Binod, P.; Pandey, A. Biological Pretreatment of Lignocellulosic Biomass – An Overview. *Bioresour. Technol.* 2015, http://dx.doi.org/10.1016/j.biortech.2015.08.030

Su, Y.; Yu, X.; Sun, Y.; Wang, G.; Chen, H.; Chen, G. Evaluation of screened Lignin-Degrading Fungi for the Biological Pretreatment of Corn Stover. *Sci. Rep.* **2018,** *8*, 1–11.

Sun, J.; Hu, Y.; Bi, Z.; Cao Y. Simultaneous Decolorization of Azo Dye and Bioelectricity Generation Using a Microfiltration Membrane Air-Cathode Single-Chamber Microbial Fuel Cell. *Bioresour. Technol.* **2009,** *100*, 3185–3192.

Thygesen, A.; Poulsen, F. W.; Angelidaki, I.; Min, B.; Bjerre A. Electricity Generation by Microbial Fuel Cells Fueled With Wheat Straw Hydrolysate. *Biomass Bioenergy.* **2011,** *35*, 4732–4739.

Vares, T.; Kalsi, M.; Hatakka, A. Lignin Peroxidases, Manganese Peroxidases, and Other Ligninolytic Enzymes Produced by Phlebia Radiate During Solid-State Fermentation of Wheat Straw. *Appl. Environ. Microbiol.* **1995,** *61*, 3515–3520.

Wang, X.; Feng, Y. J.; Lee, H. Electricity Production From Beer Brewery Wastewater Using Single Chamber Microbial Fuel Cell. *Water Sci. Technol.* **2008,** *57*, 1117–1121.

Wang, X.; Feng, Y.; Wang, H.; Qu, Y.; Yu, Y.; Ren, N.; Li, N.; Wang, E.; Lee, H.; Logan, B. E. Bioaugmentation for Electricity Generation From Corn Stover Biomass Using Microbial Fuel Cells. *Environ. Sci. Technol.* **2009,** *43*, 6088–6093.

Wang, Z.; Lee, T.; Lim, B.; Choi, C.; Park, J. Microbial Community Structures Differentiated in a Single-Chamber Air-Cathode Microbial Fuel Cell Fueled With Rice Straw Hydrolysate. *Biotechnol. Biofuels.* **2014,** *7*, 9.

Xuan, J.; Leung, M. K. H.; Leung, D. Y. C.; Ni, M. A Review of Biomass-Derived Fuel Processors for Fuel Cell Systems. *Rene. Sustain. Energy Rev.* **2009,** *13*, 1301–1313.

Yahiro, A. T.; Lee, S. M.; Kimble, D. O. Bioelectrochemistry I. Enzyme Utilizing Bio-Fuel Cell Studies. *Biochim. Biophys. Acta* **1964,** *88*, 375–383.

Yu, E. H.; Scott, K. Enzymatic Biofuel Cells – Fabrication of Enzyme Electrodes. *Energies* **2010,** *3*, 23–42.

Zhang, F.; Ge, Z.; Grimaud, J.; Hurst, J.; He, Z. Long-Term Performance of Liter-Scale Microbial Fuel Cells Treating Primary Effluent Installed in a Municipal Wastewater Treatment Facility. *Environ. Sc. Technol.* **2013,** *47*, 4941–4948.

Zhang, Y.; Min, B.; Huang, L.; Angelidaki, I. Generation of Electricity and Analysis of Microbial Communities in Wheat Straw Biomass-Powered Microbial Fuel Cells. *Appl. Environ. Microbiol.* **2009,** *75*, 3389–3395.

Zhang, J.; Zhang, B.; Tian, C.; Ye, Z.; Liu, Y.; Lei, Z.; Huang, W.; Feng, C. *Bioresour. Technol.* **2013,** *138*, 198–203.

Zhou, S.; Raouche, S.; Grisel, S.; Navarro, D.; Sigoillot, J.; Herpoel-Gimbert, I. Solid-State Fermentation in Multi-Well Plates to Assess Pretreatment Efficiency of Rot Fungi on Lignocellulosic Biomass. *Microb. Biotechnol.* **2015,** *8*, 940–949.

Zuo, Y.; Maness P.; Logan, B. E. Electricity Production From Steam-Exploded Corn Stover Biomass. *Energy Fuels.* **2006,** *20*, 1716–1721.

CHAPTER 7

Biofuel Production from Algal Biomass

RITUPARNA SAHA[1,2], DEBALINA BHATTACHARYA[3], and
MAINAK MUKHOPADHYAY[1,*]

[1]*Department of Biotechnology, JIS University, Kolkata, West Bengal, India*

[2]*Department of Biochemistry, University of Calcutta, Kolkata, West Bengal, India*

[3]*Department of Biotechnology, University of Calcutta, Kolkata, West Bengal, India*

*Corresponding author. E-mail: m.mukhopadhyay85@gmail.com

ABSTRACT

Biofuels provide a clean and green alternative to fossil fuels. With the increase in population, there is an increasing demand for energy for lighting, heating, and transportation. This has led to an overexploitation of renewable sources leaving a negative impact on the environment. This has led to the production of biofuels from a variety of sources like food crops, agri-products, and others. But the most valuable source turned out to be algae–microalgae and macroalgae. Though not much research has been made in regard to biofuel production from macroalgae, bioconversion of biofuels from microalgae has been well studied in detail.

Microalgae are mostly primitive microorganisms and require sunlight for photosynthesis. They have a high lipid and carbohydrate content, in comparison to the high lignin and hemicellulose concentration in biomass derived from terrestrial plants. Biofuels derived from microalgae are mostly third generation biofuels and includes bioethanol, biohydrogen, biogas, and biodiesel. The cultivation, harvesting and bioconversion process has been long well defined and is still under development. The biofuels have found many applications in the industry and will surely become available for other common purposes in the future.

This chapter will mainly deal with the different types of cultivation and harvesting technologies, with special attention given to the bioconversion processes that are available. The different types of biofuels and its industrial application will also be discussed. Special mention will also be made about the technologies available for biofuel production from macroalgae and its future prospects.

7.1 INTRODUCTION

Utilization of fossil hydrocarbons has played a huge part in the global economy since time immemorial—from providing fertilizers to producing plastics, and providing energy in the form of transportation, heating, and lighting. This has led to the overuse and overexcavation of prominent sites and thus has left a negative impact on the surrounding environment. Another unacceptable effect has been the CO_2 emissions in the atmosphere released for using fossil fuels, leading to global warming. With rising in average global temperature and sea level, scientists have to look for the alternatives (Hannon et al., 2010). Fossil fuels are non-renewable sources and with the rise in population, the global economy has already predicted about its increased demand for fossil fuels, which is going to eventually increase the price of the different types of fossil fuels, and therefore going to ultimately affect the world economy and its people. Over the years, the negative effects on the environment caused by the usage of fossil fuels have made the recovery of fossil fuels very hard to recover. These factors are shifting the focus from non-renewables to renewables, and driving the research and development of renewable sources and their utilization for production of biofuels (Tilman et al., 2009).

For the past few years, a wide range of renewable sources has been identified as an alternative to fossil fuels for all nations, and greatly reduce the carbon emissions into the atmosphere. A number of technologies have been developed for producing energy from the identified renewable sources. Although in some cases a series of technologies are used for energy production, studies are still going on for progression of the technologies which will likely provide a simple solution and will definitely decrease our dependence on fossil fuels. The challenge is to develop sustainable industries for these technologies which will function on renewable energy and will at the same time prove to be cost-effective (Ragauskas et al., 2006).

Fossil fuels are mostly used for generating electricity, and also as liquid fuels. Various technologies causing low pollution and producing electricity from renewable sources has now supplemented fossil fuels. They mostly include hydroelectric, nuclear, solar, geothermal and wind sources. But renewable technologies to supplement or replace liquid fossil fuels are primarily in their early developmental stages. A study by International Energy Agency estimates that by 2030 biofuel will contribute 6% of the total fuel use, but has a probability of expanding significantly if certain fundamental new fields are not developed (Hannon et al., 2010).

The most sustainable technology to replace liquid fossil fuels has been 'Biofuels.' The word is generally used to describe a diverse range of technologies which are dependent on a biological system to generate fuel. The biological systems generally vary but they have the same type of polymers from which the biofuels are extracted. These biological systems are mostly produced or derived from agricultural products. They are substantially in the form of waste that is not used or is generated after certain industrial applications (Rodionova et al., 2016). These biological systems are most often referred to as "Biomass." Biomass mainly describes plants or plant-based organic materials which are not used as food or livestock feed and is a renewable and sustainable source of energy. Biomass contains stored energy from the sun, and the chemical energy is released as heat when it is burned. All types of biomass contain the same basic constituents but in different quantities (Groom et al., 2008). A single technology or process is mainly required for the chemical to energy conversion. It all depends on the types of biomass. There are certain types which require multiple technologies or processes for the conversion as well. Recent approaches have all been made in biofuel production depending on microbial conversion, and a possibility of harvesting energy by direct conversion from microalgal and macroalgal cultivation to produce biofuel has been developed (Rodionova et al., 2016).

7.2 BIOFUEL PRODUCTION FROM MICROALGAE AND MACROALGAE

Biofuel production from renewable sources can help to reduce fossil fuel consumption and CO_2 emissions. It can also reduce the world's dependence on oil and thus help to mitigate global warming. An advantage of using biofuels is that it is made of the CO_2 used by the plants to make food during photosynthesis which is released into the air, thus keeping the net CO_2 in the atmosphere same as before (Naik et al., 2010). Also, the by-products produced during biofuel production can help to sustain and provide income and employment opportunities to people. Thus, a shift in industrial focus is required for producing alternate green processes for utilizing the feedstock and other organic materials for the production of these chemicals from renewable biomass resources.

Though not technically classified, biofuels are generally of two types based on the stages of purity and processes required for production. A first-generation biofuel, for example, biodiesel, bioethanol, and biogas, is characterized by its ability to blend with petroleum-based fuels, or other alternative technology, or in natural gas vehicles (Mohr and Raman, 2013). Though the carbon emission is less, scientists are still concerned with the source of feedstocks and the impact it can have on the land use and competition with food crops, with its impact on biodiversity. Still, the production of first-generation biofuel has been commercialized, with about 50 billion liters produced annually (Naik et al., 2010).

Second-generation biofuels are mainly produced from plant biomass and include lignocellulosic materials, which makes up the majority of the cheap, agriwastes that are available from plants. Lignocellulosic feedstock provides the production of novel biofuels like cellulosic ethanol and Fischer-Tropsch fuels (Naik et al., 2010). Plant biomass remains the most underused biological resource but can become a promising material source for fuels. There is great potential for the production of liquid fuels from this type of biomass. But the processes which are utilized for production are not cost effective and hence remains non-commercialized. When developed, second-generation biofuels could offer to give a better performance proving to be cost-effective and greatly reduce CO_2 emissions (Mohr and Raman, 2013).

Production of biofuel which is economically feasible can pave a way for relieving the dependency on fossil fuels. Deriving biofuels by cultivating algae is one such measure. It has potential over other types of biofuel production, in relation to, that it does not compete with food crops, have a rapid growth rate, can be cultivated almost anywhere even in lands not suitable for agriculture and its favorable growing conditions are found in most countries, providing a growing economy for developing countries, especially for people in rural areas (Adenle et al., 2013). Algae has an unlimited capacity and can be a potentially viable source of sustainable biofuels in the future. Algae has been found to metabolize various types of waste streams like municipal wastewater and carbon dioxide from industrial flue gas and produces products with a wide range of compositions and uses. These products are mostly used for the production of different types of biofuels and other industrial uses (Menetrez, 2012).

Biofuels from micro- and macroalgae are referred to as the third-generation biofuels and have an advantage over the other two. As previously mentioned, these marine organisms can be cultivated anywhere, be it in containment off-shore or even in brackish, saline and wastewater, without requiring any arable land, along with producing a high yield of biomass. Proper development of cultivation methods and processing has the ability to make a highly superior third generation biofuel (Daroch et al., 2013).

Microalgae are considered to be one of the most primitive life-forms and consists of both unicellular and multicellular organisms. The productivity of these organisms is very high, as they are mostly photosynthetic and converts CO_2 into carbon-rich lipids, which are few steps away from getting processed into biofuels (Brennan and Owende, 2010). However, microalgal biofuel production has not been commercialized in an industrial scale due to the high investment costs and energy required during large scale production. Research and development are being carried out for the development of technology needed for algal biofuel production to a major industrial process (Ghasemi et al., 2012).

Macroalgae, being fast-growing marine and freshwater plants, far exceeds those of terrestrial biomass. They are mostly found to inhabit an environment where there are vigorous water movement and turbulent diffusion. Unlike microalgae, macroalgae have high level of nutrient uptake as well as photosynthesis and growth (Chen et al., 2015). The absence of lignin and a low content of cellulose, makes the macroalgae to be easily convertible in biological processes as compared to land plants. Macroalgae are farmed on a massive scale in many continents but provide challenges for developing cost-effective methodologies to grow, harvest, transport, and ultimately process macroalgae for biofuel production (Kraan, 2013).

7.3 MICROALGAL CULTIVATION TECHNOLOGIES

Over the years, various types of cultivation technologies have been developed for microalgae. Though in controlled systems, but they can be further developed for large scale production as well. During culturing of microalgae, a subculture is maintained to ensure continuous cell division and growth. The water, air, and CO_2 are filtered to avoid contamination. Cultivation is mainly done in batch, semi-batch or continuous cultures. In batch culture, after inoculation of cells in the media, it is incubated for several days so that the culture reaches a desirable amount, after which it is transferred to a container with large volumes of media, where it is grown until it reaches stationary phase. In semi-batch culture, one half of the media is replenished after another half is transferred to a large-scale cultivator. In continuous culture, either new media is added after the density of the culture reaches a certain level or a steady influx of media is given whereas a part of it is continuously harvested (Ghasemi et al., 2012).

7.3.1 OPEN-AIR SYSTEMS

When mass cultivation of microalgae started, most of it was being carried out in closed culture systems, which proved to be economically unfavorable. This is why all commercial systems use open-air systems—because of its cost-effectiveness and ease of scaling up. There are mainly four types of open-air systems which are currently in use commercially—shallow big ponds, tanks, circular ponds, and raceway ponds (Borowitzka, 1999).

Open pond systems mostly constitute of shallow ponds in which nutrients are provided through run-off water from neighboring areas or sewage water/water channeled from treatment plants. The water is generally kept in motion by some rotating structures or some developed technology. Though a major disadvantage of this type of open-air system is that very few species of algae are actually able to grow in this type of systems. In most of the cases, the algal species are defined, that is, only one or more selected strains are used, and in some cases the species' remains undefined as because a mixture of strains are used (Demirbas, 2010).

A typical raceway pond is composed of a closed loop channel similar to a raceway track, kept shallow, open to the air, and generally present with a paddle wheel that keeps the water in motion, constantly mixing the nutrients and preventing sedimentation. The low depth of the pond keeps the system in low sunlight and limits light penetration through the algal broth (Slade and Bauen, 2013).

7.3.2 CLOSED PHOTOBIOREACTOR SYSTEMS

Another process by which microalgae can be cultivated is through closed reaction systems where light plays a very important role along with a bunch of other factors, which are known to influence the growth and lipid content of algae. The first and foremost important factor is lighting, which drives the conversion into stored energy by photosynthesis in the algae. Optimal light intensity and wavelength of light cause serious effects on the growth, as well as the duration till which the algae remain exposed to light. In case of closed photobioreactors, mostly LEDs (light emitting diodes) are used under strict conditions. Another factor is mixing of the algal cultures in the biomass which affects growth in two ways—one is by increasing the frequency of cell exposure to light and dark and another by increasing the mass transfer of nutrients to cells. Water consumption is also important, but unlike crops, algal growth does not require fresh water, instead, wastewater can also be used for the purpose. And closed reactor systems have an additional advantage of not losing water due to evaporation unlike open-air systems, thus water requirements remain seriously low in this type of system. CO_2 is also necessary in addition to light and water for photosynthesis. But the p_{CO_2} (partial pressure) needs to be maintained which differs from one strain of algae to another. A good source of CO_2 is flue gas as it reduces greenhouse emissions and is generated from coal-fired power plants, proving to be cost effective. Oxygen concentration is maintained by keeping the tube length limited, which restricts the scaling up process in the closed bioreactors. The major nutrient supply includes nitrogen and phosphorus as they both play a role in the growth rate and lipid production. The media in bioreactors are either supplemented from time to time in low amounts or is mostly added when half of the culture is harvested. pH is an important factor in that it has been found to affect the growth when pH varies. And last but not the least, is temperature which does not have harmful effects as such in optimal conditions, but have been found to be specific to some species of algae (Nedbal et al., 2008; Kunjapur et al., 2010; Chen et al., 2011; Zeng et al., 2011).

7.3.3 TWO-STAGE HYBRID SYSTEMS

Two-stage hybrid systems have proved to be more advantageous over time. The hybrid system forms a combination of both open ponds and closed bioreactor for yield of better results. Open ponds have proved to be an effective and profitable method for cultivating algae, but over the time becomes contaminated with redundant species very quickly. Which is why a blend of both the systems becomes the most effective choice for the cultivation of high yielding strains of algae for production of biofuels (Schenk et al., 2008; Demirbas, 2011).

In this system, open ponds are inoculated with either a singular or a mixture of strains, which were previously cultivated in a closed bioreactor, be it as simple as a plastic bag or a high tech fiber optic bioreactor. Invariably, the size of the inoculum should be large enough so that it can grow quickly as compared to some contaminated species. A way to prevent this contamination is to routinely clean or flush the ponds, which will diminish the growth of contaminating species or clear them out fully (Demirbas, 2011).

For large scale biofuel production, a series of bioreactors with increasing volume should be made to link with each other, so that the inoculum volume could increase for further subculture in open ponds. In this case, the large bioreactor should be flushed from time to time to remove the culture and inoculated with new ones. There should also be resupply of the inoculum to the open ponds to avoid contamination. For the process to work, algal species which is both fast growing in the initial inoculum stages and highly productive in the final open pond stage should be chosen. This helps to continuously flush out the cultures and avoid building up of contaminated species and simultaneously produce biofuels (Schenk et al., 2008).

7.3.4 LIPID INDUCTION TECHNIQUE

Microalgae have the ability to produce substantial amounts of triacylglycerols (TAGs) as a storage lipid under photo-oxidative stress or other adverse environmental conditions. Fatty acids form the building blocks of TAGs and whose synthesis acts as a defense mechanism under stress in algae, although little is known about its regulation in molecular and cellular level. Over the past few decades, microalgae have emerged as an alternative and renewable source for production of lipid-rich biofuels, still, the production of algal fuels has mostly remained non-viable commercially. An ongoing process that is getting developed for large scale use is the induction of lipids in algae synthetically for an increased amount of biofuel production (Hu et al., 2008).

Increasing algal oil production can lower the cost of biodiesel production. Through genetic manipulation, the production of fatty acids can be regulated through acetyl CoA carboxylase which pushes more substrate into the lipid biosynthesis pathway, producing more TAGs (Yu et al., 2011). Under normal growth conditions, a large amount of algal biomass is produced but with low lipid content. Whereas under stress conditions or non-favorable environmental conditions, many microalgae change their lipid biosynthetic pathway towards formation and accumulation of neutral fatty acids, mainly in the form of TAGs.

The high cost of biodiesel production due to low productivity of fatty acid synthesizing microalgae are what hindering the process from getting commercialized. However, to obtain increased lipid content, external stress or induction of lipid is required, but this has also become species specific. Because not all species increases the synthesis of neutral fatty acids in the form of TAGs under the same type of conditions. The stress applied needs to vary as well from species to species (Sharma et al., 2012).

7.4 MICROALGAL HARVESTING TECHNOLOGIES

For downstream processing, the separation of the microalgae from their growth medium has always posed a challenge. There is always mutual shading when a high biomass concentration is present, thus leading to a reduction in productivity. Which is why biomass concentrations in microalgal cultures are really low, as this means that a large volume of water needs to be removed for harvesting the biomass. Due to the small size of the microalgal cells, it is not possible to harvest them by means of sedimentation or simple screening. Be it for high-value products or low- value products like biofuels harvesting by centrifugation is not an option, due to the large volume of the biomass and thus proving difficult to harvest (Vandamme et al., 2010; Vandamme et al., 2013).

One of the most promising harvesting technologies that has proven cost-effective is flocculation. This is a two-step harvesting process. In the first step, the dilute suspension of microalgal culture is concentrated to a slurry. Which is dewatered using centrifugation to obtain an algal paste with a 25% dry matter content. The dewatering step is possible in this step as the volume of the biomass is high and the water content is low. Flocculation has become a widely-used technology in other industries as well, as this can be achieved in various ways. These ranges from chemical flocculation where metal salts are used for flocculation in industries such as water treatment, mining; bioflocculation which is caused by the extracellular polymers in the medium where the microalgae is growing and is thought to occur spontaneously; autoflocculation, another type of spontaneous flocculation that occurs when the pH of the medium goes beyond pH 9; and other emerging flocculation technologies involves the use of magnetic nanoparticles as well (Vandamme et al., 2013).

Among the other harvesting technologies separation of microalgae by nano- or microparticles is drawing attention in the field. This bioseparation process is mostly characterized by biocompatibility, ease in manipulating and regeneration accompanying the use of simple devices and non-destructive nature of magnetic fields. Magnetic microparticles are mainly used which adhere to the solid surface of the algae and separates it from the medium. The medium gets dewatered and the biomass gets separated. But precaution needs to be taken regarding the presence of any interfering molecule that may be present in the medium and may hinder the interaction between the biomass and the microparticles (Prochazkova et al., 2013, Wang et al., 2015).

Another efficient and cost-efficient technology includes the foam floatation system which combines dispersed air floatation with foam fractionation for harvesting, concentrating and physical separation of suspended particles. But the technology has not been commercialized into large scale yet. Most of the experiments were carried out in the laboratory and in small scale model systems. This included many other variable factors like air flow rate, foam column height, surfactant concentration and type, and also batch run time. The results revealed that highest concentration of the biomass can only be achieved depending upon certain variables and its interactions like the cationic cetyl trimethylammonium bromide (CTAB), lower surfactant concentrations and CTAB in combination with high column heights. There was also significant lipid recovery in comparison to others when CTAB was used as the surfactant. The lipid exhibited during this process by the biomass was more suited for biodiesel conversion due to the presence of the increased concentration of unsaturated and monounsaturated fatty acids. Evidence proved that CTAB actually promotes in situ lysis by solubilizing the phospholipid bilayer, thus increasing the amount of lipid content that becomes extractable. This method proves advantageous and can be used as an alternative to other technologies that are used for bulk harvesting (Coward et al., 2013; Coward et al., 2014).

7.5 MACROALGAL CULTIVATION AND HARVESTING TECHNOLOGIES

Marine macroalgae are quickly gaining importance as an attractive renewable source for producing fuels and chemicals. This type of biomass has many advantages over terrestrial plant biomass and differs from microalgae altogether. Microalgae form mainly the source of lipid- based biofuels like biodiesel, whereas macroalgae form the major source of a diverse range of carbohydrates which are used for conversion into liquid biofuels like bioethanol (Kraan, 2010).

Macroalgae or seaweed are cultivated either vegetatively or by a separate reproductive cycle. For cultivating vegetatively, algae are grown in an aquatic environment with the standard light, nutrients, temperature, salt content, and water movement. When the seaweed reaches a mature stage, it is cut or harvested by keeping a part of the seaweed for future growth. But not all seaweed can grow vegetatively, some mostly grow reproductively by alternation of generations. Cultivation through reproduction is expensive as the growth and maintenance of the seaweeds must be done in an off-shore facility (Aitken et al., 2014).

Cultivation of macroalgae is mostly done in three types of farming sites: offshore farms, nearshore coastal farms, and land-based ponds. Most have already been developed but not for use on large scale or the industry. But land-based ponds are the most favorable of the lot due to the ease in nutrient supply, and bad weather, disease, and predation could easily be avoided. For large scale production though it still has not proved to be cost effective till now (Wei et al., 2013).

For over many years seaweeds were harvested manually. But the recent demand for seaweeds to produce biofuels and other industrial products has led the development of mechanical harvesting processes. The processes differ from one species of seaweed to another. Like some seaweed that stands upright as it grows up which are mowed with rotating blades when its harvested. While others tend to float, or have weak growing forms and gets harvested through suction or dredging with a cutter and some gets transported with the whole growth structure to the shore. The mechanisms still need to improve for rapid and efficient harvesting for supplying enough biomass for biofuel production (Kirkman and Kendrick, 1997).

7.6 BIOMASS TO BIOFUEL CONVERSION TECHNOLOGIES

There are many processes available for conversion of the algal biomass after its cultivation and harvesting, to biofuels. Some are chemical and some are biological and broadly includes processes like thermochemical conversion, gasification, liquefaction, fermentation, to name a few.

7.6.1 THERMOCHEMICAL CONVERSION

Following the success of first- and second-generation biofuels which are produced from food and non-food crops, respectively, algal biomass have become an important feedstock for the production of third-generation biofuels. Its characteristic rapid growth and high carbon fixing efficiency have the potential of mass production and greenhouse gas uptake, making them highly promising for the development of biofuel production. Thermochemical conversion of biomass to biofuel is a very effective process and mainly includes a combination of torrefaction, liquefaction, pyrolysis, and gasification. Through a combination of these conversion technologies, solid, liquid, and gaseous biofuels are produced from algal feedstock for heat and power generation. While the liquid fuels can be further upgraded for production of chemicals, the gaseous biofuels can be converted to liquid ones (Chen et al., 2014; Raheem et al., 2015).

This process has been found to be appropriate for both dry and wet algal biomass. But the wet algal biomass must first undergo dewatering and then cost-intensive drying processes like sedimentation, flocculation, dissolved air floatation, centrifugation, and filtration. But hydrothermal liquefaction is mostly used to divert from the cost-intensive processes, if possible, for wet algal biomass (Kumar et al., 2016).

Though torrefaction mostly resembles pyrolysis as both take place at a high temperature, but in this process, the microalgae get thermally degraded in an inert atmosphere at a temperature of 200–300°C and the biomass gets dehydrated along with the proteins and carbohydrates, thus achieving partial carbonization. This pretreatment performance depends on the temperature and duration, making them the most important factors responsible (Chen et al., 2014).

Pyrolysis involves heating and thermal decomposition of microalgae in the absence of oxygen or air, and the pressure in the reactor mostly remains one atmosphere, where the temperature usually is between 400 and 600°C. This process mainly produces bio-oils, chars, and non-condensable gases as the major products, though they may vary depending upon the operating conditions, microalgal properties, and reaction type (Chen et al., 2015(1)).

And the conventional gasification includes the reaction of the dry microalgae with an oxidizer, such as air, water, oxygen, or steam, in a partial oxidizing environment with a temperature range of 800–1000°C and a pressure range of 1–10 bar (Khoo et al., 2013).

The process combines multiple steps for utilizing the biomass maximally, but still, the industries have faced numerous challenges which are needed to overcome for the production of biofuels in large scale.

7.6.2 *GASIFICATION*

Gasification is a process via which the carbonaceous material present in the microalgae gets converted to synthetic gas by means of partial oxidation, as mentioned before. Different types of models have been proposed over the years, like gasification of microalgae mixed with nitrogen at low temperatures or gasification of biomass at low temperature with a high moisture content. Generally, biomass with high moisture content gets converted to methane- containing gas directly, without drying. In addition to which the nitrogen in the biomass gets converted into ammonia (Amin et al., 2009).

For wet microalgal biomass, a different type of gasification commonly known as the supercritical water gasification. In this process, the aqueous phase residue after water removal is gasified at supercritical conditions of water for the production of hydrogen-rich fuel gas. The reaction conditions of which are maintained at a temperature of 600°C and a pressure of 240 bar. These conditions vary from one type of wet algal species to another (Xu et al., 2011).

Another novel method for gasification was proposed by Stucki (2009). He describes a catalytic gasification process, which utilizes dilute fossil CO_2 emissions during microalgae cultivation and converts the algal biomass to biofuel through a catalytic hydrothermal process. The products include methane in the form of clean fuel and concentrate CO_2 for sequestration as well. in addition, the process completely mineralizes nutrient-bearing organics.

7.6.3 *PYROLYSIS*

The conversion of biomass to bio-oil, synthetic gas, and charcoal at a temperature of 350–700°C in the absence of air is known as pyrolysis.

It has shown potential to be used in large scale production of biofuels from algal biomass in the industry and could replace petroleum-based liquid fuels. There are different types of pyrolysis processes which differ from one another based on their distinct characteristics. Flash pyrolysis, for example, operates at a moderate temperature of 500°C with a short hot vapor residence time. It is mostly deemed viable for the production of liquid biofuels, as this process have a 95.5% biomass to liquid fuel conversion ratio. Fast pyrolysis also takes place at the same temperature but with a higher hot vapor residence time, in comparison to flash pyrolysis. This process is deemed suitable for the formation of both liquid and gas based biofuels. Another type is slow pyrolysis which takes place at a low temperature of 400°C with very long residence time and mostly produces liquid, gas and char products in almost equal amounts (Brennan and Owende, 2010; Pragya et al., 2013).

In comparison to other conversion technologies, pyrolysis remains the most exploited process and has been well developed for commercialization purposes. Other modified methods of pyrolysis include catalytic pyrolysis which requires the presence of a catalyst for the conversion of microalgae to a liquid fuel precursor (Babich et al., 2011). Du (2011) reported a microwave assisted pyrolysis with char working as an enhancer, which gave a high yield of bio-oil.

7.6.4 LIQUEFACTION

Liquefaction is a highly efficient conversion process suitable for biomass present in wet conditions and has been collectively known as hydrothermal liquefaction (HTL). HTL is carried out in water at medium temperature of about 280–370°C and a high pressure of 10–25 MPa and produces liquid bio-crude as the main product, as well as solid, gaseous and liquid by-products. The biocrude that gets produced has energy value almost similar to that of fossil petroleum and is expected to become a suitable renewable feedstock over the years (Barreiro et al., 2013). Studies were made to quantify carbon and nitrogen as well. It was observed that both carbon and nitrogen has a tendency to accumulate in the biocrude oil derived from HTL with increase in temperature and retention time (Yu et al., 2011).

Another form of HTL is catalytic hydrothermal liquefaction in which different heterogeneous catalysts are used in presence of inert and high-pressure reducing conditions (Duan and Savage, 2011). A model system for HTL was developed in which high levels of carbon conversion to biocrude product were achieved by separating with gravity and at a relatively low temperature and in a continuous flow reactor maintaining a pressurized environment (Elliott et al., 2013). This reactor process can be used for large scale processes in the industry.

7.6.5 ANAEROBIC DIGESTION

The chemical composition of microalgae widely depends on environmental factors. It generally contains varying proportions of proteins, lipids, carbohydrates, nucleic acids, and vitamins, out of which proteins and lipids are found in much higher quantities. Lipids form an attractive substrate for anaerobic digestion due to the high methane yield. But the retention time required for the digestibility is long, sometimes as much as 20–30 days, which makes the process not so much cost-effective for large scale production (Zamalloa et al., 2011). A problem faced mostly in this process is the resistance provided by the cell wall even if the cell is dead. But the positive aspect is the high concentration of methane that is present in the biogas. Additionally, another advantage is the low concentration of the sulfur content that is responsible for corrosion in engines. Mostly anaerobic digestion has been done with fresh-water algae and not with marine algae. The process has only been done in small scale and in very few species of algae. There is much need for further development of this process (Ward et al., 2014).

7.6.6 FERMENTATION

The ability of microalgae to fix CO_2 and its ability to produce energy is supposed to be the potential measures that are required for mitigating the CO_2 in the atmosphere and ultimately reduce global warming. Along with sharing these properties microalgae has also the ability to produce ethanol from intracellular starch by self-fermentation. This system proves to be more simple and cost-effective in comparison to the other conversion technologies. Hirayama (1998) screened microalgae from seawater in search for potential strains. Out of the 200 strains that were isolated, only one strain that is Chlamydomonas sp. YA-SH-1 was selected for its excellent property of growth rate, starch content and its ability to convert starch to ethanol by fermentation.

Whereas in some cases, the biomass of multiple marine algae was treated with acid and hydrolytic enzymes which are commercially available. The hydrolysates were found to contain high concentration of a variety of carbohydrates like glucose, galactose, mannose, and so on. These were then used for fermentation with the help of a recombinant *Escherichia coli* which could easily utilize these carbohydrates and produce ethanol, which further proved to be a viable strategy to produce ethanol from the biomass of marine alga hydrolysates (Kim et al., 2011). Another significant study reported that thermochemical pre-treatment prior to algal fermentation increases the efficiency of the conversion of algal biomass to the energy in methane, although a variety of characteristics need to first determined like the pre-treatment temperature, its duration and the dosage and concentration of sodium hydroxide. This method was found to produce better methane-specific gas upon biological conversion (Chen and Oswald, 1998).

7.6.7 TRANSESTERIFICATION

The reactive extraction of lipids from algal biomass is known as transesterification and has the potential to greatly simplify and reduce the cost of algal biodiesel, due to its ability to reduce the catalysts required for the conversion of lipids into their alkyl esters (biodiesel) (Velasquez-Orta et al., 2012). Over the years, in situ transesterification has been modified and used in various ways. Like the use of ultrasound agitation during the process and combining it with co-solvent like n-pentane and dimethyl ether improves the oil to methyl esters conversion and yield, and reduces the volume of methanol required during the reaction (Ehimen et al., 2012). An enzyme catalyzed transesterification which utilized immobilized lipase was developed which increased the biodiesel yield by seven folds in comparison to alkaline based transesterification reaction process (Teo et al., 2014). Another condition is the use of the combination of ultrasound and microwave irradiation to assist biodiesel production. It was found that this combination produced a much higher concentration of fatty acid methyl esters (Ma et al., 2015). Methods were also established with direct transesterification, using 75% ethanol and a co-solvent like n-hexane, so as to reduce the energy consumption during the lipid extraction process and improve the conversion yield of the microalgal biomass (Zhang et al., 2015).

7.6.8 DIRECT COMBUSTION

Combustion takes place under a controlled oxidizing environment to convert the carbonaceous algal feed into gaseous fuels of usable heating value and also produces solid residues that can be used as products (Agrawal and Chakraborty, 2013). In a typical combustion process like that of with the *Nannochlorpsis gaditana* microalgae, it takes place in three stages. The first stage occurs in a temperature range of 30–125°C, which causes free water loss and the water that is loosely bound to biomolecules. There is a progressive destruction of the cell structure with alteration in the lipid and protein structure. The second stage where major weight loss occurs, due to the decomposition of proteins and lipids, takes place at 180–450°C, which causes char formation. And the last stage takes place at a temperature of 450–600°C which causes the oxidation of the formed char (Sanchez-Silva et al., 2013).

7.6.9 ULTRASONIFICATION TO AID BIOFUEL YIELD

Ultrasonic waves have been used in various ways to increase the biofuel yield from algal biomass. Ultrasonication can exert different effects and has been used both independently and in combination with other technologies. At low power, it exerts radiation force and acoustic streaming which are confined to occur in the cell cytoplasm whereas high power, have both chemical and physical effects, which includes free radical reactions, shock waves, microjet, and shear stress, to name a few (Kurokawa et al., 2016).

Ultrasonication for cell disruption has proved to be a cost-effective and energy-saving procedure. Cell disruption with ultrasonic waves facilitates intracellular oil release. This combined with other technologies like homogenization, centrifugation was found to have a higher lipid recovery for conversion to biofuels (Wang et al., 2015). The same has been found when Ultrasonication was used in combination with in-situ transesterification as well (Suganya et al., 2014). Continuous ultrasonication is another method that was studied with different species of microalgae. It resulted in the release of lipids from rigid walled microalgae without using any expensive dewatering steps (Nataranjan et al., 2014). Thus, proving the commercialization of the process can be well suited to the bioconversion process.

7.7 BIOETHANOL FROM ALGAL BIOMASS

Bioethanol from algal biomass can serve as an excellent alternative to bioethanol produced from agricultural food crops, and can become a green fuel for sustainability in the future. Certain species of microalgae can produce bioethanol directly by anaerobic fermentation, whereas for some, bioethanol is produced from the sugar-based biomass waste that is left behind after oil extraction, and is further hydrolyzed for bioconversion. Macroalgae can also be used for bioethanol production but there have been few studies reported and more research is required for further industrial use and commercialization (John et al., 2011).

Bioethanol from algae has been produced in different ways. One method is by hydrothermal fractionation, the likes of which was done with the microalgae *Schizocytrium* sp. This method separates sugars, lipids, and proteins by producing protein-rich solid cake, and liquid hydrolysates, which mainly contains oligomeric sugars and lipids. They get further separated by liquid-liquid fractionation, out of which the sugars are then quantified and used for bioconversion to ethanol (Kim et al., 2012). The carbohydrates from algae are derived via various disruption techniques which broadly includes physical treatment like irradiation, ultrasonication; chemical treatment like alkaline, acid hydrolysis; and enzymatic treatment (Li et al., 2014, Sirajunnisa and Surendhiran, 2016).

Macroalgae form a very suitable source for bioethanol due to its higher carbohydrate percentage. Mostly the *Enteromorpha* sp. is used as the starting material for bioethanol production because of its higher carbohydrate percentage of 70–72%. The bioconversion is mainly achieved through a series of different processes. The bioethanol produced has excellent properties and offers an alternative to conventional fuels (Nahak et al., 2011).

Bioethanol from algae has significant potential for becoming a renewable fuel in the future. It is highly coveted due to the low percentage of lignin and hemicellulose present as compared to ethanol derived from terrestrial lignocellulosic plants and thus have the ability to produce more energy (Jones and Mayfield, 2012).

7.8 BIODIESEL FROM ALGAL BIOMASS

Biodiesel from microalgae is the only biofuel which has the capability of replacing fossil fuel derived petroleum without affecting the supply of food crops and other agriproducts. Most efficient food crops do not have the ability to produce biodiesel in the amount that gets produced by algae (Chisti, 2008). In the process of bioconversion, the harvesting of microalgae and the extraction of lipids from the microalgae is the most expensive and energy consuming process. The extraction is mainly performed in a single-step production where the wet microalgal cells are mixed with solvent, methanol, and acid catalyst, after which heating is done in a pot. This method also sometimes involves an integrated in situ transesterification process for aiding in lipid extraction for its conversion to biodiesel, with a yield higher than 90% (Im et al., 2014).

Pertinent questions are still present on the commercialization aspects of large scale production of biodiesel form algal biomass. Every step needs to rightly assessed for upstream processing and critical analysis should be done of translating the findings in laboratory scale up to large scale (Rawat et al., 2013). Improvements can be made and efficiency can be increased of every step or the methods can be modified for minimalizaton of greenhouse gas emissions (Collet et al., 2014). Regardless of the slow development of the technological approaches used for conversion to biodiesel, but these solutions can be environmentally sustainable (Kim et al., 2013).

7.9 BIOGAS AND BIOHYDROGEN FROM ALGAL BIOMASS

Biogas is generally produced from maize crops through anaerobic fermentation, but microalgae have now provided a suitable alternative for renewable biogas production. Studies have demonstrated that the biogas production is generally dependant on the species and pre-treatment. Fermentation of algal biomass gives a 7–13% higher yield of methane as compared to biogas produced from maize (Mussgnug et al., 2010). Another method of obtaining biogas from algal biomass is through anaerobic digestion which produces methane-containing biogas and recycles nutrients as well, and is energy efficient as well in comparison to the series of technologies used for algal biogas production and bioconversion (Collet et al., 2011). The pre-treatment processes for efficient biogas production have also been investigated. The production gets limited in microalgae due to the complex cell wall structure present in most microalgae. Pre-treatment technologies are mostly divided into thermal, mechanical, chemical, and biological methods, which results in increasing biomass solubilization and methane yield; although most of the pre-treatment processes are species-specific (Passos et al., 2014). In spite of recent developments in the field, the scenario from cultivation to harvesting to bioconversion and its energy efficiency and cost-affectability is still yet to be determined. More research is required in order to improve the potential of the technology and bioconversion.

Biohydrogen has tremendous potential for its use as an alternate source of energy. Microalgae prove to be a much cheaper source for biohydrogen production. Generally, a two-stage process is applied—in the first stage, carbon fixation takes place and in the second, anaerobic digestion and hydrogen production occurs. But not much insight is available in these processes. And inefficient anaerobic digestion, low carbon fixation are the fundamental properties which affect hydrogen yields (Rashid et al., 2013). Studies have demonstrated valuable scenarios about the simultaneous treatment of urban wastewater using microalgae and the conversion of the biomass to obtain energy. Microalgae are first grown in the wastewater, which leads to removal of nutrients and water treatment. The biomass is then collected and kept in the bioreactor for few more days under stressed conditions leading to sugar accumulation and ultimate conversion to biohydrogen (Batista et al., 2015). The same has been achieved with different types of microalgae and has since been reported (Nobre et al., 2012).

7.10 INDUSTRIAL AND COMMERCIAL PROSPECTS OF ALGAL BIOFUELS

Algal biofuels offer great promise to contribute a significant portion of the renewable fuels that will be required in the future as an alternative to fossil fuel derived petroleum and transportation fuel. The production value is energy efficient and cost-effective and would provide an ideal alternative. Algal fuels are mainly derived from the lipids that are found in high concentrations and is a most valuable feedstock. But still, development is needed as per the technologies that are concerned in the cultivation, harvesting, and bioconversion of biofuels. Modification is required for the large-scale production and upstream processing from the laboratory scale (Pienkos and Darkins, 2009; Stephens et al., 2010).

Production of algal biofuels has a considerable advantage. Microalgae have the ability to grow in wastewaters derived from agricultural, municipal and industrial activities and provide economical and sustainable means for algal growth for biofuel production. This also has the potential of combining wastewater treatment with biofuel production. This type of hybrid technology may provide an outlook to mitigate energy efficiency, air pollution and greenhouse gas emissions (Slade and Bauen, 2013; Maity et al., 2014).

7.11 CONCLUSION

With decreasing fossil fuel resources, demand is increasing for renewable fuel sources. Algal biofuels provide an economical and sustainable resource as a green fuel. Though development and research are required for its production and utilization, still it has already established itself as an alternative to fossil fuel-based transportation fuel and others. Algal production on large scale has already been taken up by certain countries and biofuel production is underway. Though it is much costlier now, but over the years more research and further study with its conversion technologies can reduce the cost and help to use it in various industries. Algal biofuels can provide sustainable economy to the developing countries, especially to the rural people. The process is easier and energy efficient. Production of algal biofuels can become a huge industry because of its application and productivity, and hence can become a sustainable and green economy.

KEYWORDS

- **algal biomass**
- **biodiesel**
- **bioethanol**
- **biofuels**
- **macroalgae**
- **microalgae**
- **thermochemical conversion**

REFERENCES

Adenle, A. A.; Haslam, G. E.; Lee, L. Global Assessment of Research and Development for Algae Biofuel Production and its Potential Role for Sustainable Development in Developing Countries. *Energy Policy* **2013,** *61*, 182–195.

Agrawal, A.; Chakraborty, S. A Kinetic Study of Pyrolysis and Combustion of Microlagae Chlorella Vulgaris Using Thermo-gravimetric Analysis. *Bioresour. Technol.* **2013,** *128,* 72–80.

Aitken, D.; Bulboa, C.; Godoy-Faundez, A.; Turrion-Gomez, J. L.; Antizar-Ladislo, B. Life Cycle Assessment of Macroalgae Cultivation and Processing for Biofuel Production. *J. Clean. Prod.* 2014, http://dx.doi.org/10.1016/j.jclepro.2014.03.080

Amin, S. Review on Biofuel Oil and Gas Production Processes From Microalgae. *Energy Convers. Manag.* **2009,** *50,* 1834–1840.

Babich, I.V.; van der Hulst, M.; Lefferts, L.; Moulijn, J. A.; O'Connor, P.; Seshan, K. Catalytic Pyrolysis of Microalgae to High-quality Liquid Bio-fuels. *Biomass Bioener.* **2011,** *35,* 3199–3207.

Barreiro, D. L.; Prins, W.; Ronsse, F.; Brilman, W. Hydrothermal Liquefaction (HTL) of Microalgae for Biofuel Production: State of the Art Review and Future Prospects. *Biomass Bioener.* **2013,** *53,* 113–127.

Batista, A. P.; Ambrosano, L.; Graca, S.; Sousa, C.; Marques, P. A. S. S.; Ribeiro, B.; Botrel, E. P.; Neto, P. C.; Gouveia, L. Combining Urban Wastewater Treatment with Biohydrogen Production – An Integrated Microalgae-based Approach. *Bioresour. Technol.* **2015,** *184,* 230–235.

Borowitzka, M.A. Commercial Production of Microalgae: Ponds, Tanks, Tubes and Fermenters. *J. Biotechnol.* **1999,** *70,* 313–321.

Brennan, L. ; Owende, P. Biofuels From Microalgae – A Review of Technologies for Production, Processing and Extractions of Biofuels and Co-products. *Renew. Sustain. Energy Rev.* **2010,** *14,* 557–577.

Chen, C.; Yeh, K.; Aisyah, R.; Lee, D.; Chang, J. Cultivation, Photobioreactor Design and Harvesting of Microalgae for Biodiesel Production: A Critical Review. *Bioresour. Technol.* **2011,** *102,* 71–81.

Chen, P. H.; Oswald, W. J. Thermochemical Treatment for Algal Fermentation. *Environ. Int.* **1998,** *24,* 889–897.

Chen, W.; Lin, B.; Huang, M.; Chang, J. Thermochemical Conversion of Microalgal Biomass into Biofuels: A Review. *Bioresour. Technol.* 2014, http://dx.doi.org/10.1016/j.biortech.2014.11.050

Chen, H.; Zhou, D.; Luo, G.; Zhang, S.; Chen, J. Macroalgae for Biofuels Production: Progress and Perspectives. *Renew. Sustain. Energy Rev.* **2015,** *47,* 427–437.

Chen, Y.; Wu, Y.; Hua, D.; Li, C.; Harold, M. P.; Wang, J.; Yang, M. Thermochemical Conversion of Low-lipid Microalgae for the Production of Liquid Fuels. *RSC Adv.* 2015, doi: 10.1039/C4RA13359E (1)

Chisti, Y. Biodiesel From Microalgae Beats Bioethanol. *Trends in Biotechnology* **2008,** *26,* 126-131.

Collet, P.; Helias, A.; Lardon, L.; Ras, M.; Goy, R.; Steyer, J. Life-cycle Assessment of Microalgae Culture Coupled to Biogas Production. *Bioresour. Technol.* **2011,** *102,* 207–214.

Collet, P.; Lardon, L.; Helias, A.; Bricout, S.; Lombaert-Valot, I.; Perrier, B.; Lepine, O.; Steyer, J.; Bernard, O. Biodiesel From Microalgae – Life Cycle Assessment and Recommendations for Potential Improvements. *Renew. Energy* **2014,** *71,* 525–533.

Coward, T.; Lee, J. G. M.; Caldwell, G. S. Development of a Foam Floatation System for Harvesting Microalgae Biomass. *Algal Res.* **2013,** *2,* 135–144.

Coward, T.; Lee, J. G. M.; Caldwell, G. S. Harvesting Microalgae by CTAB-aided Foam Floatation Increases Lipid Recovery and Improves Fatty Acid Methyl Ester Characteristics. *Biomass Bioener.* **2014,** *67*, 354–362.

Daroch, M.; Geng, S.; Wang, G. Recent Advances in Liquid Biofuel Production From Algal Feedstocks. *Appl. Energy* **2013,** *102*, 1371–1381.

Demirbas, A. Use of Algae as Biofuel Sources. *Energy Convers. Manag.* **2010,** *51*, 2738–2749.

Demirbas, M. F. Biofuels From Algae for Sustainable Development. *Appl. Energy* **2011,** *88*, 3473–3480.

Du, Z.; Li, Y.; Wang, X.; Wan, Y.; Chen, Q.; Wang, C.; Lin, X.; Liu, Y.; Chen, P.; Ruan, R. Microwave-assisted Pyrolysis of Microalgae for Biofuel Production. *Bioresour. Technol.* **2011,** *102*, 4890–4896.

Duan, P.; Savage, P. E. Hydrothermal Liquefaction of a Microalga With Heterogeneous Catalysts. *Ind. Eng. Chem. Res.* **2011,** *50*, 52–61.

Ehimen, E.A.; Sun, Z.; Carrington, G.C. Use of Ultrasound and Co-Solvents to Improve the In-situ Transesterification of Microalgae Biomass. *Procedia Environ. Sci.* **2012,** *15*, 47–55.

Elliott, D. C.; Hart, T. R.; Schmidt, A. J.; Neuenschwander, G. G.; Rotness, L. J.; Olarte, M. V.; Zacher, A. H.; Albrecht, K. O.; Hallen, R. T.; Holladay, J. E. Process Development for Hydrothermal Liquefaction of Algae Feedstocks in a Continuous-flow Reactor. *Algal Res.* **2013,** *2*, 445–454.

Ghasemi, Y.; Rasoul-Amini, S.; Naseri, A. T.; Montazeri-Najafabady, N.; Mobasher, M. A.; Dabbagh, F. Microalgae Biofuel Potentials (Review). *Appl. Biochem. Microbiol.* **2012,** *48*, 126–144.

Groom, M. J.; Gray, E. M.; Townsend, P. A. Biofuels and Biodiversity: Principles for Creating Better Policies for Biofuel Production. *Conserv. Biol.* **2008,** *22*, 602–609.

Hannon, M.; Gimpel, J.; Tran, M.; Rasala, B.; Mayfield, S. Biofuels From Algae: Challenges and Potential. *Biofuels* **2010,** *1*, 763–784.

Hirayama, S.; Ueda, R.; Ogushi, Y.; Hirano, A.; Samejima, Y.; Hon-Nami, K.; Kunito, S. Ethanol Production From Carbon Dioxide by Fermentative Microalgae. *Adv. Chem. Convers. Mitigating Carbon Dioxide* **1998,** *114*, 657–660.

Hu, Q.; Sommerfels, M.; Jarvis, E.; Ghiradi, M.; Posewitz, M.; Seibert, M.; Darzins, A. Microalgal Triacylglycerols as Feedstocks for Biofuel Production: Perspectives and Advances. *Plant J.* **2008,** *54*, 621–639.

Im, H.; Lee, H.; Park, M. S.; Yang, J.; Lee, J. W. Concurrent Extraction and Reaction for the Production of Biodiesel From Wet Microalgae. *Bioresour. Technol.* **2014,** *152*, 534–537.

John, R. P.; Anisha, G. S.; Nampoothiri, K. M.; Pandey, A. Micro and Macroalgal Biomass: A Renewable Source for Bioethanol. *Bioresour. Technol.* **2011,** *102*, 186–193.

Jones, C. S.; Mayfield, S. P. Algae Biofuels: Versatility for the Future of Bioenergy. *Curr. Opin. Biotechnol.* **2012,** *23*, 346–351.

Khoo, H. H.; Koh, C. Y.; Shaik, M. S.; Sharatt, P. N. Bioenergy Co-products Derived From Microalgae Biomass Via Thermochemical Conversion – Life Cycle Energy Balances and CO_2 Emissions. *Bioresour. Technol.* **2013,** *143*, 298–307.

Kim, J.; Yoo, G.; Lee, H.; Lim, J.; Kim, K.; Kim, C. W.; Park, M. S.; Yang, J. Methods of Downstream Processing for the Production of Biodiesel From Microalgae. *Biotechnol. Adv.* **2013,** *31*, 862–876.

Kim, J. K.; Um, B.; Kim, T. H. Bioethanol Production From Micro-algae, *Schizocytrium* sp., Using Hydrothermal Treatment and Biological Conversion. *Korean J. Chem. Eng.* **2012,** *29*, 209–214.

Kim, N.; Li, H.; Jung, K.; Chang, H. N.; Lee, P. C. Ethanol Production From Marine Algal Hydrolysates Using *Escherichia coli* KO11. *Bioresour. Technol.* **2011,** *102*, 7466–7469.

Kirkman, H.; Kendrick, G. A. Ecological Significance and Commercial Harvesting of Drifting and Beach-cast Macro-algae and Seagrasses in Australia: A Review. *J. Appl. Phycol.* **1997,** *9*, 311–326.

Kraan, S. Mass-cultivation of Carbohydrate Rich Macroalgae, A Possible Solution for Sustainable Biofuel Production. *Mitig. Adapt. Strateg. Glob. Change* **2013,** *18*, 27–46.

Kumar, G.; Shobana, S.; Chen, W.; Bach, Q.; Kim, S.; Atabani, A. E.; Chang, J. A Review of Thermochemical Conversion of Microalgal Biomass for Biofuels: Chemistry and Processes. *Green Chem.* **2016,** doi: 10.1039/C6GC01937D

Kunjapur, A. M.; Eldridge, R. Photobioreactor Design for Commercial Biofuel Production From Microalgae. *Ind. Eng. Chem. Res.* **2010,** *49*, 3516–3526.

Kurokawa, M.; King, P. M.; Wu, X.; Joyce, E. M.; Mason, T. J.; Yamamoto, K. Effect of Sonication Frequency on the Disruption of Algae. *Ultrason. Sonochem.* **2016,** *31*, 157–162.

Li, K.; Liu, S.; Liu, X. An Overview of Alga Bioethanol Production. *Int. J. Energy Res.* **2014,** doi: 10.1002/er.3164

Ma, G.; Hu, W.; Pei, H.; Jiang, L.; Song, M.; Mu, R. *In situ* Heterogeneous Transesterification of Microalgae Using Combined Ultrasound and Microwave Irradiation. *Energy Convers. Manag.* **2015,** *90*, 41–46.

Maity, J. P.; Bundschuh, J.; Chen, C.; Bhattacharya, P. Microalgae for Third Generation Biofuel Production, Mitigation of Greenhouse Emissions and Wastewater Treatment: Present and Future Prospectives – A Mini Review. Energy 2014, http://dx.doi.org/10.1016/j.energy.2014.04.003

Menetrez, M. Y. An Overview of Algae Biofuel Production and Potential Environmental Impact. *Environ. Sci. Technol.* **2012,** *46*, 7073–7085.

Mohr, A.; Raman, S. Lessons From First Generation Biofuels and Implications for the Sustainability Appraisal of Second Generation Biofuels. *Energy Policy* **2013,** *63*, 114–122.

Mussgnug, J. H.; Klassen, V.; Schluter, A.; Kruse, O. Microalgae as Substrates for Fermentative Biogas Production in a Combined Biorefinery Concept. *J. Biotechnol.* **2010,** *150*, 51–56.

Nahak, S.; Nahak, G.; Pradhan, I.; Sahu, R. K. Bioethanol From Marine Algae: A Solution to Global Warming Problem. *J. Appl. Environ. Biol. Sci.* **2011,** *1*, 74–80.

Naik, S. N.; Goud, V. V.; Rout, P. K.; Dalai, A. K. Production of First and Second Generation Biofuels: A Comprehensive Review. *Renew. Sustain. Energy Rev.* **2010,** *14*, 578–597.

Nataranjan, R.; Ang, W. M. R.; Chen, X.; Voigtmann, M.; Lau, R. Lipid Releasing Chracteristics of Microalgae Species Through Continuous Ultrasonication. *Bioresour. Technol.* **2014,** *158*, 7–11.

Nedbal, L.; Trtilek, M.; Cerveny, J.; Komarek, O.; Pakrasi, H. B. A Photobioreactor System for Precision Cultivation of Photoautotrophic Microorganisms and for High-content Analysis of Suspension Dynamics. *Biotechnol. Bioeng.* **2008,** *100*, 902–910.

Nobre, B. P.; Villalobos, F.; Barragan, B. E.; Oliviera, A. C.; Batista, A. P.; Marques, P. A. S. S.; Mendes, R. L.; Sovova, H.; Palavra, A. F.; Gouveia, L. A Biorefinery From *Nannochloropsis* sp. Microalga – Extraction of Oils and Pigments. Production of Biohydrogen from the Leftover Biomass. *Bioresour. Technol.* 2012, http://dx.doi.org/10.1016/j.biortech.2012.11.084

Passos, F.; Uggetti, E.; Carrere, H.; Ferrer, I. Pretreatment of Microalgae to Improve Biogas Production: A Review. *Bioresour. Technol.* **2014,** *172*, 403–412.

Pienkos, P. T.; Darzins, A. The Promise and Challenges of Microalgal-derived Biofuels. *Biofuels Bioprod. Bioref.* **2009,** *3*, 431–440.

Pragya, N.; Pandey, K. K.; Sahoo, P. K. A Review on Harvesting, Oil Extraction and Biofuels Production Technologies From Microalgae. *Renew. Sustain. Energy Rev.* **2013,** *24*, 159–171.

Prochazkova, G.; Safarik, I.; Branyik, T. Harvesting Microalgae with Microwave Synthesized Magnetic Microparticles. *Bioresour. Technol.* **2013,** *130*, 472–477.

Ragauskas, A. J.; Williams, C. K.; Davison, B. H.; Britovsek, G.; Cairney, J.; Eckert, C. A.; Frederick, W. J.; Hallett, J. P.; Leak, D. J.; Liotta, C. L.; Mielenz, J. R.; Murphy, R.; Templer, R.; Tschaplinski, T. The Path Forward for Biofuels and Biomaterials. *Science* **2006,** *311*, 484–489.

Raheem, A.; Azlina, W. A. K. G. W.; Yap, Y. H. T.; Danquah, M. K. Thermochemical Conversion of Microalgal Biomass for Biofuel Production. *Renew. Sustain. Energy Rev.* **2015,** *49*, 990–999.

Rashid, N.; Rehman, M. S. U.; Memon, S.; Rahman, Z. U.; Lee, K.; Han, J. Current Status, Barriers and Developments in Biohydrogen Production by Microalgae. *Renew. Sustain. Energy Rev.* **2013,** *22*, 571–579.

Rawat, I.; Kumar, R. R.; Mutanda, T.; Bux, F. Biodiesel From Microalgae: A Critical Evaluation from Laboratory to Large Scale Production. *Appl. Energy* **2013,** *103*, 444–467.

Rodionova, M.V. et al. Biofuel Production: Challenges and Opportunities. *Int. J. Hydrogen Energy* **2016,** *11*, 1–12.

Sanchez-Silva, L.; Lopez-Gonzalez, D.; Garcia-Minguillan, A. M.; Valverde, J. L. Pyrolysis, Combustion and Gasification Characteristics of *Nannochlorpsis gaditana* Microalgae. *Bioresour. Technol.* **2013,** *130*, 321–331.

Schenk, P. M.; Thomas-Hall, S. R.; Stephens, E.; Marx, U. C.; Mussgnug, J. H.; Posten, C.; Kruse, O.; Hankamer, B. Second Generation Biofuels: High-Efficiency Microalgae for Biodiesel Production. *Bioenerg. Res.* **2008,** *1*, 20–43.

Sharma, K. K.; Schumann, H.; Schenk, P. M. High Lipid Induction in Microalgae for Biodiesel Production. *Energies* **2012,** *5*, 1532–1553.

Sirajunnisa, A. R.; Surendhiran, D. Algae – A Quintessential and Positive Resource of Bioethanol Production: A Comprehensive Review. *Renew. Sustain. Energy Rev.* **2016,** *66*, 248–267.

Slade, R.; Bauen, A. Micro-algae Cultivation for Biofuels: Cost, Energy Balance, Environmental Impacts and Future Prospects. *Biomass Bioener.* **2013,** *53*, 29–38.

Stephens, E.; Ross, I. L.; Mussgnug, J. H.; Wagner, L. D.; Borowitzka, M. A.; Posten, C.; Kruse, O.; Hankamer, B. Future Prospects of Microalgal Biofuel Production Systems. *Trends Plant Sci.* **2010,** *15*, 554–564.

Stucki, S.; Vogel, F.; Ludwig, C.; Haiduc, A. G.; Brandenberger, M. Catalytic Gasification of Algae in Superficial Water for Biofuel Production and Carbon Capture. *Energy Environ. Sci.* **2009,** *2*, 535–541.

Suganya, T.; Kasirajan, R.; Renganathan, S. Ultrasound-enhanced Rapid in situ Transesterification of Marine Macroalgae *Enteromorphia compressa* for Biodiesel Production. *Bioresour. Technol.* **2014,** *156*, 283–290.

Teo, C. L.; Jamaluddin, H.; Zain, N. A. M.; Idris, A. Biodiesel Production via Lipase Catalyzed Transesterification of Microlagae Lipids from *Tetraselmis* sp. *Renew. Energy* **2014,** *68*, 1–5.

Tilman, D.; Socolow, R.; Foley, J. A.; Hill, J.; Larson, E.; Lynd, L.; Pacala, S.; Reilly, J.; Searchinger, T.; Somerville, C.; Williams, R. Beneficial Biofuels – The Food, Energy, and Environment Trilemma. *Science* **2009,** *325*, 270–271.

Vandamme, D.; Foubert, I.; Meesschaert, B.; Muylaert, K. Flocculation of Microalgae Using Cationic Starch. *J. Appl. Phycol.* **2010,** *22*, 525–530.

Vandamme, D.; Foubert, I.; Muylaert, K. Flocculation as a Low-cost Method for Harvesting Microalgae for Bulk Biomass Production. *Trends Biotechnol.* **2013,** *31*, 233–239.

Velasquez-Orta, S. B.; Lee, J. G. M.; Harvey, A. P. Evaluation of FAME Production from Wet Marine and Freshwater Microalgae by *in situ* Transesterification. *Biochem. Eng. J.* **2013,** *76*, 83–89.

Wang, D.; Li, Y.; Hu, X.; Su, W.; Zhong, M. Combined Enzymatic and Mechanical Cell Disruption and Lipid Extraction of Green Alga *Neochloris oleoabundans*. *Int. J. Mol. Sci.* **2015,** *16*, 7707–7722.

Wang, S.; Stiles, A. R.; Guo, C.; Liu, C. Harvesting Microalgae by Magnetic Separation: A Review. *Algal Res.* **2015,** *9*, 178–185.

Ward, A. J.; Lewis, D. M.; Green, F. B. Anaerobic Digestion of Algae Biomass: A Review. *Algal Res.* **2014,** *5*, 204–214.

Wei, N.; Quarterman, J.; Jin, Y. Marine Macroalgae: An Untapped Resource for Producing Fuels and Chemicals. *Trends Biotechnol.* **2013,** *31*, 70–77.

Xu, L.; Brilman, D. W. F.; Withag, J. A. M.; Brem, G.; Kersten, S. Assessment of a Dry and a Wet Route for the Production of Biofuels from Microalgae: Energy Balance Analysis. *Bioresour. Technol.* **2011,** *102*, 5113–5122.

Yu, G.; Zhang, Y.; Schideman, L.; Funk, T.; Wang, Z. Distributions of Carbon and Nitrogen in the Products from Hydrothermal Liquefaction of Low-lipid Microalgae. *Energy Environ. Sci.* **2011,** *4*, 4587–4595.

Yu, W.; Ansari, W.; Schoepp, N. G.; Hannon, M. J.; Mayfield, S. P.; Burkart, M. D. Modifications of the Metabolic Pathways of Lipid and Triacylglycerol Production in Microalgae. *Microbial Cell Fact.* **2011,** *10*, 1–11.

Zamalloa, C.; Vulsteke, E.; Albrecht, J.; Verstraete, W. The Techno-economic Potential of Renewable Energy Through the Anaerobic Digestion of Microalgae. *Bioresour. Technol.* **2011,** *102*, 1149–1158.

Zeng, X.; Danquah, M. K., Chen, X. D.; Lu, Y. Microalgae Bioengineering: From CO_2 Fixation to Biofuel Production. *Renew. Sustain. Energy Rev.* **2011,** *15*, 3252–3260.

Zhang, Y.; Li, Y.; Zhang, X.; Tan, T. Biodiesel Production by Direct Transesterification of Microalgal Biomass with Co-Solvent. *Bioresour. Technol.* 2015, http://dx.doi.org/10.1016/j.biortech.2015.07.052

CHAPTER 8

Potential Feedstocks for Second-Generation Ethanol Production in Brazil

LUIZA HELENA DA SILVA MARTINS[1], JOÃO MOREIRA NETO[2], PAULO WESLEM PORTAL GOMES[1], JOHNATT ALLAN ROCHA DE OLIVEIRA[3], EDUARDO DELLOSSO PENTEADO[4], and ANDREA KOMESU[4,*]

[1]*Centro de Ciências Naturais e Tecnologia, State University of Pará (UEPA), Belém, PA, Brazil*

[2]*Departamento de Engenharia (DEG), University Federal of Lavras (UFLA), Lavras, MG, Brazil*

[3]*Instituto de Ciências da Saúde, Faculdade de Nutrição, Federal University of Pará (UFPA), Belém, PA, Brazil*

[4]*Departamento de Ciências do Mar (DCMar), Federal University of São Paulo (UNIFESP), Santos, SP, Brazil*

[*]*Corresponding author. E-mail: andrea_komesu@hotmail.com.*

ABSTRACT

An increasing interest for discovering new alternative fuel sources has been observed due to the current concerns related to the petroleum-derived fuels, such as reduction in petroleum reserves, price volatility, as well as environmental issues. One important alternative fuel is the ethanol or bioethanol. It is the most widely used biofuel in transportation all around the world and can be produced from different feedstock types, which are classified in three agricultural categories: simple sugars, starch, and lignocellulose. The feedstock price is highly volatile, which directly affect ethanol production costs, and other problem is the availability. The use of lignocellulosic wastes to ethanol production has become a focus of research in several parts of the world. Each region of the world focused on a specific waste from its local industry. Brazil is a country of immense agro-industrial production and with different residues that can be explored to ethanol production. This work will summarize information about the potential raw materials for ethanol production in Brazil: sugarcane bagasse, leaves and straw; cassava residues, peach palm branches; tucumã; palm oil residues; and cocoa. In addition, challenges in second-generation ethanol production will also be pointed out.

8.1 INTRODUCTION

It is very important to emphasize that the world population has been increasing at an alarming rate and with it the demand for liquid fuels in the transport sector is also increasing. Factors such as global warming, depletion of fossil fuels, and rising prices of petroleum-based fuels are gaining major concern, and the demand for the situation has forced the research for sustainable, renewable, efficient, and low-cost alternative energy from sources with lower greenhouse emissions (Nigam and Singh, 2011; Martinez, 2016).

The world energy matrix presents significant structural changes over the years. Regional factors, such as the level of development, geography, and natural resources, as well as international agreements and fluctuating prices of fossil fuels, directly influence the use of a certain energy source (MME, 2017; Margon et al., 2018). In this context, ethanol or bioethanol appears as a cleaner and renewable alternative. Of all biofuels, ethanol is the only one currently produced on a large scale. The world production of ethanol is about 58 billion liters, of which 70% (38.5 billion liters) are produced in Brazil and in the United States (Perdices et al., 2012).

Ethanol can be produced from different raw materials and conversion technologies, which can be first, second, or third generation. "First-generation" technologies are based on the alcoholic fermentation of the carbohydrates present, for example, in sugarcane juice or corn starch enzymatic hydrolysates. The "second-generation" technologies use agricultural and agro-industrial residues, and the fermentation process is based on the carbohydrates released from the vegetal biomass by hydrolysis of cellulose and hemicelluloses (Pitarelo et al., 2012; Margon et al., 2018). In the "third generation," the biofuel is derived from marine biomasses. It is important to note that the characteristics of the biofuel itself may not change between these "generations," but rather, the source from which the fuel is derived changes (Alaswad et al., 2015). About 90% of the ethanol is derived from the fermentation of the sugar or starch, the remainder is produced synthetically (Perdices et al., 2012).

The food versus fuel competition turned out to be a dilemma nowadays and has been widely focused and discussed. Therefore, the production of biofuels derived from nonedible biomass residues is gaining attention as "second-generation" ethanol (Derman et al., 2018).

An alternative and promising route that can minimize the environmental impacts caused by fossil fuels, so as not to affect food supply, is undoubtedly the use of so-called "second-generation" biofuels, which are those obtained from agroindustry waste, different from "first-generation" biofuels, which are those obtained from edible materials, such as beets, corn, etc. "Second-generation" ethanol and biobutanol from cellulose are being considered as liquid biofuels suitable for achieving global targets for reducing greenhouse gas emissions, diversifying the energy mix, and alleviating the demand for nonrenewable energy sources. Biofuels are expected to be able to supply about a quarter of the world's projected energy needs (heat, electricity, transport) by 2035 (Cardona et al., 2018).

One of the biggest benefits of using "second-generation" technology is that it can reduce carbon dioxide emissions by up to 90%, as well as the emission of other pollutants such as NO_X and SO_X compared to petroleum fuels (Carvalho et al., 2017).

Lignocellulosic biomasses consist of three main components, namely cellulose, hemicelluloses, and lignin, which are polymers of sugars from carbon C_5 and C_6 that can be used in the production of ethanol. Cellulose is the main component of these biomasses, which can be converted into ethanol through processes such as saccharification followed by fermentation. If the cellulose content is higher in the sample, higher glucose content can be produced. In relation to lignin, it can produce a compound known as phenol. The hemicelluloses can also be hydrolyzed and used in the production of xylose (Derman et al., 2018).

"Second-generation" bioethanol or lignocellulosic ethanol in Brazil is more consolidated from sugarcane bagasse biomass (Carvalho et al., 2017). However, the country is well known for its large agribusiness sector, of which fruit growing stands out, and many of the waste generated does not have an adequate end. These residues are mostly lignocellulosic in their structure, which could be potential raw materials for the production of "second-generation bioethanol," in addition, an alternative capable of better targeting these wastes, valuing them as a by-product. In the last decade, new efficient and low-cost technologies have been developed for the production of "second-generation" ethanol.

In this context, this work aims to present the potential lignocellulosic feedstocks for bioethanol production in Brazil, which is a country well known for possessing a wealth of biodiversity, as well as great potential in the agro-industry sector.

8.2 LIGNOCELLULOSIC MATERIALS

In order to add economy to the conservation of the environment, in recent years several different countries have been strategically organized to obtain new energy sources, with lignocellulosic residues being an alternative because they are available in significant quantities, with an emphasis on natural plant fibers due to the diversity of species to be researched (Fernandes, 2013).

In Brazil, tons of fiber are produced and can be found in nature, grown as an agricultural activity or waste generated by the agroindustry, which has many sources that are not always properly used. Thus, numerous processes arise to transform them into chemical compounds and products with high-value added, such as alcohol, enzymes, organic acids, amino acids, and others (Fernandes, 2013). Ethanol produced from lignocellulosic materials ("second-generation" technology) can increase production per hectare in three times, and unlike the raw materials of "first-generation" technologies, do not compete with food production. In addition, lignocellulosic materials are more abundant in nature and contain a high polymerized sugar content in the form of cellulose and hemicellulose. The bioethanol production from different energy crops is summarized in Table 8.1.

The lignocellulosic biomass is composed of carbohydrates polymers (cellulose and hemicellulose), lignin, and a remaining smaller part (extractives, acids, salts, and minerals). The cellulose and hemicellulose, which typically comprise two-thirds of the dry mass, are polysaccharides that can be hydrolyzed to sugars and eventually be fermented to ethanol. The

TABLE 8.1 Comparison of Bioethanol Production from Different Energy Crops.

Crops	Yield ($t\ ha^{-1}\ yr^{-1}$)	Conversion rate to bioethanol ($L\ t^{-1}$)	Bioethanol yield ($L\ ha^{-1}\ yr^{-1}$)
Sugarcane	70	70	4900
Cassava	40	150	6000
Sweet sorghum	35	80	2800
Maize	5	410	2050
Wheat	4	390	1560
Rice	5	450	2250

Source: Reprinted with permission from Jansson et al., 2009. © 2009 Elsevier.

lignin cannot be used for ethanol production (Hamelinck et al., 2005). Hamelinck et al. (2005) describes the biomass structure:

- Cellulose (40–60% of the dry biomass) is a linear polymer of cellobiose (glucose–glucose dimer); the orientation of the linkages and additional hydrogen bonding make the polymer rigid and difficult to break. In hydrolysis, the polysaccharide is broken down to free sugar molecules by the addition of water. This is also called saccharification. The product, glucose, is a six-carbon sugar or hexose (Hamelinck et al., 2005).
- Hemicellulose (20–40%) consists of short highly branched chains of various sugars: mainly xylose (five-carbon), and further arabinose (five-carbon), galactose, glucose, and mannose (all six-carbon). It also contains smaller amounts of non-sugars such as acetyl groups. Hemicellulose, because of its branched, amorphous nature, is relatively easy to hydrolyze (Hamelinck et al., 2005).
- Lignin (10–25%) is present in all lignocellulosic biomass. Therefore, any ethanol production process will have lignin as a residue. It is a large complex polymer of phenylpropane and methoxy groups, a non-carbohydrate polyphenolic substance that encrusts the cell walls and cements the cells together. It is degradable by only a few organisms, into higher value products such as organic acids, phenols, and vanillin. Via chemical processes, valuable fuel additives may be produced (Hamelinck et al., 2005).

At present, several technologies are in use for converting cellulosic feedstocks into ethanol. However, all these technologies can be grouped into two broad categories, namely, hydrolysis and thermochemical conversion. In hydrolysis, the polysaccharides (cellulose and hemicellulose) present in a feedstock are broken down to free sugar molecules (glucose, mannose, galactose, xylose, and arabinose). These free sugar molecules are then fermented to produce ethanol. As lignin cannot be used for ethanol production, it is removed during the conversion process and is generally utilized to meet electricity or heat requirement of an ethanol mill. In the thermochemical conversion process, the feedstock is gasified to produce syngas (a mixture of carbon monoxide, hydrogen, CO_2, methane, and nitrogen) and then syngas is either fermented or catalytically converted to obtain ethanol (Dwivedi et al., 2009). Details of specific technologies under each broad category of conversion technology, i.e., hydrolysis and thermochemical conversion are discussed by Dwivedi et al. (2009). Lignocellulosic biomass ethanol production demands a good knowledge of the material structure used in the biotechnological transformation (Oliveira et al., 2016).

Bearing the importance of "second generation" of ethanol production, this chapter will summarize information about the potential feedstocks that can be better explored in Brazil (Fig. 8.1) as follows.

8.2.1 SUGARCANE BAGASSE, LEAVES, AND STRAW

The sugarcane is named after the scientist of *Saccharum officinarum* and belongs to the botanical family Poaceae, being a grass of the class Liliopsida and the order Ciperales, represented by plants that present small flowers, practically devoid of perianth and protected by dry bracts, in typical inflorescences (Aranha and Yahn, 1987). It is considered to originate in Southeast Asia, the Greater New Guinea, and Indonesia (Heinz, 2013).

Sugarcane bagasse and straw are usually burnt in the industries to provide the required energy demand for the biorefinery. If, on the other hand, both were used for the production of ethanol, much more ethanol would be produced from each hectare of processed cane (Canilha et al., 2012).

Due to the importance of sugarcane bagasse as an industrial byproduct, there is great interest in the development of chemical methods for its transformation into fuels and chemicals that offer strategic economic and environmental advantages. This biomass presents approximately 40–45% of cellulose, 30–35% of hemicelluloses, and 20–30% of lignin (Table 8.2),

FIGURE 8.1 **(See color insert.)** Potential feedstocks for ethanol production in Brazil.

the latter is not easily separated and rapidly solubilized due to its natural recalcitrance (Peng et al., 2009).

Sugarcane is among the main agricultural crops grown in tropical countries. Annual world production of sugarcane is 1.6 billion tons and generates 279 million metric tons of biomass waste (Chandel et al., 2012).

Sugarcane bagasse, leaves, and straw contain an appreciable amount of cellulose and hemicelluloses, which can be polymerized by chemical

TABLE 8.2 Chemical Composition (% w/w, Dry Basis) of Brazilian Sugarcane Bagasse Reported in the Literature.

Component (%)	Author						
	Pitarelo (2007)	Da Silva et al. (2011)	Canilha et al. (2008)	Rocha et al. (2011)	Brienzo et al. (2009)	Rabelo et al. (2011)	Martins et al. (2015)
Cellulose	41.1	38.8	45.0	45.5	42.4	38.4	39.49
Hemicelluloses	22.7	26.0	25.8	27.0	25.2	23.2	26.07
Lignin	31.4	32.4	19.1	21.1	19.6	25.0	23.33
Ash	2.4	2.8	1.0	2.2	1.6	1.5	3.24
Extractives	6.8	—	9.1	4.6	—	—	
Others	—	—	—	—	—	—	

Source: Reprinted from Canilha et al. (2012).

or enzymatic treatments in sugar monomers (glucose, xylose, arabinose, mannose, and galactose). Sugar obtained from these biomasses can be converted into bioethanol and value-added products of significance, which has joint economic importance (Chandel et al., 2012). These biomasses for industrial purposes could provide a sustainable and economical solution for the production of biologically based products as well as products with high added value such as ethanol, xylitol, organic acids, industrial enzymes, and other products.

The efficient use of biomass of cane in products of economic importance, both biological and non-biological, will help to improve the socioeconomic status of developing countries, create employment opportunities, and improve the environment. Among sugarcane producing countries, Brazil is the main producer, with 625 million tons of sugarcane in 2011, followed by India and China. Sugarcane bagasse has long been considered in industries for the production of "Second-Generation Ethanol" (Chandel et al., 2012).

Lignocellulosic biomasses, such as sugarcane biomass, represent an abundant resource where carbohydrates can be extracted and converted into simple sugars and fermented in ethanol, but all these involved stages end up making this process complex due to the complexity of the material lignocellulosic. The process generally involves the following steps: size reduction; pre-treatment; enzymatic hydrolysis to extract sugars; liquid-solid separation of pentose-rich hydrolysates (hemicellulose); purification/detoxification of hemicellulosic sugars; combined or separate fermentation of hexoses and pentose hydrolysates; distillation of bioethanol; and utilization of residual lignin-rich solids as boiler fuel. In order to increase the economic viability of biofuel production, improvements and simplification of processes are necessary to reduce production costs and the variation of yields of the final product (Wang et al., 2018).

Sugarcane biomasses are chemically composed of cellulose, hemicelluloses, and lignin. The fractions of cellulose and hemicellulose are composed of a mixture of polymers of carbohydrates. Several different strategies have been studied for the conversion of polysaccharides to fermentable sugars. The hemicellulosic fraction can be hydrolyzed with dilute acids, followed by cellulose hydrolysis by cellulase to produce fermentable sugars to ethanol fermentation (Canilha et al., 2012).

Generally, the biological process of converting sugarcane biomass into fuel ethanol involves: (1) pretreatment to remove lignin or hemicelluloses to release cellulose; (2) depolymerization of carbohydrate polymers to produce free sugars by cellulase mediated action; (3) fermentation of hexose and/or pentoses sugars to produce ethanol; (4) distillation of ethanol. Ethanol produced from sugarcane wastes is one of the most suitable alternatives for partial replacements of fossil fuels because it provides renewable and less carbon-intensive energy than gasoline. Bioethanol reduces air pollution and also helps mitigate climate change by reducing greenhouse gas emissions (Canilha et al., 2012).

8.2.2 CASSAVA RESIDUE

Cassava was originated in South America and was domesticated for more than 5000 years. It was derived from the wild ancestral *M. esculenta ssp. flabellifolia*. The selection that occurred during the domestication of this crop resulting in many morphological, physiological, and biochemical differences between the cassava we know today and its wild ancestor. Characteristics such as the increase in the size of the tuberous roots, the higher starch content in the tuberous roots (Fig. 8.2) and the vegetative propagation through stem cuttings are the results of human selection in the cassava crop. Interestingly, some progenies in the second generation (Fig. 8.2) of a cross between cassava and *M. esculenta ssp. Flabellifolia* showed tuberous root weight considerably higher than that of cassava (Jansson et al., 2009).

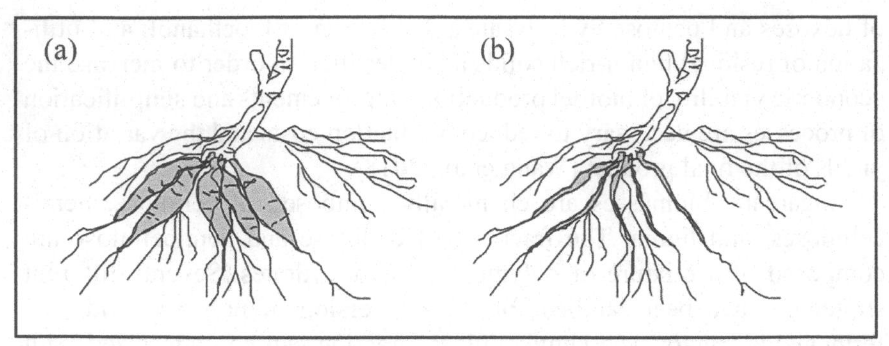

FIGURE 8.2 (See color insert.) (a) Sheme of cassava root and (b) cassava wild ancestor *M. esculenta ssp. Flabellifolia.* Source: Reprinted with permission from Jansson et al., 2009. © 2009 Elsevier.

Many characteristics of cassava, such as high dryness and heat tolerance, as well as low demands on the use of agricultural fertilizers and also their high starch content, end up making this crop one of the most attractive for the production of starch in the future. We know that as population pressure grows along with climate change, cassava production is expected to increase in the coming decades. Thus, cassava crop improvement has been the subject of much research, although cassava is cultivated mainly by small farmers, its use for agro-industrial processing in Asia has been increasing. In addition to its traditional role as a food crop, cassava can increase its value by becoming an important biofuel crop. High starch yield and total dry matter, despite poor soil and dry conditions coupled with low agrochemical requirements, result in an energy input that represents only 5–6% of the final energy content of the total cassava biomass. This translates into a 95% energy yield assuming the full utilization of energy content in the total biomass, according to Janssem et al. (2009).

According to Veiga et al. (2016), since the energy production, projects may involve large amounts of investment, a project to utilize cassava residues or any other biomass as a source of energy cannot be justified by the application of national productivity because subsistence production will always have an economically prohibitive yield. In addition, in order to properly evaluate productivity in the field, the essential characteristics of cassava residues as raw material for energy production must be determined in each case.

Considering the main types of cassava roots processing in Brazil, such as the cassava flour manufacture and the starch extraction, the by-products generated may be solid or liquid. Some of the solid by-products are brown peel, discard of sub-peel, fiber, bran, and sweep. Among the liquid residues is quenched with manueira, which is extremely toxic and polluting because it has hydrocyanic acid (Sagrilo et al., 2003). Bran is the main solid waste produced in stalls and, in general, it is thrown in waterways or left in ditches that drain and carry a great deal of organic load for them.

The root culture is very rich in carbohydrates, presenting about 30%, of which 25–40% is the starch (Adeoti, 2010; Martinez, 2016). In the starch, for each ton of processed root, about 930 kg of bran is produced with 85% moisture. After drying, this residue presents on average 75% of starch and 11.5% of fibers. Considering its composition and the considerable amount generated, several researches have been carried out in order to take advantage of this residue (Cereda, 1996; Martinez, 2016).

Veiga et al. (2016) studied the yield of roots and crop residues from different cassava crops and found satisfactory results for all crops studied. Table 8.3 shows a comparison with the typical productivity of sugarcane, the most important energy crop in Brazil. The sugarcane presents dry matter yield about 50% higher than the cassava studied in Veiga's work. Although the processes involved in producing ethanol using these plants are slightly different because of different feedstocks (sugars or starches), the ethanol yield in both cases is of the same order of magnitude. Since the beginning of the Brazilian Ethanol National program in the 1970s, intensive research has been conducted on genetically improved cane and new treatments, resulting in a productivity increase of about 50 t ha^{-1} $year^{-1}$ in 1980 to the current 85 t ha^{-1} $year^{-1}$, according to the AgriEnergy 2012 Annual report (Brazil, 2012).

TABLE 8.3 Productivity Comparison Between Cassava and Sugarcane in Brazil (Veiga et al., 2016).

	Cassava roots (t ha^{-1} yr^{-1})	**Sugarcane (t ha^{-1} yr^{-1})**
Productivity (wb)	30.1	85
Drymatter	13.1	36
Waste (db)	9.2	21.2
Raw material for ethanol production	9.8	13.17
Ethanol production	4.9	5.36

wb, water basis; db, dry basis

8.2.3 PEACH PALM (Bactris gasipaes) AND TUCUMÃ (Astrocaryum aculeatum)

The lignocellulosic residues found in the Brazil Northern cities fairs, such as peach palm peel (*Bactris gasipaes*) and tucumã (*Astrocaryum aculeatum*) bark, are abundant and have a low cost of cellulose, hemicellulose among other polymers, which often by not being employed become garbage, causing pollution to the city. It is noteworthy that there are no studies using the potential of these residues for the production of ethanol (Silva et al., 2017).

Silva et al. (2017) studied the potential of residues of peach palm and tucumã for the production of biofuels since such organic waste constitutes a sanitary inconvenience to the city of Manaus located in the north region of Brazil. These authors studied the enzymatic hydrolysis of these residues using the following enzymes: celluclast® 1.5 L (10 U/g sample), pectinex ultra SP-L (10 U/g), and glycosidase (10 U/g alpha-glucosidase). The enzymes were used independently and in combination. Reactions were incubated at 40°C for 48 h with shaking at 200 rpm. This stage is important for the release of reducing sugars. Among the residues studied, peach palm was the most satisfactory in relation to the release of fermentable sugars of potential for the production of "second-generation" bioethanol, with total reducing sugar values (RSVs) of 26.8 g/L for the residue of peach palm and 3.63 g/L for the tucumã.

8.2.4 PALM OIL RESIDUES

The Palm Oil industry can generate a huge amount of solid waste in the form of lignocellulosic waste. Part of these lignocellulosic biomasses is available in the form of bunches of fruit. Due to its physicochemical characteristics and its abundant supply, it emerged as potential biomass in the production of biofuels. About 23% of empty fruit bunches are produced by tons of processed fresh fruit. In 2016, the yield of palm oil processing was 15.91 tons per hectare; thus, the annual production of waste generated (bunches) was estimated at 3.66 tons per hectare palm oil production. Thus, this residue emerged as a potential raw material for the production of biofuels, such as bioethanol, through a series of pre-treatments, hydrolysis, saccharification, and fermentation. Palm oil wastes are one of the cheapest raw materials found as an alternative energy source when compared to edible sources, such as corn, sweet potatoes, and other foods used for the production of "first-generation" bioethanol. This is because these residues have high levels of cellulose that can be converted into fermentable sugars, being a potential raw material for conversion into bioethanol through alcoholic fermentation (Derman et al., 2018).

The use of açaí seeds in northern of Brazil for fuel ethanol production purposes would be a way to add economic value for this material and remove it from the public roads of these cities (Oliveira et al., 2014).

The açaí tree (*Euterpe oleracea*) is an Amazonian palm that has the pulp obtained from its fruits widely consumed by the population of the north of Brazil and that has had its consumption expanded to the international market with the increase of the exportation of the pulp of fruit (Galotta and Boaventura, 2005).

Açaí seeds in recent years have been shown to be a processing waste, for which few technological alternatives of use have been developed. This waste has been presented as an environmental disorder for the cities of the north of Brazil, mainly Belém and Manaus (Silva and Rocha, 2003) and still continues today.

FIGURE 8.3 (See color insert.) Açaí seeds after pulp extraction, washed, dried, and grinded (adapted from Oliveira, 2014).

The state of Pará is currently the largest producer of açaí in Brazil with 120.890 tons/year of the fruit; of this, total 93.521 tons/year is waste (core), that is, about 83% (IBGE, 2008). There are only 3000 establishments in the city of Belém that commercialize the processed açaí, which represents a daily consumption of 440 thousand kg of the fruit (IBGE, 2008), which generates a surplus of 365 tons per day of organic waste.

Oliveira et al. (2014) observed that the açaí seeds contained mainly glucan (44.14%), xylan (18.70%), and lignin (18.37%) values that approximate amount of 60%, which would justify the use of this biomass to ethanol production (Kim et al., 2005), since associated with methods to optimize enzymatic attack such as the removal of hemicellulose and lignin and modification of cellulose structure (Rabelo et al., 2011) could lead to suitable process. The lignin content of seeds of açaí biomass is in the range reported for other agricultural residues such as corn stover (17–19%) (Kim et al., 2005).

Ash and extractives were present in concentrations higher (1.26% and 8.15%) than that reported to others biomass such as sugarcane bagasse (Gomez et al., 2010) that were 1.79% and 3.25%, respectively. Açaí seeds also contain a relatively low amount of protein (4.20%) when was compared to soybean straw (Cassales, 2011).

The efficient utilization of lignocellulosic material for ethanol production requires pre-treatments that can act in various forms on the lignocellulosic material, improving enzymatic hydrolysis, and fermentation rates (Heredia-Olea et al., 2013).

The treatment of açaí seeds with sulfuric acid dilute was performed by Oliveira et al., (2014) with a set experiment varying the concentration of solids (5, 10, and 15% weight/volume of solution) and sulfuric acid concentration (0.5, 1.0, and 1.5% w/v). It was observed that the recoveries achieved of material treated are ranged from 56.87% to 93.87%, depending on the pretreatment conditions. The lowest recovery was obtained at 1% acid concentration and 1.6% solid loading for 60 min, which was probably to decomposition of the resulting cellulose. In this study, the values obtained for the total concentration of furfural and hydroxymethylfurfural ranged from 0.22 g/L to 1.182 g/L, which was lower than the concentration considered able to inhibit fermentation (Nichols et al., 2008).

The combination between delignification with NaOH 1% and sulfuric acid dilute treatment has shown great results to reduce the percentages of hemicellulose and lignin from açaí seeds according to Oliveira et al. (2015). Remove until 99% of hemicellulose and until 31.56% of lignin from açaí seeds . However, in this same condition was observed a high removal of cellulose (63.5%), which represents an undesirable condition. In Figure 8.4, from the left to right is possible to observe açaí seeds after treatment with H_2SO_4 dilute and delignified with NaOH 1%.

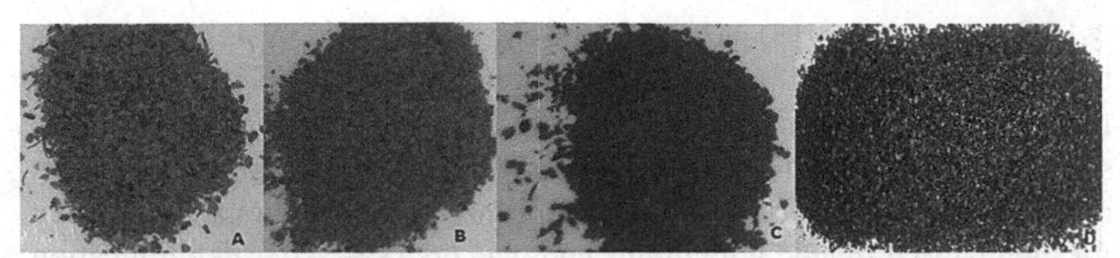

FIGURE 8.4 (See color insert.) From the left to right: açaí seeds after treatment with 0.2% (a), 1.0% (b), 1.84% (c) de H_2SO_4 and treated with 1.84% of H_2SO_4/delignified with 1% of NaOH (adapted from Oliveira, 2014).

Great disorders of biomass structure were observed in Figure 8.5c after delignification of H_2SO_4-treated açaí seeds. It becomes visibly more porous and amorphous in comparison to untreated material (a) or H_2SO_4 treated (b) (Oliveira et al., 2016). As also verified by Zúñiga (2010), this material heterogeneity is a complicating factor to more efficient Scanning Electron Microscope biomass structure evaluation. In Figure 8.5a–c, photomicrographies of treated and untreated açaí seeds are presented.

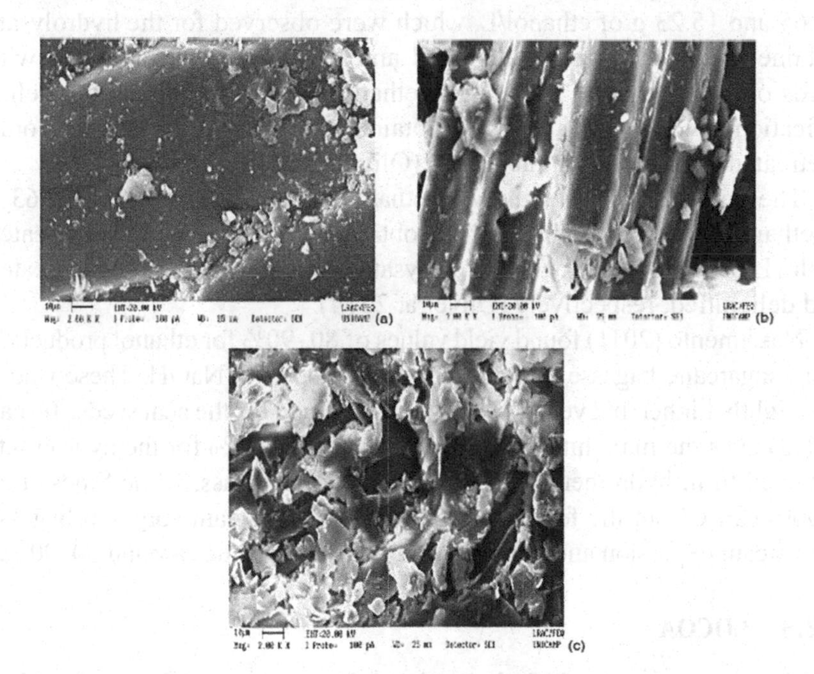

FIGURE 8.5 Untreated açaí seed (a), after H_2SO_4 dilute treatment (b), and treated with H_2SO_4 dilute and delignified with NaOH 1%. Source: Reprinted from Oliveira et al., 2016. Open access.

Oliveira et al. (2015) performed the enzymatic hydrolysis of açaí seeds treated with dilute sulfuric acid and delignified with NaOH and observed for glucose concentrations yields ranging from 0.26 to 0.47 g glucose/g of seed pretreated with dilute acid. For enzymatic conversion, found values ranged from 62.40% to 94.76%. The optimum conversion value obtained for the material treated with dilute acid was 66.48%. After delignification for the same condition it was possible to obtain a maximum increase up to a conversion of 42.53%, which demonstrated the value conversion of 94.76%.

The fermentation of enzymatic hydrolysates obtained with açaí seeds treated with H_2SO_4 diluted and delignified was studied by Oliveira (2014) and showed fermentation yield values of 73.79% (pre-treated with diluted H_2SO_4 only), 79.69% (pre-treated and delignified) and 89.31% (standard glucose). In the same study, the glycerol values reached the maximum levels of 5.12 and 4.99% after the 24 h of fermentation for the hydrolysates obtained with the same material.

After 24 h of fermentation, the maximum ethanol contents was of 21.65 and 15.23 g of ethanol/L, which were observed for the hydrolysate obtained with açaí seeds delignified and for hydrolysate obtained with seeds only treated with H_2SO_4. The ethanol content obtained after delignification is 42% higher than that obtained for the hydrolysate with only pretreated treated with dilute H_2SO_4 (Oliveira, 2014).

The volumetric productivity of ethanol obtained was 0.9 and 0.63 g of ethanol L^{-1} h^{-1} for the hydrolysate obtained with açaí seeds only treated with H_2SO_4 dilute and for the hydrolysate obtained with the seeds treated and delignified, respectively (Oliveira, 2014).

Nascimento (2011) found yield values of 80–90% for ethanol production from sugarcane bagasse hydrolysates pretreated with NaOH. These values are slightly higher, but very close to those obtained for the açaí seeds. Toscan (2013) cites the maximum conversion value of 76.72% for the hydrolysate obtained from hydrothermally pretreated elephant grass, while Sipos et al. (2009) carried out the fermentation of pre-treated steam sorghum bagasse with steam explosion and obtained again the yield values around 80–90%.

8.2.5 COCOA

Cocoa, known worldwide for being the main raw material for the chocolate industry, the fruit of cocoa comes from the tree species popularly known as cacaueiro (cacaotree), scientifically known as *Theobroma cacao L,* which means food of the gods, has a life expectancy of more than 100 years and can reach 4–8 m in height, producing fruits that measure up to 25 cm in length containing from 30 to 40 seeds of 2–3 cm in length each, surrounded by a white mucilaginous pouch, weighing in average 500 g (Pinheiro and Silva, 2017).

In Brazil, cocoa was first cultivated in 1679, but mainly due to the poverty of the soils of the region that covers the State of Pará, some producers were not successful in implanting the cocoa crop, thus not consolidating the cultivation in this region. On the other hand, the state of Bahia had a positive experience with the planting of cacao. The adaptation of cacao to the edaphoclimatic characteristics of the southern region of Bahia was so significant that the region produced 95% of all Brazilian cacao. The development of the cacao culture extended to the south of the Bahian reconceive, extreme south of Bahia to the state of Espirito Santo, taking this region to receive the name of "Cocoa Region" (Pinheiro and Silva, 2017).

Cacao shell residues have cellulose of 31.45%, lignin of 15.8%, and crude fiber of 50.2% in its composition, according to Pinheiro and Silva (2017). Therefore, this feedstock presents promisor to "second-generation" bioethanol production.

8.3 CHALLENGES IN SECOND-GENERATION ETHANOL PRODUCTION

Several limitations and drawbacks of bioethanol production from lignocellulosic biomass have been reported. To bioethanol production, the challenges are as follows.

- Land and water use competition (Leong et al., 2018).
- Requires sophisticated downstream processing technologies (Leong et al., 2018).
- Recalcitrant structures of the feedstock (Jambo et al., 2016).
- Pretreatment steps and enzymatic hydrolysis for saccharification optimized for the production of "second-generation" ethanol and operating at high solids loading (Komesu et al., 2018).
- Isolation and development of powerful strains with high substrate tolerance (Komesu et al., 2018).

- Reduce the generation of by-products, such as lactic acid, acetic acid, carbon dioxide, and glycerol during ethanol fermentation (Komesu et al., 2018).
- Development of strains able to utilize xylose or mixed sugars in lignocellulosic biomass (Komesu et al., 2018)
- Production cost of bioethanol from lignocellulosic materials is relatively high when based on current technologies.

8.4 CONCLUDING REMARKS

The potential of "second-generation" feedstock for bioethanol production has been evaluated in this chapter. Bioethanol from lignocellulosic biomass is a fuel that can provide considerable environmental, economic, and strategic benefits. Great efforts have been made to try to develop technologies that allow them to compete with petroleum-derived fuels. Bioethanol can be produced from different types of feedstock. These raw materials are classified into three agricultural categories: simple sugars, starch, and lignocellulose. The price of the raw material is highly volatile, which directly affects the costs of producing bioethanol. Brazil is a country of immense agro-industrial production and with different residues that can be explored to ethanol production. Therefore, potential raw materials for ethanol production in Brazil are sugarcane and straw bagasse; cassava residues, peach palm branches; tucumã; palm oil residues; and cocoa. To produce bioethanol at low cost, it is necessary to continue to advance the knowledge of the process steps (pre-treatment, hydrolysis, fermentation, and distillation) and integrate them into a single process developing the concept of biorefinery.

KEYWORDS

- **ethanol**
- **biofuel**
- **lignocellulosic biomasses**
- **Brazil feedstocks**
- **second-generation ethanol**

REFERENCES

Adeoti, O. Water Use Impact of Ethanol at a Gasoline Substitution Ratio of 5% from Cassava in Nigeria. *Biomass Bioenergy*. **2010,** *34,* 985–992.

Alaswad, A.; Dassisti, M.; Prescott, T.; Olabi, A.G. Technologies and Developments of Third Generation Biofuel Production. *Renewable Sustain. Energy Rev.* **2015,** *51,* 1446–1460.

Altman, F. F. A. The Açaí Seed (Euterpe olaracea Mart.). Belém, Brasil, Instituto Agronômico. Boletim Técnico, 31, 1956.

Aranha, C.; Yahn, C. Botânica da cana-de-açúcar. PARANHOS, SB Cana-de-açúcar: cultivo e utilização. Campinas: Fundação Cargill, v. 1, 1987.

Brazil, Ministério da Agricultura, Pecuária e Abastecimento, in: StatisticalYearbookofAgroenergy 2012, 2013 (Brasília).

Brienzo, M.; Siqueira, A. F.; Milagres, A. M. F. Search for Optimum Conditions of Sugarcane Bagasse Hemicellulose Extraction. *Biochem. Eng. J.* 2009, *46*(2), 199–204.

Canilha, L.; Carvalho, W.; Giulietti, M.; Felipe, M. D. G. A., Almeida E Silva, J. B. Clarification of a Wheat Straw-Derived Medium with Ion-Exchange Resins for Xylitol Crystallization. *J. Chem. Technol. Biotechnol.* **2008,** *83* (5), 715–721.

Canilha, L.; Chandel, A. K.; Suzane dos Santos Milessi, T.; Antunes, F. A. F.; Luiz da Costa Freitas, W.; das Graças Almeida Felipe, M.; da Silva, S.S. Bioconversion of Sugarcane Biomass into Ethanol: An Overview About Composition, Pretreatment Methods, Detoxification of Hydrolysates, Enzymatic Saccharification, and Ethanol Fermentation. *BioMed Res. Int.* 2012, 1-15.

Cardona, E.; Llano, B.; Peñuela, M.; Peña, J.; Rios, L.A. Liquid-Hot-Water Pretreatment of Palm-Oil Residues for Ethanol Production: An Economic Approach to the Selection of the Processing Conditions. *Energy* **2018,** *160,* 441–451.

Carvalho, J. A.; Borges, W. M. S.; Carvalho, M. Z.; Arantes, A. C. C.; Bianchi, M. L. Bagaço de Cana-de-Açúcar Como Fonte de Glicose: Pré-tratamento Corona. *Revista Virtual de Química*. **2017,** *9,* 97–106.

Cassales, A.; Souza-Cruz, P. B. De.; Rech, R.; Ayub, M. A. Z. Optimization of Soybean Hull Acid Hydrolysis and Its Characterization as a Potential Substrate for Bioprocessing. *Biomass Bioenergy*. **2011,** *35,* 4675–4683.

Cereda, M.P. Caracterização, Usos e Tratamentos de Resíduos da Industrialização da Mandioca. Botucatu: Centro de Raízes Tropicais, 1996.

Chandel, A. K.; da Silva, S. S.; Carvalho, W.; Singh, O.V. Sugarcane Bagasse Andleaves: Foreseeable Biomass of Biofuel and Bio-Products. *J. Chem. Technol. Biotechnol.* **2012,** *87*(1), 11–20.

da Silva, A. S. A.; Inoue, H.; Endo, T.; Yano, S.; Bon, E. P. Milling Pretreatment of Sugarcane Bagasse and Straw for Enzymatic Hydrolysis and Ethanol Fermentation. *Bioresource Technol.* **2010,** 101 (19), 7402–7409.

Derman, E.; Abdulla, R.; Marbawi, H.; Sabullah, M. K. Oil Palm Empty Fruit Bunches as a Promising Feedstock for Bioethanol Production in Malaysia. *Renewable Energy*. **2018,** *129,* 285–298.

Dwivedi, P.; Alavalapati, J. R. R.; Lal, P. Cellulosic Ethanol Production in the United States: Conversion Technologies, Current Production Status, Economics, and Emerging Developments. *Energy Sustain. Develop.* **2009,** *13,* 174–182.

Fernandes, S. M. Prospecção de Micro-Organismos Lignolíticos da Microbiota Amazônica Para a Produção de Biocombustível de Segunda Geração e Compostagem. Master Dissertation, Universidade do Estado do Amazonas, AM, Brazil, 2013.

Galotta, A. L. Q. A.; Boaventura, M. A. D. Chemical Constituents of the Root and Stem of açaí leaf (Euterpe precatoriamart.,Arecaceae). *Química Nova.* **2005,** *28*, 610–613.

Gomez-Rueda, S. M. Pretreatment and Enzymatic Hydrolysis of Sugarcane Bagasse. Master Dissertation, UniversidadeEstadual de Campinas, Campinas, SP, Brazil, 2010.

Hamelinck, C. N.; Hooijdonk, G. V.; Faaij, A. P. C. Ethanol from Lignocellulosic Biomass: Techno-Economic Performance in Short-, Middle- and Long-Term. *Biomass Bioenergy.* **2005,** *28*, 384–410.

Heinz, D. J. *Sugarcane Improvement Through Breeding*; Elsevier: New York, NY, 2013.

Heredia-Olea, E., Pérez-Carrillo, E., Serna-Saldívar, S.O. Production of Ethanol from Sweet Sorghum Bagasse Pretreated with Different Chemical and Physical Processes and Saccharified with Fiber Degrading Enzymes. *Bioresour Technol.* **2013,** *134*, 386–390.

IBGE. Production of Vegetable Extraction and Forestry 2007. Comunicação Social, IBGE, [Online] 2008. http://www.ibge.gov.br/home/presidencia/noticias/noticia_impressao.php?id_noticia=1270 (accessed Apr 21, 2013).

Jambo, S. A.; Abdulla, R.; Azhar, S. H. M.; Marbawi, H.; Gansau, J. A.; Ravindra, P. A Review on Third Generation Bioethanol Feedstock. *Renewable Sustain. Energy Rev.* **2016,** *65*, 756–769.

Jansson, C.; Westerbergh, A.; Zhang, J.; Hu, X.; Sun, C. Cassava, a Potential Biofuel Crop in (the) People's Republic of China. *Appl. Energy.* **2009,** *86*, S95–S99.

Kim, S.; Holtzapple, M.T. Lime Pretreatment and Enzymatic Hydrolysis of Corn Stover. *Bioresour. Technol.* **2005,** *96*, 1994–2006.

Komesu, A.; Oliveira, J.; Martins, L. H. S.; Wolf Maciel, M. R.; MacielFilho, R. Lactic Acid and Ethanol: Promising Bio-Based Chemicals from Fermentation. In *Principles and Applications of Fermentation Technology*; Kuila, A.; Sharma, V., Eds.; Scrivener Publishing LLC: Beverly, MA; **2018;** p 87–116.

Leong, W.-H.; Lim, J.-W.; Lam, M.-K.; Uemura, Y.; Ho, Y.-C. Third Generation Biofuels: A Nutritional Perspective in Enhancing Microbial Lipid Production. *Renewable Sustain. Energy Rev.* **2018,** *91*, 950–961.

Margon, R. A., Pinotti, L. M., Freitas, R. R. Enzymatic Hydrolysis of Eucalyptus Biomass for Bioethanol Production: A Bibliometric Analysis. *Res. Soc. Develop.* **2018,** *7* (4), 1474301.

Martinez, D. G. Ethanol Production from Cassava Processing Waste. Master Dissertation, Universidade Estadual do Oeste do Paraná, Cascavel, Brazil, 2016.

Martins, L. H.S.; Rabelo, S. C.; da Costa, A. C. Effects of the Pretreatment Method on High Solids Enzymatic Hydrolysis and Ethanol Fermentation of the Cellulosic Fraction of Sugarcane Bagasse. *Biores. Technol.* **2015,** *191*, 312–321.

Ministério De Minas E Energia. Resenha Energética Brasileira: exercício de 2016. Brasília: Secretaria de Planejamento e Desenvolvimento Energético, [Online] 2017. http://www.mme.gov.br/documents/10584/3580498/02+-+Resenha+Energ%C3%A9tica+Brasileira+2017+-+ano+ref.+2016+%28PDF%29/13d8d958-de50-4691-96e3-3ccf53f8e1e4?version=1.0 (accessed Jan 02, 2018).

Nascimento, V. M. Alkalinepre-treatment (NAOH) of Sugar Cane Baga for Ethanol Production and Xylooligomer Production. Master Dissertation, Universidade Federal de São Carlos, São Carlos, Brazil, 2011.

Nichols, N. N.; Sharma, L. N.; Mowery, R. A.; Chambliss, C. K.; Walsum, G. P. V.; Dien, B. S.; Iten L. B. Fungal Metabolism of Fermentation Inhibitors Present in Corn Stover Dilute Acid Hydrolysate. *Enzyme Microb. Technol.* **2008,** *42*, 624–630.

Nigam, P. S.; Singh, A. Production of Liquid Biofuels from Renewable Resources. *Progress Energy Combustion Sci.* **2011,** *37*, 52–68.

Oliveira, J. A. R. Investigation of the Steps for the Second Generation Ethanol Production Process from the Biomass of Açaí Seeds (euterpeoleracea). PhD Dissertation, UniversidadeEstadual de Campinas, Campinas, Brazil, 2014.

Oliveira, J. A. R; Komesu, A.; Martins, L. H. S; MacielFilho, R. Evaluation of Microstructure of Açaí Seeds Biomass Untreated and Treated with H_2SO_4 and NaOH by SEM, RDX and FTIR. *Chem. Eng. Trans.* **2016,** *50*, 379–384.

Oliveira, J. A. R., Martins, L. H. S.; Komesu, A., Maciel Filho, R. Evaluation of Alkaline Delignification (NaOH) of Açaí Seeds (EutherpeOleracea) treated with H_2SO_4 Dilute and Effect on Enzymatic Hydrolysis. *Chem. Eng. Trans.* **2015,** *43*, 499–504.

Oliveira, J. A. R.; Komesu, A.; Maciel Filho, R. Hydrothermal Pretreatment for Enhancing Enzymatic Hydrolysis of Seeds of Açaí (Euterpe oleracea) and Sugar Recovery. *Chem. Eng. Trans.* **2014,** *37*, 787–792.

Peng, F.; Ren, J. L.; Xu, F.; Bian, J.; Peng, P.; Sun, R. C. Comparative Study of Hemicelluloses Obtained by Graded Ethanol Precipitation from Sugarcane Bagasse. *J. Agri. Food Chem.* **2009,** 57 (14), 6305–6317.

Perdices, M. B.; Arcaya, G. E. A.; Coral, D. S. O.; Mendoza, M. A. G.; Nunes, D. C. L. Bioetanol a Partir de Materiais Lignocelulósicos Pela Rota da Hidrólise. In *Biocombustíveis*; Lora, E. E. S.; Venturini, O. J., Eds.; Interciência: Rio de Janeiro, Brazil, **2012;** Vol. 1; p 536.

Pinheiro, I. R.; Silva, R. O. Reaproveitamento dos Resíduos Sólidos da Indústria Cacaueira. *Blucher Chem. Eng. Proc.* **2017,** 4 (1), 95–99.

Pitarelo, A. P. Avaliação da Susceptibilidade do Bagaço e da Palha de cana-de-açúcar à Bioconversão via Pré-Tratamento a Vapor e Hidrólise Enzimática.UFPR, Curitiba, 2007.

Pitarelo, A. P.; Silva, T. A.; Peralta-Zamora, P. G.; Ramos, L. P. Efeito do Teor de Umidade Sobre o Pré-tratamento a Vapor e a Hidrólise Enzimática do Bagaço de Cana-de-açúcar. *Química Nova.* **2012,** *35* (8), 1–8.

Rabelo S. C.; Carrere H.; MacielFilho, R.; Costa, A. C. Production of Bioethanol, Methane and Heat from Sugarcane Bagasse in A Biorefinery Concept, *Bioresource Technol.* **2011,** *102*, 7887–7895.

Rocha, M. G. J.; Martin, C.; Soares, I. B.; Maior, A. M. S.; Baudel, H. M.; De Abreu, C. A. M. Dilute Mixed-Acid Pretreatment of Sugarcane Bagasse for Ethanol Production. *Biomass Bioenergy.* **2011,** *35* (1), 663–670.

Rodríguez-Zúñiga, U. F.; Lemo, V.; Farinas, C. S.; Bertucci Neto, V.; Couri, S. Evaluation of Agroindustrial Residues as Substrates for Cellulolytic Enzymes Production Under Solid State Fermentation. In *Encontro Da Sociedade Brasileira De Pesquisa Em Materiais - SBPMat*, 7, 2008, Guarujá; BRAZILIAN MRS MEETING, 7., 2008, Guarujá. Abstracts. Rio de Janeiro: SBPMat, 2008. non-paged. 1CD-ROM.

Sagrilo, E.; VidigalFilho, P. S.; Pequeno, M. G.; Scapim, C. A.; Vidigal, M. C. G.; Diniz, S. P. S. S.; Modesto, E. C.; Kvitschal, M.V. Effect of Harvest Period on the Quality of Storage Roots and Protein Content of the Leaves in Five Cassava Cultivars (Manihotesculenta, Crantz). *Brazilian Arch. Biol. Technol.* **2003,** *46*, 295–305.

Silva Fernandes, F.; Farias, A. V.; dos Santos, R. A.; de Souza, É. S.; de Souza, J. V. B. Determinação da Quantidade de Açúcares Redutores dos Hidrolisados Enzimáticos de Resíduos Amazônicos Para Obtenção de Bioetanol. *J. Eng. Exact Sci.* **2017,** *3* (4), 0608–0613.

Silva, I. T.; Rocha, B. R. P. Biomass Energy, Family Agriculture and Social Insertion in Contribution to Sustainable Development in the Isolated Communities of the State of Pará. Anais Do Simpósio Amazônia, Cidades E Geopolíticas Das Águas. Projeto MEGAM, **2003,** 172–173.

Sipos, B.; Réczey, J.; Somorai, Z.; Kádár, Z.; Dienes, D.; Réczey, K. Sweet Sorghum as feedstock for Ethanol Production: Enzymatic Hydrolysis of Steam-Pretreated Bagasse. *Appl. Biochem. Biotechnol.* **2009,** *153*, 151–162.

Toscan, A. Effect of Hydrothermal Pre-treatment on the Yield of Enzymatic Hydrolysis of Elephantgrass. Master Dissertation, Centro de Ciências Agrárias e Biológicas, Universidade de Caxias do Sul, Caxias do Sul, Brazil, 2013.

Veiga, J. P. S.; Valle, T. L.; Feltran, J. C.; Bizzo, W.A. Characterization and Productivity of Cassava Waste and Its use as an Energy Source. *Renewable Energy*. **2016,** *93*, 691–-699.

Wang, Z.; Dien, B. S.; Rausch, K. D.; Tumbleson, M. E.; Singh, V. Fermentation of Undetoxified Sugarcane Bagasse Hydrolyzates Using a Two Stage Hydrothermal and Mechanical Refining Pretreatment. *Bioresource Technol.* **2018,** *261*, 313–321.

Zúñiga, U. F. R. Bioprocess Development for Production of Specific Cellulases in the Production Chain of Second-Generation Ethanol. Ph.D. Thesis, University of São Paulo, São Paulo, Brazil, 2010.

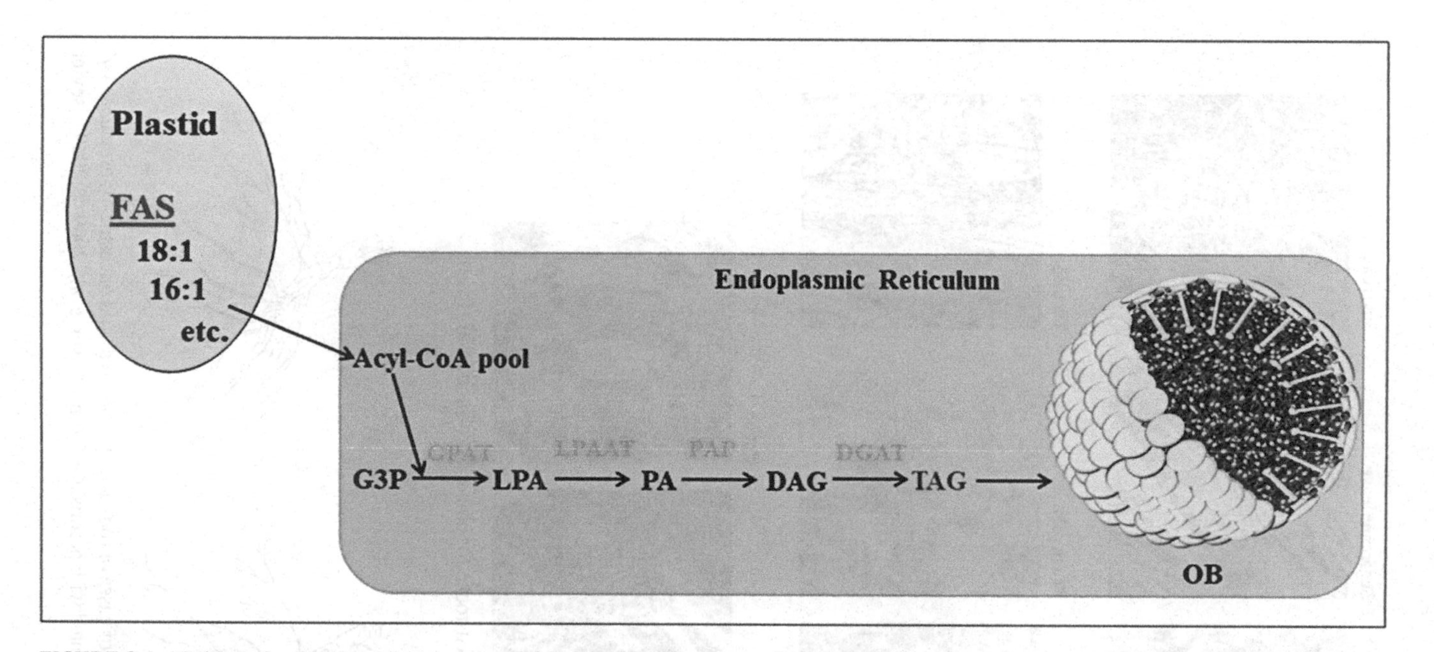

FIGURE 3.1 Schematic diagram of triacylglycerol biosynthesis in plants. Green compartment represents plastid and orange compartment represents ER. DAG, tiacylglycerol; DGAT, diacylglycerol acyltransferase; FAS, fatty acid synthase; G-3-P, glycerol-3-phosphate; GPAT, glycerol-3-phosphate acyltransferase; LPA, lysophosphatidic acid; LPAAT, lysophosphatidic acid acyltransferase enzyme; PA, phosphatidic acid; TAG, triacylglycerol.

FIGURE 8.1 Potential feedstocks for ethanol production in Brazil.

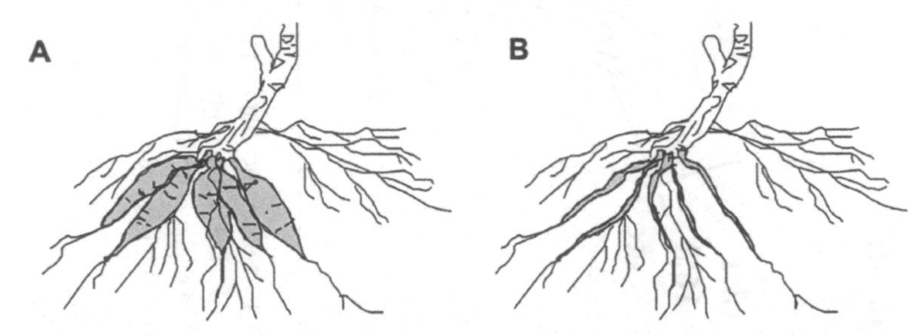

FIGURE 8.2 (A) Sheme of cassava root and (B) cassava wild ancestor *M. esculenta ssp. flabellifolia*. Source: Reprinted with permission from Jansson et al., 2009. © 2009 Elsevier.

FIGURE 8.3 Açaí seeds after pulp extraction, washed, dried, and grinded (adapted from Oliveira, 2014).

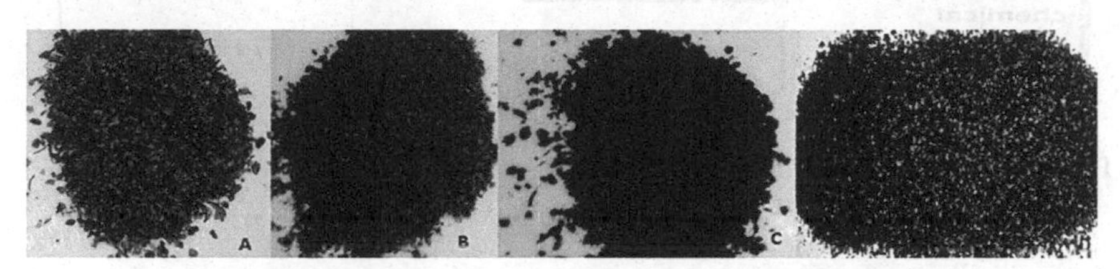

FIGURE 8.4 From the left to right: açaí seeds after treatment with 0.2% (a); 1.0% (b), 1.84% (c) de H_2SO_4 and treated with 1.84% of H_2SO_4 /delignified with 1% of NaOH (adapted from Oliveira, 2014).

FIGURE 9.2 (A) *Pistia stratiotes* mat growth on a water pond and (B) rosette structure of the plant.

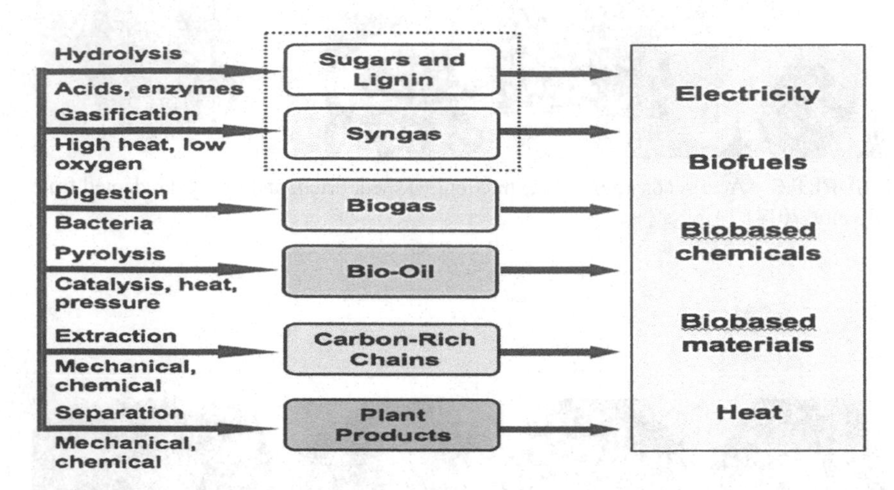

FIGURE 17.1 Bioenergy formation process.

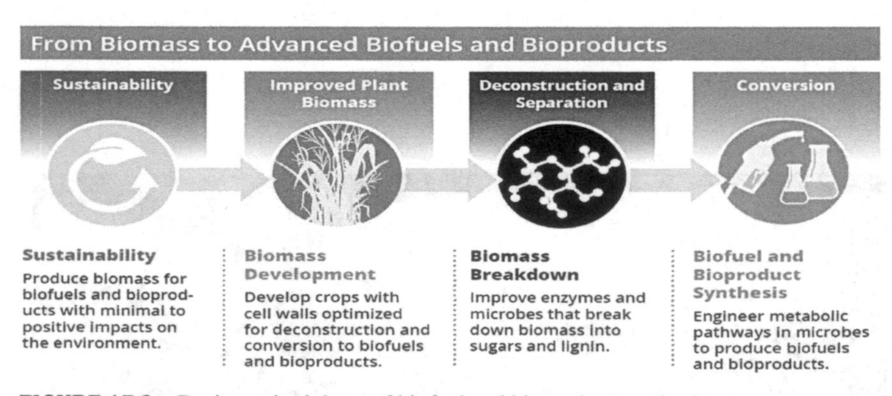

FIGURE 17.2 Basic methodology of biofuel and bioproduct synthesis.

CHAPTER 9

Perspective of Liquid and Gaseous Fuel Production from Aquatic Energy Crops

GUNASEKARAN RAJESWARI[1], SAMUEL JACOB[1,*], and RINTU BANERJEE[2]

[1]Department of Biotechnology, School of Bioengineering, SRM Institute of Science and Technology, Kattankulathur 603203, Tamil Nadu, India

[2]Department of Agricultural and Food Engineering, Indian Institute of Technology, Kharagpur 721302, West Bengal, India

[*]Corresponding author. E-mail: samueljacob.b@ktr.srmuniv.ac.in

ABSTRACT

Aquatic plants, otherwise termed as weeds, are fast-growing plants that complete their life cycle in water bodies. Because of its rapid growth and multiplication, it causes deleterious effects to aquatic environment and public health both directly and indirectly. Therefore, considering the harmful effects and economical loss it is essential to control the population of aquatic weeds in water bodies. Because of their high productivity and organic content, aquatic weeds have attracted interest to utilize it for biofuel production (ethanol, methane, and hydrogen) and green leaf manure. This chapter is envisaged to have an insight into variety of aquatic weeds, macrophytes, and dedicated energy crops that have significant impact on lignocellulosic biofuel.

9.1 INTRODUCTION

Economic growth of a country relies on many factors among them energy plays a pivotal role. A report by International Energy Agency

(IEA, 2011) suggests that between 2010 and 2035 global primary energy demand will increase by one-third. Courtney et al. (2003) reported that the present crude oil reserves will get extinguish within 40–70 years, whereas natural gas will run off by next 50 years. In addition, CO_2 emissions due to energy generation from fossil fuels will rise by 20% and it has been predicted that average global temperature could rise between 1.4°C and 5.8°C by 2100 (Dow et al., 2006). Fossil fuels are finite in nature and are considered as depleting resources. Thus, strong efforts are oriented to search for new energy sources, such as renewable fuels, where the resources from which they are derived are not significantly depleted by their utilization.

Recently, renewable energy is progressively substituting conventional fuels in four different sectors such as power generation, heating, transportation fuels, and off-grid rural energy services (Appels, 2011). According to REN21's (2014) report, about 19% to energy utilization and 22% to power generation in 2012 and 2013, respectively, have been shared through renewable sources. Further, this energy consumption is segregated as conventional biomass: 9%, nonbiomass-based heat energy: 4.2%, hydroelectricity: 3.8% and 2% electricity from conventional renewable such as solar (photo-voltaic), geothermal, wind, and biomass (REN21, 2014). For a country like India with a growing urban population and massive agricultural area coverage, fuels derived from biomass and waste could be a suitable sustainable management option.

Biofuels include liquid and gaseous fuels derived from biomass conversion by microbial intervention. The sources of biomass include biodegradable portion of municipal wastes, industrial solid/liquid effluents, forestry, and agricultural energy crops, that is, lignocellulosic biomass (Angenent et al., 2004).

The biofuel should meet the following criteria to be justified as sustainable and alternative:

- Provide a net energy gain.
- Environmental friendly.
- Cost effective and economically feasible.
- Large production capacities and should not hamper the food chain/web.

Upon combustion, biofuels reduce net emission of carbon dioxide and other associated greenhouse gases attributed to global environmental climate change. In general, biomass-derived fuels are considered as CO_2 neutral since the CO_2 emission during combustion of these biofuel is the same CO_2 that the plants have metabolized during photosynthesis to synthesize organic materials that forms the biomass (Jorgensen, 2009). There are mainly four types of biofuels: (1) bioethanol, (2) biodiesel, (3) biohydrogen, and (4) biomethane. These biofuels have marked differences and similarities in technological processes as well as their usages.

Several feedstocks that include corn, sugar beet, and other dedicated energy crops were widely utilized for bioethanol production. However, pertaining to the food versus fuel conflict, the food crops are being replaced by agro-residues and unutilized lignocellulosic biomass such as poplars, switch grass, bamboo, *Lantana camara*, *Ricinus*, etc. (Kuila et al., 2011). In a developing country such as India where bioethanol mandate has been fixed at E10 by 2020, relaying on terrestrial-based energy crops leads to deficient supply of raw material for ethanol production. However, another viable feedstock could be aquatic plants obtained from constructed wetlands. This chapter is focused on recent developments of using aquatic plants, such as cattails, water lettuce, water hyacinth, duckweeds (*Azolla*), marine macrophytes, and microalgae, as feedstock for liquid biofuel production.

9.2 AQUATIC WEEDS AS ENERGY CROPS

Aquatic weeds are fast-growing plants that complete their life cycle in water bodies. Because of its rapid growth and multiplication, it causes deleterious effects to aquatic environment and public health both directly and indirectly. Some of the deleterious effects of aquatic weeds are listed below:

- It reduces the effectiveness of ponds for fish growth and production by reducing the oxygen levels present in it.
- The flow capacity of irrigation channels are adversely reduced thereby limiting the water accessibility to the farmers.
- Dense growth of aquatic weeds provides suitable environment for propagation of malaria and filarial parasites.

Among different countries India has the largest water canal network, where the flow velocity is significantly deterred by about 30–40% because of the dense growth of aquatic weeds (http://agriwaterpedia.info/wiki/Water_related weeds). Therefore, considering the harmful effects and economical loss it is essential to control the population of aquatic weeds in water bodies. Different seaweeds that are exploited for liquid biofuel production have been represented in Figure 9.1.

FIGURE 9.1 Different marine macrophytes (sea weeds) employed for biofuel production.

9.3 DEDICATED AQUATIC ENERGY CROPS

Several varieties of aquatic biomass, for example, water weeds such as giant kelp, algae, seaweed (marine macroalgae), etc., can constitute a potential feedstock for bioethanol production. Macroalgae (seaweeds) that occupy regions close to the shore in an aquatic environment are a potential source of value-added chemicals with varied applications. The potential application with commercial viability of aquatic biomass is mainly dependent on their types and biochemical compositions. Conventionally, food industry is the dominant market for macroalgae and its derived value-added chemicals. Recent developments in bioprocessing of macroalgae increased their interest in other product development such as pharmaceutical additives, nutraceutical/functional food preparation, cosmetics, and healthcare products. As world energy demands continue to rise and fossil fuel resources are alarmingly declining, macroalgae have attracted attention for the production of biofuels, such as bioethanol and biogas, as a potential renewable feedstock because of their considerable carbohydrates composition. Macroalgae have several advantages over terrestrial biomass, primarily due to their high biomass productivity, noncompetition with food crops for the usage of cultivable land, consumption of fresh water resources for irrigation, and uptake of CO_2 as the predominant carbon input. Gao and McKinley (1993) estimated the average yield of macroalgae in the world and it was found to be 730,000 kg ha^{-1} $year^{-1}$ which is comparatively higher than the other terrestrial lignocellulosic feedstocks. Yokoyama et al. (2007) reported that after the oil crisis in the United States and Japan in 1970s the utilization of marine algae biomass as a source of bioenergy is widespread. The other major aquatic crops that have been exploited for liquid biofuel production are discussed in the subsequent sections.

9.3.1 *MARINE MACROPHYTES (SEA WEEDS)*

Johna et al. (2011) proposed that multicellular macroalgae (seaweed) are rich source of carbohydrates and lipids suitable for the production of bioethanol. Three types of sea weeds, namely, *Gelidium elegans* Kuetzing (red seaweed), *Ulva pertusa* Kjellman (green seaweed), and *Alaria crassifolia* Kjellman (brown seaweed) were effectively utilized for bioethanol production with a maximum yield of 55 g/L (Yanagisawa et al., 2011). Though the cellulose content in seaweeds is relatively less (~46.08%) compared to other lignocellulosic biomass (45–55%), the lignin content is negligible (~1.83%) that helps in easy hydrolysis of holocellulose. *G. elegans* Kuetzing contains both glucan and galactan which can be hydrolyzed to fermentable sugars by the adoption of successive and combined saccharification of *Saccharomyces cerevisiae* IAM 4178. The combined saccharification was effective in case of red seaweed as galactans are hydrolyzed by acids which contains 70.9 g/L glucose and 53.2 g/L galactose and successively subjected to fermentation by yeast which leads to the production of ethanol as high as 5.5% (55.0 g/L).

Trivedi et al. (2013) reported *Ulva fasciata* Delile (marine macroalgae) as a potent feedstock for the bioethanol production having high amount of polysaccharides which can be grown worldwide regardless of geographical barriers. By pretreating the biomass with aqueous solution at 120°C for 1 h followed by enzymatic hydrolysis with cellulase 2,2119 at 45°C, 2%

(v/v) of enzyme for 36 h sugar yield of about 206.82 ± 14.96 mg/g was obtained. This study investigated the growth rate of *U. fasciata* at 25 ± 1°C (Provasoli, 1968), which showed a gradual increase in the daily growth rate (DGR%) from 19.15 ± 0.91% to 24.25 ± 1.62% which is relatively higher than the genus *Gracilaria* and *Kappaphycus* (3–8%) as reported by other researchers (Cirik et al., 2010; Mantri et al., 2009; Padhi et al., 2011). Bruhn et al. (2011) estimated production capacity of *Ulva* sp., as 45 t DW ha^{-1} y^{-1} which is 2–20 times higher than the convention terrestrial bioenergy crops. Kumar et al. (2011) studied different tropical seaweeds and reported 46–57% of higher carbohydrate concentration for different taxa of *Ulva*. Siddhanta et al. (2009) reported that the cellulose content of both red and brown seaweed are relatively low on dry weight basis 2–10% but in case of green seaweed (*U. fasciata*) it was about 15 ± 2.3% of dry weight.

Dahlia et al. (2013) evaluated the lipid content and fatty acid composition for the biodiesel production from different marine macroalgae such as *Ulva linza* (Chlorophyceae), *Jania rubens* (Rhodophyceae), and *Padina pavonica* (Phaeophyceae) which are abundant near coastal environments of Abu Qir Bay coast, Alexandria, Egypt. The variation in the total lipid contents and fatty acid composition of the macroalgae during spring, summer, and autumn were estimated to determine the most favorable conditions for the growth of seaweeds. The seawater parameters such as pH, temperature, and salinity were found to significantly vary during different seasons; however, they showed no impact on fatty acid profiles. *U. linza* resulted in high amount of total lipid content with 4.14% dry weight during spring season. This study concluded that the fatty acid methyl ester profiles analyzed by gas chromatography indicate that brown and green seaweeds are the suggested feedstock for the biodiesel production.

Hayward et al. (1996) proposed that brown weeds which include kelps such as *Laminaria saccharina* has the ability to grow up to 4 m whereas *Saccorhiza polyschides* can reach up to 4.5 m in length. Adams et al. (2009) reported ethanol fermentation from *Saccharina latissima* (brown algae) containing mannitol and laminarin that constitute 55% of dry weight by considering the various pretreatment and hydrolysis strategies. This study eliminates the need for pretreatment process in both fresh and defrosted macroalgae substrates and the ethanol yield was found to be higher (0.45% (v/v)) than those with altered pH and temperature pretreatment by using 1 U kg^{-1} laminarinase and *S. cerevisiae*.

Enteromorpha sp. (macroalgae) is a very good potential source for the bioethanol production as it contains 70–72% of carbohydrates in its biomass. Nahak et al. (2011) performed various pretreatments on *Enteromorpha* sp. (dry and fresh algae) as it is abundantly available in Indian East Coastal marine ecosystem. The pretreated biomass with dilute sulfuric acid, nitric acid, and steam flashing are subjected to fermentation by industrial enzymes resulted in 0.–0.94 mg/g and 0.21–0.52 mg/g of fermentable sugars in dry and fresh algae, respectively.

Kumar et al. (2013) studied the third largest genus of class Rhodophyta (*Gracilaria verrucosa)* in which pulp contains 62–68% holocellulose which is subjected to the enzymatic hydrolysis to yield sugars. The biochemical properties of *G. verrucosa* contain total carbohydrate concentration (%, w/w) that ranges from 62.17 ± 1.26 to 67.50 ± 1.50, while the reducing sugar content of pulp varied from 34.83 ± 0.76 to 37.67 ± 0.58 and glucose ranges from 21.67 ± 0.58 to 24.80 ± 0.52. The hydrolysate when fermented with *S. cerevisiae* resulted in 86% theoretical yield (0.43 g/g sugars) which is better than the other algae biomass such as *A. crassifolia* (0.281 g/g), *G. elegans* (0.376 g/g), *U. pertusa* (0.381 g/g), and *Sargassum sagamianum* (0.386 g/g) as reported by Yanagisawa et al. (2011). Some of the aquatic biomass utilized for liquid biofuel production have been tabulated in Table 9.1.

TABLE 9.1 Reducing Sugar and Bioethanol Yield from Aquatic Biomass.

Marine macrophytes	Reducing sugars	Ethanol yield	References
Green sea weeds			
Ulva fasciata Delile	206.82 ± 14.96 mg/g	0.45 g/g	Trivedi et al. (2013)
Ulva pertusa	NA	0.381 g/g	Yanagisawa et al. (2011)
Red sea weeds			
Gracilaria verrucosa	0.87 g/g cellulose	0.43 g/g	Kumar et al. (2013)
Kappaphycus alvarezii	0.306 g/g	0.39 g/g	Khambhaty et al.(2012)
Gelidium elegans Kuetzing	70.9 g/L glucose and 53.2 g/L galactose	0.376 g/g	Yanagisawa et al. (2011)
Gelidium amansii	0.422 g/g	0.38 g/g	Park et al. (2012)
Gracilaria salicornia	16.6 g/g	NA	Wang et al. (2011)
Brown sea weed			
Saccharina japonica	0.456 g/g	0.169 g/g	Jang et al. (2012)
Laminaria japonica	0.376 g/g	0.41 g/g	Kim et al. (2011)
Alaria crassifolia	NA	0.281 g/g	Yanagisawa et al. (2011)

9.3.2 *Pistia stratiotes* (WATER LETTUCE)

Pistia stratiotes commonly known as water lettuce and locally as Jalkhumbi is a free floating invasive aquatic weed. It is a stoloniferous plant and is frequently observed in the form of dense mat as shown in Figure 9.2 in water bodies which are rich in lime content. *P. stratiotes* has a wide distribution in eastern part of India (i.e., Assam, Odisha, and West Bengal). Aquaculture ponds and tanks, deep water rice fields, and lakes are severely inhabited by this aquatic weed. Among global countries India, South America, Australia, and South Africa were found to be infested with this aquatic weed (Fig. 9.3). Like most of the floating weeds *P. stratiotes* is also removed from water bodies by mechanical and hand removal. *P. stratiotes* has a significant ability to multiply the biomass rapidly (Reddy and De Busk, 1984). Because of its high productivity and organic content, *P. stratiotes* has attracted interest to utilize it for bioethanol (Mishima et al., 2008), methane production (Abbasi et al., 1991), and green leaf manure (Raju and Gangwar, 2004). However, very little research work has been done on utilizing this green biomass as an energy crop candidate.

9.3.3 AZOLLA

The ponds and lakes receiving wastewater enhance the colonization of aquatic plant biomass because of the increased nutrient availability. *Azolla*, commonly termed as duck weed fern, fairy moss, or water fern, belongs to the family Salviniaceae and is found to infest the flooding regions of paddy fields. It is cocultivated with paddy crops as it has the tendency to fix the atmospheric nitrogen with the help of symbiotic cyanobacterium *Anabaena azollae (Miranda et al., 2016)*. Hasan et al. (2009) reported that *Azolla pinnata* is a fast-growing plant which has a doubling time of 3–10 days resulting in average fresh biomass yield of 37.8 t/ha. Chemical composition of *Azolla* cell wall such as cellulose/hemicellulose, starch, and lipids resembles as that of other conventional terrestrial crops that are being utilized as a feedstock for biofuel production. Thus, the scope of utilizing or substituting the conventional biomass for fuel synthesis is highly recommended.

Azolla pinnata and *Azolla filiculoides* are two predominant species that were used as feedstock for the synthesis of value-added chemicals through hydrothermal liquefaction of gaseous fuels such as biohydrogen, biomethane, and bioethanol.

FIGURE 9.2 (See color insert.) (A) *Pistia stratiotes* mat growth on a water pond and (B) rosette structure of the plant.

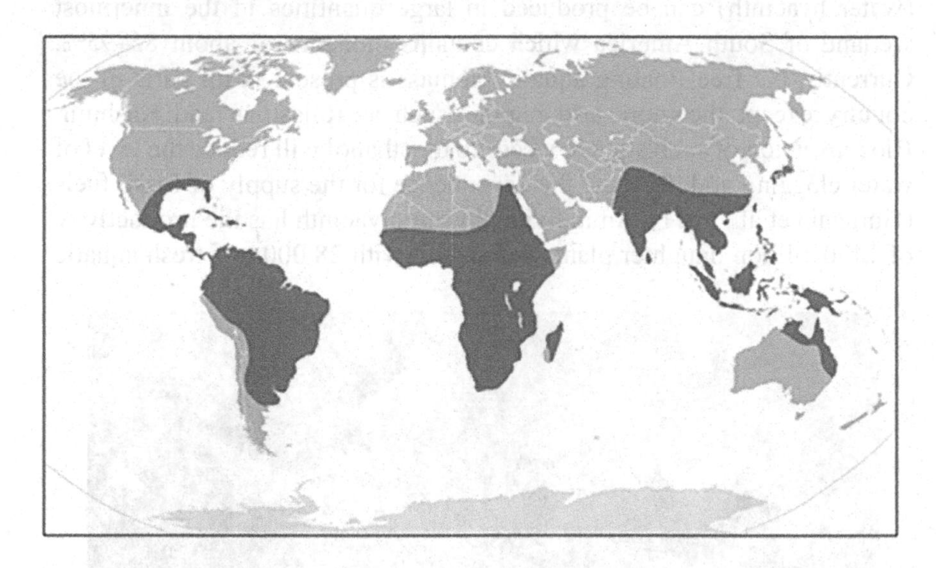

FIGURE 9.3 Worldwide distribution of *Pistia stratiotes* (majority of Asian and South American countries got infested with this weed).

The eutrophication caused by wastewater discharge in water bodies results in rapid colonization of aquatic flora. In general, *Azolla* species have faster growth rate that produces considerable amount of biomass in a shorter duration by utilizing organic components present in the wastewater. It is also found that *Azolla* grows along with its evolutional symbiont *Anabaena azollae* (blue green algae) that helps in nitrogen fixation. Usually, *Azolla* is found to infest in the flooding waters of paddy fields and helps in ensuring nitrogen availability to the paddy crops. The lignocellulosic composition of *Azolla* leaves resembles as that of terrestrial crops and microalgae which substantiate it as a potential feedstock for bioethanol production. Figure 9.4 represents a species of *Azolla* (*Azolla pinnata*).

9.3.4 WATER HYACINTH

Water hyacinth (*Eichhornia crassipes* Martius), an aquatic free-floating weed was first familiarized from South Africa to China in 1901 and extensively utilized as pig feed in 1950s (Xieet al., 2001; Chu et al., 2006). Bergier et al. (2012) identified that *E. crassipes* and *Eichhornia azurea* (water hyacinth) can be produced in large quantities in the innermost wetland of South America which contains moisture of about 85–95%. Currently, the free-floating aquatic biomass is present in all parts of the country except the more arid regions such as Rajasthan and Kashmir. Bioconversion of such aquatic weeds to bioethanol will reduce the level of water clogging and decrease the dependence for the supply of fossil fuels (Guragain et al., 2011). An individual water hyacinth has the productivity of 1490 million daughter plants per annum with 28,000 t of fresh aquatic

FIGURE 9.4 *Azolla pinnata.*

biomass by covering an area of about 1.40 km^2. *E. crassipes* Martius has the ability to live in both fresh water and wastewater and the reproduction cycle was found to be 8 days (Lu et al., 2007; Bayrakcin and Koçar, 2014). The cell wall components of water hyacinth are tabulated in Table 9.2 which shows that biomass can serve as a potential feedstock for fuel production as compared to that of terrestrial biomass.

The total amount of C6 sugars (glucose and galactose) and C5 sugars (xylose and arabinose) are estimated to be 19.8%, 6.5%, 11.5%, and 9.0% (w/w), respectively, which makes them a sustainable source for bioethanol production by yeast or bacterial fermentation process (Girisuta et al., 2008). An effective pretreatment to enhance the hydrolysis efficiency in water hyacinth was carried out by combining the biological and chemical pretreatment method. The cotreated water hyacinth by a white rot fungus (*Echinodontium taxodii*) with 0.25% (v/w) of sulfuric acid showed maximum yield of reducing sugar from 208.8 to 366.0 mg/g dry matter which is comparatively higher than the sole acid pretreatment (112.9–322.3 mg/g dry matter) which constitute 1.13–2.11-fold increase at 100°C for 60 min. The ethanol yield from the cotreated water hyacinth (0.192 g/g of dry matter) showed 1.34–fold increase than the acid-treated sample (0.146 g/g of dry matter) by the fermentation with industrial yeast (Ma et al., 2010).

FIGURE 9.5 Water hyacinth (*Eichhornia crassipes* Martius).

TABLE 9.2 Cell Wall Composition of Water Hyacinth.

Cellulose (%w/w)	Hemicellulose (%w/w)	Lignin (%w/w)	References
18.2	29.3	2.8	Ma et al. (2010)
24.5	34.1	8.6	Ruan et al. (2016)
40.2 (holocellulose)		6.5	Harun et al. (2011)
19.5	33.4	9.27	Gunnarsson and Petersen (2007)
15–25	30–55	3–4	Bayrakcin and Koçar (2014)
24.7 ± 0.4	32.2 ± 0.3	3.2 ± 0.2.	Das et al. (2016)

A new pretreatment method for water hyacinth like crude glycerol pretreatment was performed by Guragainet al. (2011) and compared with the conventional dilute acid treatment (DAT). As there was no significant difference found between crude glycerol and DAT in terms of enzymatic hydrolysis yield and fermentation yield of pretreated sample, based on the availability and cost, the crude glycerol pretreatment was considered as an alternative route for the simultaneous production of biodiesel and bioethanol.

Ruan et al. (2016) studied the alkali (NaOH) pretreatment of biomass (water hyacinth) and evaluated the pretreatment efficiency by the hydrolysis of pretreated feedstock with cellulase from *Trichoderma reesei* at optimized condition 50°C, pH 4.8, and 60 g/L of substrate for 48 h with almost 100% cellulose conversion efficiency. The NaOH pretreated water hyacinth biomass resulted in increase of cellulose composition from 24.5% to 59.9%, corresponding decrease in the hemicellulose and lignin content from 34.1% and 8.6% to 18.2% and 2.1% on the basis of dry weight, respectively. Structural change with water hyacinth biomass was examined by using Fourier transform infrared (FTIR) and scanning electron microscope (SEM) before and after pretreatment and hydrolysis.

The dilute sulfuric acid (2% v/v) pretreatment of water hyacinth biomass at high temperature and pressure was investigated by Das et al. (2016), which resulted in the change in the composition of cellulose (24.7–35.4% w/w), hemicellulose (32.2–19.6% w/w), and lignin (3.2–0.9% w/w). The maximum sugar yield of about 425.6 mg/g was obtained by enzymatic saccharification at 5% solid content, 50°C, pH 5.5, and 30 filter paper units (FPU) of cellulase concentration for 1 day. Central composite design was used for the optimization of mixed ethanol fermentation by *S. cerevisiae*

(MTCC 173) and *Zymomonas mobilis* (MTCC 2428). The optimized conditions for maximum ethanol fermentation (13.6 mg/mL) by coculturing were pH 6.41, 37.7 h, and 1:1 ratio of *Saccharomyces* to *Zymomonas*.

E. crassipes Martius native of Brazil and Ecuador region was genetically manipulated to upregulate the cellulose biosynthesis pathway enzymes which enhance the production of high amount of polysaccharides in the biomass that have synergic effects in the yield of biofuel. Similarly, genetic alternation in the lignin biosynthetic pathway can effectively reduce the cost involved in the pretreatment process. The production of essential cell wall-degrading enzymes such as cellulase and hemicellulase within the plant by means of bioengineering will effectively decrease the dependency of bioreactor for enzyme production (Sticklen, 2008). Wymelenberg et al. (2006) proposed a transgenic approach in *Phanerochaete chrysosporium* genome (fungus) which contains large number of genes involved in the breakdown of lignin in such a way that it facilitates the extraction of saccharides from the water hyacinth and reduces the cost for the production of biofuels. The fermentation process for bioethanol production in water hyacinth can be efficiently carried out by microbes such as *Klebsiella oxytoca* (0.96 g/L/h), *Escherichia coli* (0.85 g/L/h), and *Zymomonas mobilis* (0.61 g/L/h), suggested by other researchers (Dien et al., 2003; Yomano et al., 1998; Ohta et al., 1991; Mohagheghi et al., 2002). The various strategies that have been widely employed for the conversion of aquatic biomass to liquid biofuel are schematically represented in Figure 9.6.

9.4 CONCLUSION

The weeds that infest the aquatic bodies might clog the water streams which can turn to be a challenging task for the municipalities to remove. These lead to different health issues by harboring mosquito larvae, destruction of aquatic ecosystem as the dissolved oxygen start to deplete, and reduction in sunlight percolation into the water. On the other, scientific removal of these weeds and their further processing for biofuel production can be a viable option for their efficient control and management. Therefore, scientific interventions and its advancements are in need of time for removal of weeds from the national river, tributaries, lakes, etc., and further adoption of biorefinery technologies to utilize it as feedstock alternative to terrestrial energy.

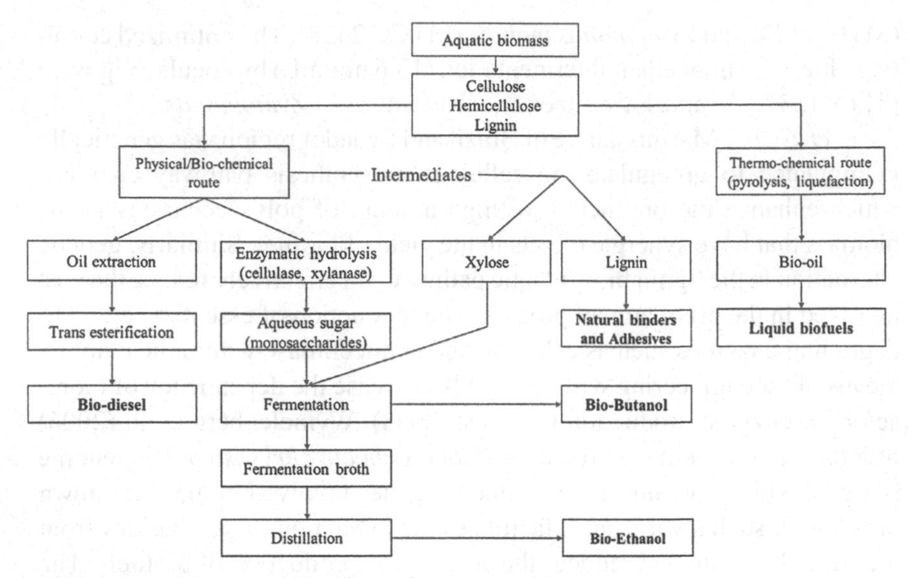

FIGURE 9.6 Whole crop biorefinery from aquatic biomass.

KEYWORDS

- **aquatic energy crop**
- **liquid biofuel**
- **lignocellulosics**
- **energy**
- **cost-effective**

REFERENCES

Abbasi, S. A.; et al. Biogas Production from the Aquatic Weed Pistia (*Pistia stratiotes*). *Bioresour Technol.* **1991,** *37*, 211–214.

Adams, J. M.; et al. Fermentation Study on *Saccharina latissima* for Bioethanol Production Considering Variable Pre-treatments. *J. Appl. Phycol.* **2009,** *21*, 569–574.

Angenent, L. T.; et al. Production of Bioenergy and Biochemicals from Industrial and Agricultural Wastewater. *Trends Biotechnol.* **2004,** *22*, 477–485.

Appels, L.; et al. Anaerobic Digestion in Global Bio-energy Production: Potential and Research Challenges. *Renew. Sustain. Energy Rev.* **2011,** *15*, 4295–4301.

Bayrakcin, A. G.; Koçar, G. Second-generation Bioethanol Production from Water Hyacinth and Duckweed in Izmir: A Case Study. *Renew. Sustain. Energy Rev.* **2014,** *30*, 306–316.

Bergier, I.; et al. Biofuel Production from Water Hyacinth in the Pantanal Wetland. *Ecohydrol. Hydrobiol.* **2012,** *12*, 77–84.

Chu, J. J.; et al. Invasion and Control of Water Hyacinth (*Eichhornia crassipes*) in China. *J. Zhejiang Univ. Sci. B.* **2006,** *7* (8), 623–626.

Cirik, S.; et al. Greenhouse Cultivation of *Gracilaria verrucosa* (Hudson) Papenfuss and Determination of Chemical Composition. *Turk. J. Fish Aquat. Sci.* 2010, *10*, 559–564.

Courtney, B.; Dorman, D. Worldwide Fossil Fuels; Department of Chemistry, Louisiana State University, Louisiana, USA, 2003.

Dahlia, M.; et al. Lipid Content and Fatty Acid Composition of Mediterranean Macroalgae as Dynamic Factors for Biodiesel Production. *Oceanologia* 2014. http://dx.doi.org/10.1016/j.oceano.2014.08.001 (accessed Mar 8, 2019).

Dien, B. S.; et al. Bacteria Engineered for Fuel Ethanol Production: Current Status. *Appl. Microbiol. Biotech.* **2003,** *63*, 258–266.

Dow, K.; Downing , T. *The Atlas of Climate Change: Mapping the World's Greatest Challenge*; University of California Press: Los Angeles, USA, 2011. https://www.ucpress.edu/book/9780520268234/the-atlas-of-climate-change (accessed Mar 8, 2019).

Girisuta, B.; et al. Experimental and Kinetic Modelling Studies on the Acid-catalysed Hydrolysis of the Water Hyacinth Plant to Levulinic Acid. *Bioresour. Technol.* **2008,** *99* (83), 67–75.

Gunnarsson, C. C.; Petersen, C. M. Water Hyacinths as a Resource in Agriculture and Energy Production: A Literature Review. *Waste Manag.* **2007,** *27*, 117–129.

Guragain, Y. N.; et al. Comparison of Some New Pretreatment Methods for Second Generation Bioethanol Production from Wheat Straw and Water Hyacinth. *Bioresour. Technol.* **2011,** *102*, 4416–4424.

Harun, M. Y.; et al. Effect of Physical Pretreatment on Dilute Acid Hydrolysis of water Hyacinth (*Eichhornia crassipes*). *Bioresour. Technol.* **2011,** *102*, 5193–5199.

Hayward, P.; Nelson-Smith, T.; Shields, C. *Sea Shore of Britain and Northern Europe*; Collins: London, 1996.

IEA. *Technology Roadmap, Biofuels for Transport*; 2011. http://www.iea.org/publications/freepublications/publication/biofuels_roadmap.pdf (accessed Mar 8, 2019).

Jang, J. S.; et al. Optimization of Saccharification and Ethanol Production by Simultaneous Saccharification and Fermentation (SSF) from Seaweed, *Saccharina japonica*. *Bioprocess Biosyst. Eng.* **2012,** *35*, 11–18.

Johna, R. P.; et al. Micro and Macroalgal Biomass: A Renewable Source for Bioethanol. *Bioresour. Technol.* **2011,** *102*, 186–193.

Jorgensen, P. J. *Biogas: Green Energy*, 2nd ed.; 2009 (Plant Energy and Researcher for a Day: Faculty of Agricultural Sciences, Aarhus University). http://dca.au.dk/fileadmin/DJF/Kontakt/Besog_DJF/Oevelsesvejledning_og_baggrundsmateriale/Biogas_-_Green_Energy_2009_AU.pdf (accessed Mar 8, 2019).

Khambathy, Y.; et al. *Kappaphycus alvarezii* as a Source of Bioethanol. *Bioresour. Technol.* **2012,** *103*, 180–185.

Kim, N. J.; et al. Ethanol Production from Marine Algal Hydrolysates Using *Escherichia coli* KO11. *Bioresour. Technol.* 2011, *102*, 7466–7469.

Kuila, A.; et al. Production of Ethanol from Lignocellulosics: An Enzymatic Venture. *EXCLI J.* **2011,** *10*, 85–96.

Kumar, M.; et al. Minerals, PUFAs and Antioxidant Properties of Some Tropical Seaweeds from Saurashtra Coast of India. *J. Appl. Phycol.* **2011,** *23*, 797–810.

Kumar, S.; et al. Bioethanol Production from *Gracilaria verrucosa*, a Red Alga, in a Biorefinery Approach. *Bioresour. Technol.* 2013. http://dx.doi.org/10.1016/j.biortech.2012.10.120 (accessed Mar 8, 2019).

Lu, J.; et al. Water Hyacinth in China: Sustainability Science-based Management Framework. *J. Environ. Manag.* **2007,** *40*, 823–830.

Ma, F.; et al. Combination of Biological Pretreatment with Mild Acid Pretreatment for Enzymatic Hydrolysis and Ethanol Production from Water Hyacinth. *Bioresour. Technol.* **2010,** *101*, 9600–9604.

Mantri, V. A.; et al. The Carpospore Culture of Industrially Important Red Alga *Gracilaria dura* (Gracilariales, Rhodophyta). *Aquaculture* **2009,** *297*, 85–90.

Mohagheghi, A.; et al. Co-fermentation of Glucose, Xylose, and Arabinose by Genomic DNA Integrated Xylose/Arabinose Fermenting Strain of *Zymomonas mobilis* AX101. *Appl. Biochem. Biotechnol.* **2002,** *98*, 885–898.

Nahak, S.; et al. Bioethanol from Marine Algae: A Solution to Global Warming Problem. *J. Appl. Environ. Biol. Sci.* **2011,** *1* (4), 74–80.

Padhi, S.; et al. Cultivation of *Gracilaria verrucosa* (Huds) Papenfuss in Chilika Lake for Livelihood Generation in Coastal Areas of Orissa State. *J. Appl. Phycol.* **2011,** *23*, 151–155.

Park, J. H.; et al. Use of *Gelidium amansii* as a Promising Resource for Bioethanol: A Practical Approach for Continuous Dilute-acid Hydrolysis and Fermentation. *Bioresour. Technol.* **2012,** *108*, 83–88.

Provasoli, L. Media and Products for the Cultivation of Marine Algae. In *Culture and Collection of Algae*; Watanabe, B. S., Hattori, A., Eds.; Japan Society of Plant Physiology: Tokyo, **1968;** pp 63–75.

Raju, R. A.; Gangwar, B. Utilization of Potassium Rich Green Leaf Manure for Rice (*Oryza sativa*) Nursery and Their Effects on Crop Productivity. *Indian J. Agron.* **2004,** *49*, 244–247.

Reddy, K. R.; De Busk, W. F. Growth Characteristics of Aquatic Macrophytes Cultured in Nutrient-enriched Water: Water Hyacinth, Water Lettuce and Pennywort. *Econ. Bot.* **1984,** *38*, 229–239.

REN21. *Renewables 2014*, Global Status Report; REN21 Secretariat: Paris, 2014; pp 30–31. http://www.ren21.net/Portals/0/documents/Resources/GSR/2014/GSR2014_full%20report_low%20res.pdf (accessed Mar 8, 2019).

Ruan, T.; et al. Water Hyacinth *Eichhornia crassipes* Biomass as a Biofuel Feedstock by Enzymatic Hydrolysis. *Bioresour. Technol.* **2016,** *11* (1), 2372–2380.

Siddhanta, A. K.; et al. The Cellulose Contents of Indian Seaweeds. *J. Appl. Phycol.* **2009,** *23*, 919–923.

Sticklen, M. B. Plant Genetic Engineering for Biofuel Production: Towards Affordable Cellulosic Ethanol. *Nat. Rev. Genet.* **2008,** *9*, 433–443.

Trivedi, N.; et al. Enzymatic Hydrolysis and Production of Bioethanol from Common Macrophytic Green Alga *Ulva fasciata* Delile. *Bioresour. Technol.* 2013. DOI: http://dx.doi.org/10.1016/j.biortech.2013.09.103.

Wymelenberg, V. A.; et al. Structure, Organization, and Transcriptional Regulation of a Family of Copper Radical Oxidase Genes in the Lignin-degrading Basidiomycete *Phanerochaete chrysosporium*. *Appl. Environ. Microbiol.* **2006,** *72*, 4871–4877.

Xie, Y.; et al. Invasive Species in China: An Overview. *Biodivers. Conserv.* 2001, *10*, 1317–1341.

Yanagisawa, M.; et al. Production of High Concentrations of Bioethanol from Seaweeds that Contain Easily Hydrolysable Polysaccharides. *Proc. Biochem.* **2011,** *46*, 2111–2116.

Yomano, L. P.; et al. Isolation and Characterization of Ethanol-tolerant Mutants of *Escherichia coli* KO11 for Fuel Ethanol Production. *J. Ind. Microbiol. Biotechnol.* **1998,** *20*, 132–138.

CHAPTER 10

Metabolic Engineering for Liquid Biofuels Generations from Lignocellulosic Biomass

DIPANKAR GHOSH*

Department of Biotechnology, JIS University, 81 Nilgunj Road, Agarpara, Kolkata 700109, West Bengal, India

E-mail: d.ghosh@jisuniversity.ac.in; dghosh.jisuniversity@gmail.com

ABSTRACT

Lignocellulosic biomass has been considered as renewable feedstock for liquid biofuels generations. Lignocellulosic residues are inedible, inexpensive, and immensely available in nature which makes it attractive alternative toward belittling fossil fuels reserves. Liquid biofuels generations from lignocellulosic residues have positive socioeconomic impacts likely reduction in agricultural land usage, water utilization, and greenhouse gases emissions. Liquid biofuels have higher octane and cetane numbers, which makes it better fuels with higher combustion speed and ignition quality. Naive microorganisms involve in lignocellulosic bioconversions toward liquid biofuels, however, processes are limited due to lower titer, molar yield, and productivity. It seems naive microorganisms are inefficient to degrade lignocellulosic biomass into accessible monomers. Competing metabolic pathways in native microbes are redirecting flux toward by-products accumulations. Hence, current chapter focuses on metabolic engineering and synthetic biology approaches to resolve these bottlenecks by knocking out competing pathways and introducing novel pathways for ameliorating liquid biofuels generations.

10.1 INTRODUCTION

Biofuels are the sustainable renewable alternatives of depleting fossil fuel reserves. It does not impair with the environmental balance, food security, cultivable land usage, and ecological diversity. It contributes to the social and economical development of local, rural, indigenous people, and communities through initiation of occupational opportunities. It pertains predominantly in various forms like gaseous, liquid, and solid. It is broadly categorized as primary and secondary biofuels. Primary biofuels are the natural unprocessed organic material enriched biomass (i.e., firewood, wood chips, wood pellets, animal wastes, litter of crops, and forest residues). These organic residues are directly combusted to supply cooking fuel, heating, or electricity generations. Secondary biofuels are processed in solid form (i.e., charcoal), or liquid form (i.e., bioethanol, biodiesel or bioesters, and bio-oil), or gaseous form (i.e., biogas, biomethane, and biohydrogen) (Nigam and Singh, 2011). Secondary biofuels have been applied in transportation and high temperature industrial processes. Secondary biofuels have been sub-classified into first, second, third, and fourth generations (Sikarwar et al.; 2017), considering feedstock usage, catalytic processes, and targeted biofuels (Fig. 10.1). The key benefits of biofuels generations are (1) energy security (i.e., domestic energy source, locally distributed, well connected supply-demand chain, and higher reliability); (2) economic stability (i.e., stability of price, generate employment, rural development, lower inter-fuels competition, reduce demand-supply gap, open new industrial dimensions, and control on monopoly of fossil enriched nations); (3) environmental gain (i.e., betterment in waste utilization, local pollution minimization, lower greenhouse gases (GHGs) emission from energy consumption, and landfill sites reduction). However, first-generation biofuels create agnosticism to scientific networks. Stultification of environment and carbon imbalance restrict continuation of first-generation biofuel production. Hence, an efficient alternative renewable technology requires reprieving GHGs. On the other hand, "food-versus-fuel" moot is another key concern upon first-generation biofuel productions (Laursen, 2006). In this current global scenario, non-food-based feedstock likely lignocellulosic biomass (LB) can offer the potency to produce second-generation biofuels. It could significantly minimize carbon dioxide generations, do not compete with food crops, and offer better engine performance. Moreover, it is a biorefinery approach which deals with biomass production, biomass transformation, and production of target biofuels (Fernando et al., 2006). 100 billion tons organic dry residues of land biomass per annum and 50 billion tones of aquatic biomass are generated on earth. Where, only 1.25% of the entire land biomass is utilized as food. The rest of the biomass is unutilized or recycled back into the earth's system, which can further be channelized as renewable cheaper feedstock for biofuels generations (Osamu and Carl, 1989). Third- and fourth-generation biofuel are derived through atmospheric carbon dioxide sequestration using green algae or blue-green algal biocatalysts (Chen et al., 2015). Algal biocatalysis has been accelerated using metabolic engineering and synthetic biology strategies to enhance the desired hydrocarbon yields along with creating an artificial carbon sink to make it a carbon-neutral energy source (Lu and Fu, 2011; Anandarajah et al., 2012). Even though, this technology is at a very preliminary phase of scientific investigations.

10.2 LIGNOCELLULOSIC BIOMASS COMPOSITION

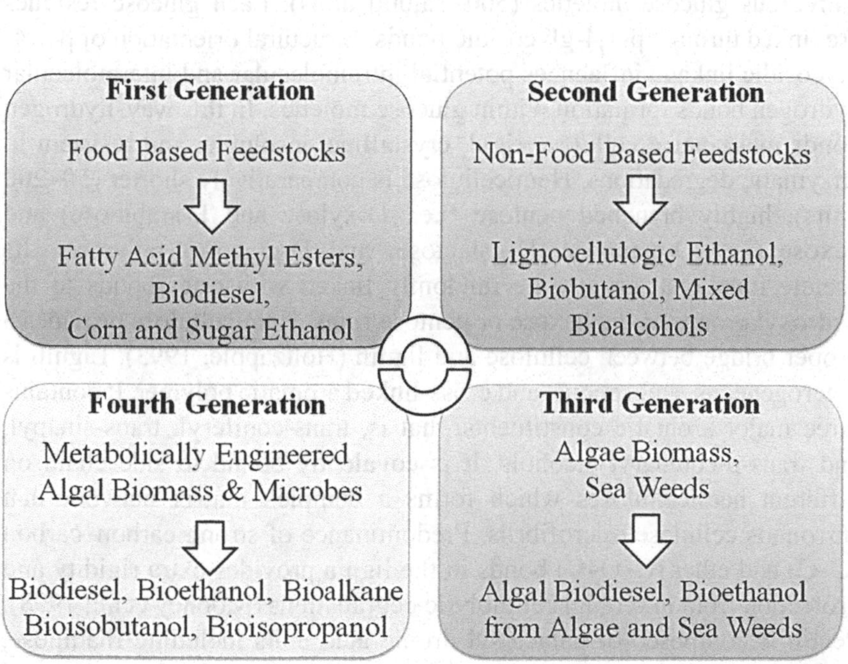

FIGURE 10.1 Trends of generations for advanced biofuel productions.

LB has been referred to be one of the enriched value-added feedstock as an alternative to fossil fuel resources (Zhou and Lu, 2008; Huber et al., 2006). LB-derived residues are assimilated from accessible atmospheric carbon dioxide, sunlight (photon energy), and water (terminal electron acceptor) through photosynthesis. Lignocellulosic feedstocks have essential benefits over other biomass reserves. LBs are non-edible portion of the plant and do not compete with food supplies (Sun and Cheng, 2002). However, agricultural, forestry, and agro-industrial LB wastes are annually accumulated in extensive quantities. LBs disposal from waste streams of industries such as food, agricultural, forestry, paper pulp, and timber to the soil or landfill poses severe environmental issues. But these LBs could be utilized toward value-added biomolecule production (Taherzadeh and Karimi, 2008). In general, LBs can be classified into four categories such as hardwood, softwood, agricultural wastes, grasses, and microbes (green algae, fungi, and few bacteria). The major building blocks of LBs are cellulose, hemicelluloses, lignin including minor constituents, that is, pectin compounds and proteins (Saini and Tewari, 2015). Cellulose is a linear, longer, unbranched homopolysaccharide which consists of anhydrous glucose moieties (500–15,000 units). Each glucose residues are linked through β-1,4-glycosidic bonds. Structural orientation of β-1,4-glycosidic linkage influences potential intramolecular and intermolecular hydrogen bonds formation within glucose moieties. In this way, hydrogen bonds make naive cellulose rigid, crystalline, insoluble, and resistant to enzymatic degradations. Hemicellulose is comparatively shorter (50–200 units), highly branched pentose (i.e., D-xylose and L-arabinose) and hexose (i.e., D-mannose, D-galactose, and D-glucose) polymers. Its acetate functional groups are randomly linked with ester bonds to the hydroxyl groups of the hexose or pentose rings. Hemicellulose provides a proper bridge between cellulose and lignin (Holtzapple, 1993). Lignin is heterogeneous, amorphous, and cross-linked aromatic polymer. It contains three major aromatic constituents, that is, trans-coniferyl, trans-sinapyl, and trans-p-coumaryl alcohols. It is covalently bound to side chain on different hemicelluloses which forms a complex matrix network that surrounds cellulose microfibrils. Predominance of strong carbon–carbon (C–C) and ether (C–O–C) bonds in the lignin provides extra rigidity and protection from microbial cellulolytic degradations (Mooney et al., 1998). Pectin is composed of galactosyl uronic acid units including rhamnose, arabinose, and galactose residues. Pectin is hydrophilic in nature and provides some adhesive properties to LBs. Some proteinaceous materials are minor constituents of LBs which may be covalently cross-linked with pectin and lignin (Chesson and Forsberg, 1998).

On the other hand, LBs contain elevated fraction of oxygen and lower amount of hydrogen and carbon in comparison to natural petroleum resources. A diverse group of valued biofuels can be generated from LBs due to its biochemical compositional variety. At present, bioethanol production from LBs have only well established as mature technology (Cherubini and Strømman, 2011; Altaf et al., 2007). Higher oxygen level in LB-derived biofuel reduces heat content and prevents its blending with existing fossil fuels (Lange, 2007). To this end LBs need to be deoxygenated and depolymerized to produce transportation fuels and biochemicals. However, oxygenation and/or deoxygenation of LBs refer to completely different terminal end products. Beside these, efficiency of depolymerizations of diverse groups of LBs can also be varied considering variable proportions of cellulose, hemicellulose, and lignin for the production of value-added biofuels and biomolecules (Isikgor and Becer, 2015).

10.3 CHALLENGES IN LIGNOCELLULOSIC BIOMASS ACCESSIBILITY

Most native LBs are inherently resistant to degradation to accessible simple sugars. Biochemical nature of LBs makes it recalcitrant to degradation that has evolved over aeons to protect the plant from attack by insects and microbes. Typical biochemical features of LBs are crystallinity, low accessible surface area, and a high percentage of lignin (natural hard cement for cellulose and hemicelluloses) and hemicelluloses (physical protector of cellulose fibers). To this end, improvement of LB accessibility is one of the key challenges of biorefinery approach to fractionate LBs into its three major constituents, that is, cellulose, hemicellulose, and lignin even further into simple sugars. Pretreatment processes are the possible way out for deconstruction of LBs. In practice, different kinds of pretreatment processes may be necessary to modify raw LB feedstocks to a state where fermentable sugars can be extracted and transformed into value-added biofuels. There are diverse classes of pretreatment methods that have been under application including physical pretreatments (milling, irradiation, hydrothermal high pressure, and pyrolysis), physicochemical treatments (explosion, alkali lysis, acid hydrolysis, oxidizing agents, gaseous agents, and solvent extraction of lignin), and finally biological pretreatment (using fungi, actinomycetes, and few bacterial species) (Fernando et al., 2006; Ghosh and Hallenbeck, 2012). In theory, pretreatment approaches alter the natural binding characteristics of LBs by modifying the supramolecular structure of the cellulose–hemicellulose–lignin matrix. Pretreatment of LB has a great potential for the improvement of efficiency and lowering the cost of production (Mosier et al., 2005). The integration of various biomass pretreatment methods with other processes like enzymatic saccharification, detoxification, fermentation of the hydrolysates, and recovery of products will greatly reduce the overall cost of using lignocellulose for practical purposes (Saha, 2005). However, these above mentioned pretreatment approaches have several bottlenecks (Fig. 10.2) concerning higher operation cost, expensive pretreatment processes, higher amount of water usage, unwanted by-products formations, toxicity issues, lower product recovery, etc. These current limitations practically de-accelerate

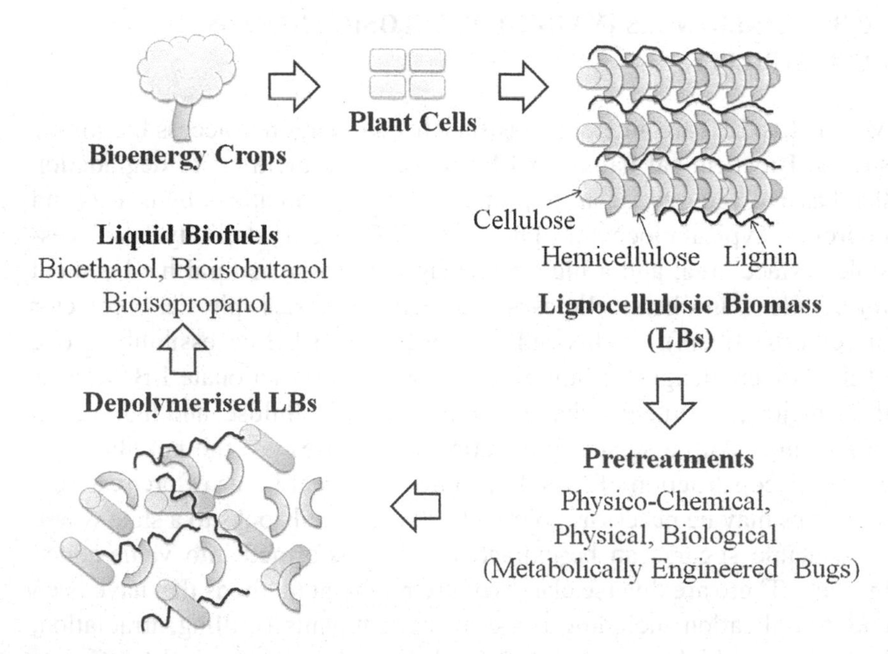

FIGURE 10.2 Lignocellulosic biomass accessibility for liquid biofuel production.

the economic viability, feasibility, and sustainability of large-scale biofuels productions from LBs feedstock.

Cost of LBs feedstock influences the cost of value-added biofuel generations. Bioenergy crops are the major source of LBs feedstocks. Bioenergy crops are broadly grouped into two major classes, namely, gymnosperms and angiosperms. Gymnosperms are soft woods like pine, spruce, fir, and cedar, whereas angiosperms are monocots (all perennial grasses such as switch grass, sorghum, sugarcane, and bamboo); herbaceous species (e.g., corn, wheat, and rice); flowering plants (e.g., alfalfa, soybean, and tobacco); and hardwoods (e.g., poplar, willow, and black locust) (Balan et al., 2011; Tao et al., 2012). As per current global records, about 70% of total LB feedstocks are derived from agricultural residues and 30% from forest residues (Sokhansanj et al., 2009; David and Ragauskas, 2010; Sannigrahi and Ragauskas, 2010; Xu and Cheng, 2011; Zegada-Lizarazu and Monti, 2012; Zub and Brancourt-Hulmel, 2010; Dohleman and Long, 2009). However, the great challenge is to convince farmers to cultivate bioenergy crops for LB feedstocks generation rather than grains production for food supply. Lack of assurance of buyback guarantees to surpass their reluctance and risk of failure to sell their cultivated energy crop products are major rate-limiting stages of LB feedstock productions at present.

Biomass harvesting for LB feedstocks generations is another burning issue considering expensive machinery requirements. Several kinds of machineries are required for harvesting different types of LB biomasses depending on the texture of biomasses. There are a few factors which influence biomass harvesting such as soil contaminations of biomass, local weather conditions, and moisture content of agricultural residues (Thorsell et al., 2004; Sokhansanj and Hess, 2009; Hoskinson et al., 2007; Hess et al., 2009; Huggins et al., 2011). Another cost-intensive aspect of LB feedstock generation is biomass yield which occupies one-third of biofuel generation costs. Soil topology, soil fertility, location, environmental conditions, and genetics of plants are the determinants of biomass yield (Duffy and Nanhou, 2002).

LB densifications, transportations of biomass, and storage are a few other factors influencing LB-based feedstock generation and biofuel generation costs. Biomass densification is LB feedstock packing process through elastic and plastic deformations. Different technologies are currently in application for densification of biomass. However, these approaches are highly expensive and energy-consuming processes

(Tumuluru et al., 2011). Lower densities of biomass, seasonal availability, texture variation, and distance have influenced transportation costs. LBs with lower density occupy more transportation carrier space and volume (Sultana and Harfield, 2010; Caputo et al., 2005; Kumar and Flynn, 2006; Badger, 2003). Preservation of harvested LB feedstocks from rain or moisture has been carried out through covering of polythene wraps before shipping to biorefinery. However, it requires protecting LB feedstock from fungal degradation while stored under aerobic environments. It could probably reduce the accessible sugar contents of LB biomass. Therefore, it is always recommended to store LB feedstock at around lower than 10% controlled moisture environment (per dry weight basis) to extend longevity (Rentizelas and Tatsiopoulos, 2009; Johansson, 2006). Moreover, preventive measures are also essential to minimize self-ignition of LB feedstock during long-term storage (Gray and Hasko, 1984).

To this end, regional biomass processing depots (RBPDs) needs to be developed to take care of these issues. But there are a few gainsays in building RBPDs. It includes (1) identification of energy and cost requirements for LB feedstock size reduction, pretreatment, densification operational units; (2) installation of economically viable transportation facilities (rail and heavy duty trucks facilities, etc.) near RBPDs; and (3) establishment of small technology transfer laboratories in RBPDs (Kurian et al., 2013; Carolan and Dale, 2007; Sharma et al., 2013). However, implementation of these innovations demands skilled manpower, proper planning, advanced biochemical technologies, medium of exchange, time, and routine inspections.

10.4 METABOLIC ENGINEERING FOR IMPROVING LIGNOCELLULOSIC BIOMASS YIELD AND DECOMPOSITION

In nature, diverse microbial communities secrete enzymes to break down LBs for generating simple sugar monomers toward survival. LB feedstock decomposition has been carried out by microbes following two ways in the environment, namely, (1) cellulosomal enzyme complex and (2) free enzyme mechanisms. Cellulosomal enzyme system is a complex mixture of enzymes that are docked to the cohesive and doctrine domains through anchoring on the microbial cell surface. In free enzyme system, enzymes are molecular scissors and released as indigenous component for degrading target LB feedstocks (Bayer and Lamed, 2013; Hatakka and Hammel, 2010). Free enzyme system is comparatively convenient and easy to duplicate for LB feedstock degradation in biorefineries. Free enzyme system has been classified into four major categories: (1) cellulase, (2) hemicellulase, (3) pectinase, and (4) ligninase (Zhang and Ma, X., 2012; Gao et al., 2010; Gao et al., 2011; Banerjee et al., 2010; Van-Dyk and Pletschke, 2012). These diverse groups of free enzyme systems are required to break down different kinds of chemical bonds via hydrolysis of LB feedstock into its simple monomeric subunits toward value-added biofuel generation. However, free enzyme systems efficiency is restricted considering unavailability of efficient enzymes, cost of the active enzyme productions, enzymatic hydrolysis time, enzyme recycling, and solid loading (Hong et al., 2013; Zeng et al., 2014; Jeoh et al., 2007; Hall et al., 2010; Jin et al., 2012; Weiss et al., 2013; Gao et al., 2011; Cruz et al., 2013; Lu et al., 2010). Several approaches have already been applied to minimize the cost of the active enzymes. These approaches define (1) screening and identification of novel multifunctional enzyme which can degrade complex LB feedstocks; (2) identification of novel hydrolytic enzymes bearing higher specific activity and turnover number; (3) modifications of enzyme catalytic site by protein engineering; (4) direct evolution of industrial enzymes; and (5) expression of effective enzymes in plant to accelerate direct extraction of fermentable sugars from LB feedstocks (Rizk and Elleuche, 2012; Horn et al., 2012; Dashtban and Qin, 2009; Song et al., 2012; Egelkrout and Howard, 2012). Enzymatic hydrolysis time reduction depends on a few factors like efficacy of pretreatments, enzymatic activities, and substrate concentrations. Dilute phosphoric acid pretreatment; catalyzed stream explosion that reduces the hydrolysis time through elevating intractability of LB feedstock for quicker sugar bioconversion rates (Hall et al., 2010). But this process is not effective as it downregulates microbial catalytic efficiencies for biofuel generation. In addendum, enzyme recycling also minimizes processing costs of biorefineries. However, this approach is not industrially sustainable. The main reasons are deactivation of expensive enzymes due to thermal denaturation, chemical denaturation, and shear stress. The enzyme immobilization on nanoparticles or polymeric matrices (Ansari and Husain, 2012), ion exchange absorptions (Wu et al., 2010), and ultracentrifugation (Qi et al., 2011) are the most common methods of enzyme recycling approach. Solid loading is another major concern over sugar biotransformation of

LB feedstocks. Presence of different biochemical enzyme inhibitors in pretreated LB feedstock brings down the catalytic efficiencies of degradative enzymes. Therefore, it seems existing technologies are not sustainable and noneconomic in its current state considering industrial demands. In this current global scenario, metabolic engineering and synthetic system biology approaches could be the potential gap filler which initiates the new concept of consolidated bioprocessing (CBP) (Datta, 2016).

CBP is the single-step process where LBs have been decomposed by potential degradative microbial communities without adding externally purified enzymes toward value-added biofuel generation. Those natural microbes need to be enriched with a few basic features such as inhibitor tolerance, efficacy to simultaneously metabolize broad range of sugar moieties, adoptability on cheaper media, acidic pH tolerance, and resistance to higher temperature. A possible avenue to evolve such microscopic bugs is to metabolically engineer naive LB-degrading microorganisms. An alternative way is the heterologous expression of LB hydrolytic enzymes in non-LB-degrading microbes for improving decomposition of LB feedstocks. However, lower LB yield influences the CBP in its current state. Metabolic engineering has been implemented to tackle this issue. Biomass yield of different energy crops has been improved through genetic modifications in following genes and metabolic networks such as photosynthetic genes, transcription factors, cell cycle regulators, hormone metabolism, lignin alterations, and micro-RNA (Vanholme et al., 2012; Rojas et al., 2010). LB yield has also been improved through metabolic pathway manipulations in both abiotic and biotic stress factors (Sticklen, 2006; Mathur and Sharma, 2008). Moreover, a combined approach of transgenesis, classical plant-based breeding, and advance agricultural routine practices has been shown to accelerate the yield of LBs.

LB decomposition needs to attain next level importance toward the utilization of accumulated higher LB to achieve green bioprocess. *Clostridium thermocellum*, *Geobacillus* sp., *Saccharomyces* sp., and *Escherichia coli* strains have been recognized as the most promising microbial platform for metabolic engineering toward increasing LB decompositions. *C. thermocellum* and *Geobacillus* sp. have been manipulated for the use in industrial LB conversion (Lynd et al., 2005; Akinosho et al., 2014). The endoglucanase (BsCel5 gene) homologous over expression in *Bacillus subtilis* enables it to grow on the solid cellulosic materials as the sole carbon source and then evolve the enzyme to increase specific activity on regenerated amorphous cellulose (Zhang et al., 2011). *E. coli* has been metabolically engineered to grow on cellulose and hemicellulose via expressing cellulase, xylanase, glucosidase, and xylobiosidase enzymes under control of native *E. coli* promoters (Bokinsky et al., 2011). Metabolically engineered *Saccharomyces* sp. has been constructed following potential cellulase display on the yeast cell surface toward better utilization of LB (Matano and Kondo, 2013). Cellulose degradation has improved the engineering of cellulosomal complex in *C. thermocellum*, which includes some bacterial and fungal glycoside hydrolases that show multimodular architecture comprising a carbohydrate-binding module (CBM), a linker, and a catalytic domain. Firm binding to the cellulose and the CBM helps promote its degradation by increasing its local concentration and bringing the enzyme in close and prolonged association with its recalcitrant lignocellulosic feedstocks (Schubot et al., 2004; Payne et al., 2013). Metabolic engineering and synthetic biology have provided an extra pace to produce large amounts of cellulases in heterologous hosts by choosing one among the several mechanisms, at either transcriptional or translational level, that regulate gene expression in microbial bugs (Mazzoli and Pessione, 2012; Ilmén et al., 2011). In other cases, efficient cellulase secretion in *Clostridium acetobutylicum* and *Lactobacillus plantarum* have been elevated by the replacement of native signal peptides by either signal peptides of efficiently secreted autologous proteins or optimized synthetic sequences (Wieczorek and Martin, 2010; Hyeon et al., 2011). However, Cel48F/Cel9G engineering with CBM3a and X2 modules of the *C. cellulolyticum* CipC scaffoldin has prevented toxic effects and triggered enzyme secretion in *C. acetobutylicum* (Chanal et al., 2011). In order to avoid the hydrolysis of the heterologously expressed cellulases, utilization of protease-inactivated microbial strains, such as *B. subtilis* WB800 and *Lactococcus lactis* HtrA mutants, may be demanded (Wieczorek and Martin, 2010; Anderson et al., 2011).

Lignin biosynthetic pathway and its enzymes are well characterized. Lignin reduction remains a challenging task. This problem stems from a lack of specificity in traditional lignin-reduction methods, which usually compromise plant growth or impair the plant defense system. Emerging strategies like genome bioediting and transgene regulation provide new options to achieve controlled lignin manipulations in targeted plant tissues when applies in conjunction with tissue-type-specific or cell-type-specific promoters. It will finally give the opportunity to design crops with optimized lignin composition and distribution while retaining all other traits related to the phenylpropanoid pathway. This new trend for lignin engineering focuses on the redirection of carbon flux to the production of related phenolic compounds and on the replacement of monolignols with novel lignin monomers to improve biophysical and chemical properties of lignins such as recalcitrance or industrial use. Concerning lignin decomposition, downregulation of caffeic acid O-methyltransferase or COMB (EC 2.1.1.68) and cinnamoyl-CoA reductase or CCR (EC1.2.1.44) enzymes in switch grass trigger moderate lignin reduction, along with a decrease in the syringyl and guaiacyl units (S/G) ratio. This results in an increase in lignin decomposition (Fu et al., 2011). Another possible way to lower lignin content is by blocking the free parahydroxyl groups in monolignols responsible for the creation of lignin subunits via oxidative coupling (notably 4-O-methylation of monolignols). This shows no detrimental effect on transport across the membrane or on growth and development of the plant (Zhang et al., 2012).

A recent work showed that expressing the bacterial hydroxycinnamoyl-CoA hydratase-lyase (HCHL) in *Arabidopsis* allows the overproduction of such C6C1 aromatics, which are incorporated into the lignin and reduce its molecular weight (Eudes et al., 2012). The study focuses on the in-planta expression of a bacterial phenylpropanoid side-chain cleavage enzyme for the production of C6C1 lignin monomers. Lignin from the engineered plants incorporates higher amount of C6C1 monomers and has a lower degree of polymerization, which results in higher biomass saccharification yields. Efforts are made to engineer only lignifying tissues by using a secondary cell wall-specific promoter. Rerouting of naive lignin metabolic network is another promising avenue to improve the quality of LB toward improving decomposition efficacy and quality of feedstock into its simple monomeric conformations. In lignin biosynthetic pathways, type III polyketide synthases enzymes reroutes hydroxycinnamoyl-CoA away from lignin biogenesis in presence of co-substrate malonyl-CoA. Type III polyketide synthases involves the biosynthesis of lignans and neolignans, which could be metabolized to reroute coniferyl alcohol away from lignin production (Satake and Murata, 2013). In this process, lignin precursor phenyl alanine, cinnamate, and coniferyl alcohol could be converted into benzenoid/phenylpropanoid volatile at the expense of lignin biogenesis (Colquhoun and Clark, 2011).

D-galacturonic acid is the main constituent of pectin complex polymer of LB. D-galacturonic acid oxidative hydrolysis generates a value-added precursor, that is, dicarboxylic acid and meso-galactaric acid. In filamentous fungus *Aspergillus niger* and *Hypocrea jecorina*, the biochemical reaction is carried out by uronate dehydrogenase (UDH). However, this hydrolysis process has been inhibited due to the presence of competing pathway which produces D-galacturonic acid with the advent of D-galacturonic acid reductase (*gaaA*). This D-galacturonic acid metabolic bypass pathway has been knocked out through CRISPR-Cas9 gene knock out tool kit toward improving pectin hydrolysis into value-added biomolecules (Kuivanen et al., 2016; Mojzita et al., 2010). In another approach, D-galacturonic acid is catabolized into L-ascorbic acid value-added product where EgALase (L-galL aldonolactonase from *Euglena gracilis*), Smp30 (aldonolactonase from rat functioning in the animal L-AA pathway), and MgGALDH (L-galL dehydrogenase from *Malpighia glabra*) over express in L-galacturonic acid dehydratase (*gaa*B) knocked out *A. niger* strain (Kuivanen et al., 2015).

10.5 METABOLIC ENGINEERING FOR IMPROVING LIQUID BIOFUELS GENERATIONS FROM LIGNOCELLULOSIC BIOMASS

Transportation sector consumes one-fourth of global primary energy reserve per annum. It not only depletes the natural energy reserves as well as accelerates greenhouse gas emissions. At present, primary source of energy transportation section has been deluged through petroleum-derived liquid fuels (Bhutto et al., 2016). Although there is always a big debate that which forms of fuels is comparatively better considering liquid or gaseous form. Gaseous fuels include biogas, producer gas, and methane having high content of carbon dioxide, carbon monoxide, and hydrogen. Whereas, liquid fuels represent liquid petroleum gas (LPG), diesel fuels, and gasoline. Gaseous fuels are always user friendly concerning current internal combustible engine mechanism. Gaseous fuels have a higher octane number which is advantageous in terms of high compression ratio and better ignition properties. Moreover, gaseous fuels generate less-contaminating pollutants and lower amount of waste sludge. However, it is not considered as the most effective transportation fuels. Higher cost of storage actually limits the gaseous fuels economy. Even gaseous fuels critical value for liquefaction is not reasonable at all. In contrary, liquid fuels are promising because it has comparatively a better critical value of liquefaction, less compression pressure, comparatively reasonable octane number, higher calorific values, easy storage, and less quantity of carbon monoxide as far composition (Ortiz-Canavate, 1994). To this end, liquid fuels could be the most viable fuels in the transport sector and crucial substitute to combat greenhouse gas emissions as compared to gaseous forms. But current global scenario has clearly showed an increase in trend of drastic usage of natural petroleum-based energy storage along with global climate change. This is the greatest challenge facing our communities in upcoming decades. To this end, liquid biofuel production from LB could be one of the most promising avenues at current stage to create more sustainable and balanced economy. Major benefits in LB-to-liquid biofuels include higher abundance and accessibility of LB and low procurement cost. However, this CBP suffers due to unavailability of high-yielding native microbial biocatalysts, deficiency in infrastructure, and lack of social and political supports. Metabolic engineering and synthetic biology approach is a key advanced technology for transforming naive microbial bugs toward efficient microbial cell factories toward liquid biofuel generation from lignocellulosic feedstocks. The most promising established microbial platform which is involve in liquid biofuel generation from LB are facultative mesophilic bacteria (*E. coli*, *Klebsiella oxytoca*, and *B. subtilis*), thermophilic bacteria (*Clostridium* sp., *Thermoanaerobacter* sp., and *Geobacillus thermoglucosidasius*) and yeast (*Saccharomyces cerevisiae*, *Zymomonas mobilis*, and *Pichia stipitis*). Metabolically engineered microbial communities biosynthesize a few most value-added liquid biofuels such as bioethanol, bioisobutanol, and bioisopropanol from LB.

Bioethanol has been established as a promising clean renewable biofuel, principally as a fuel additive which could improve the engine performance and reduce air pollution, and is the first biofuel produced on a large scale. Bioethanol derived from LB definitely has a benefit over bioethanol derived from food crops, since it would not compete with human food demands. Another beneficial aspect includes the fact that it could use cheap by-products from farms and municipal wastes (Ghosh and Hallenbeck, 2012). LB hydrolysate contains not only glucose as sugar residues it also releases xylose which is 40% of the total fermentable sugars (Jönsson and Nilvebrant, 2013; Casey et al., 2010). *Z. mobilis* is the most potential native bioethanol producer (Ajit and Chisti, 2017; Scully and Orlygsson, 2014).

Z. mobilis utilizes naturally hexose sugar (i.e., glucose) to generate higher bioethanol yield following Entner–Doudoroff (ED) metabolic pathway (Akponah and Ubogu, 2013) and end up with less cellular biomass production. *Z. mobilis* cannot naturally ferment LB or LB hydrolysate pentose sugar moiety, that is, xylose. *Z. mobilis* is engineered to utilize xylose by incorporating the xylose-utilizing pathway from *E. coli* which includes xylose isomerase (*xyl*A)-, xylulose kinase (*xyl*B)-, transketolase (*tkt*A)-, and transaldolase *(tal*B) operon (Zhang et al., 1995). However, *Z. mobilis* has shown a very poor efficacy not having an active xylose transport mechanism (Parker et al., 1995). Currently *Z. mobilis* ZM4 complete genome sequence, transcriptomics, and metabolomics profiling have been completed under aerobic and anaerobic fermentative mode (Seo et al., 2005; Yang et al., 2009). Information achieved through these biological technologies should accelerate engineering of a new ethanologenic *Z. mobilis* strain with industrial potential in near future. In addition, *met*B/*yfd*Z and *tes*A operon has been over expressed in *Z. mobilis* for enhancing amino acid biosynthesis and free fatty acid biosynthesis to improve acid stress tolerance, heat stress tolerance, and ethanol tolerance, respectively (Wang et al., 2016). Another potential bioethanol-producing yeast is *S. cerevisiae* which uses Embden–Meyerhof–Parnas (EMP) for hexose utilization. It reveals competence for pentose sugar, that is, xylose assimilation through addition of xylose metabolic network, that is, xylose reductase (XR) and xylulose dehydrogenase (XDH) from yeast *P. stipitis* (Jeffries et al., 2007). The generated xylulose can be transformed into xylulose-5-phosphate by xylulose kinase (XK) of *S. cerevisiae* to enter into pentose phosphate pathway (Jeffries and Jin, 2004). But this metabolically engineered *S. cerevisiae* is not industrially viable due to intracellular redox (NADPH, NADH) imbalance (Kotter and Ciriacy, 1993). Intracellular redox has also been influenced by the ammonia assimilation. These current issues have been resolved through deletion of NADPH-dependent glutamate dehydrogenase (GDH1) and overexpression of NADH-dependent glutamate dehydrogenase (GDH2) (Eliasson et al., 2000). Furthermore, additional metabolic engineering efforts have been carried out to facilitate xylose uptake to enhance ethanol yield and productivities from complete utilization of pentose sugar xylose (Saloheimo et al., 2007; Rintala et al., 2008; Hector et al., 2008).

E. coli, naturally adept at utilizing a wide spectrum of hexoses and pentoses, which is rendered more competent for ethanol production by the incorporation of the pyruvate to ethanol pathway from *Z. mobilis* and by knocking out side pathways that would otherwise divert carbon flow from bioethanol production. Initially, pyruvate decarboxylase (PDC) of *Z. mobilis* has been expressed in *E. coli* to maintain the redox balance of the NADH pool. It has not been successful because the native alcohol dehydrogenase (ADH) activity of *E. coli* is insufficient to achieve higher ethanol yields. Moreover, *E. coli* FMJ39 strain has been transformed with plasmid pLOI0297 bearing the PET operons, *Z. mobilis* alcohol dehydrogenase (*adh*B), and PDC under the control of the *E. coli lac*-promoter. *E. coli* FMJ39 has been chosen due to its carrying deletions in lactate dehydrogenase (*ldh*A) and pyruvate formate lyase (PFL); consequently, this strain is unable to grow anaerobically, since it is incapable of fermenting pyruvate to regenerate the NAD^+ needed for glycolysis. As an added feature, this indirectly selects for *E. coli* FMJ39 carrying pLOI0297 as this restores anaerobic growth on glucose. In this strain, most of the pyruvate is funneled to ethanol production (Beall and Ingram, 1991). However, plasmid-bearing strains are too genetically unstable for industrial applications. Hence, *E. coli* KO11 has been engineered with *Z. mobilis* PDC and alcohol dehydrogenase (*adh*B) integrated into its chromosome behind the *pfl*B promoter (Ohta et al., 1991). *E. coli* FBR5 strain which carries pLOI297 (PET operon) and mutations in *pfl* and *ldh*A genes has a high ethanol yield. Introduction of additional mutations can also improve the yields of *E. coli* strains. Deleting *frd* gene (fumarate reductase) blocks the succinate pathway and increases the ethanol yield (Dien et al., 2000). *E. coli* strains have been mutated in their phosphor-enol-pyruvate-glucose phosphotransferase system (ptsG) that shows an increased ability to utilize arabinose, xylose, and glucose, simultaneously (Nicholas and Bothart, 2001). Two more effective approaches have been attempted to correct possible metabolic imbalances. In one approach, citrate synthase (*cit*Z) from *B. subtilis* was expressed in *E. coli* KO11. Unlike that of *E. coli*, the Gram positive citrate synthase is not repressed by elevated concentrations of NADH and expression of this citrate synthase in *E. coli* KO11 increased the growth as well as ethanol yield by ~75% in corn steep liquor medium (Underwood et al., 2002). Very recently, an ethanol-tolerant *E. coli* ET1bc carrying *Z. mobilis pdc* and *adh*B has shown to have an enhanced ethanol yield (~1.3-fold compared to ethanol-sensitive strain, *E. coli* JMbc) from xylose as sole carbon source (Wang et al., 2008). Furthermore, few *E. coli* strains with increased tolerance to ethanol have been evolved via adaptive laboratory evolution (ALE), and strains with increased resistance to furfural, the inhibitor released during lignocellulose pretreatment, have been metabolically engineered (Miller et al., 2002). Metabolically engineered *Klebsiella oxytoca* strain has been developed to improve bioethanol production from cellulose. *K. oxytoca* P2 has been designed chromosomally with integrated PDC and alcohol dehydrogenase (*adh*B) genes from *Z. mobilis*. This metabolically engineered *Klebsiella* strain has around 30% ethanol yield using microcrystalline cellulose as a substrate (Wood and Ingram, 1992; Golias et al., 2002).

Mesophilic microorganisms are efficient; however, these bugs suffer from poor efficiency to hydrolyse LB feedstock, poor tolerance to extreme pH, and higher salt concentrations. As a result mesophilic microbial cultures are easily contaminated with inhibitory microbes to reduce LB or LB hydrolysate flux toward value-added liquid biofuel generation (Jin et al., 2014). Thermophilic bacteria are tractable to metabolic engineering for enhanced bioethanol yield from LB, namely, *C. thermocellum*, *G. thermoglucosidasius*, *Thermoanaerobacter ethanolicus*, *Thermoanaerobacter mathranii*, and *Thermoanaerobacterium saccharolyticum*. Thermophilic microorganisms would have a number of potential advantages over mesophilic microorganisms. Thermophilic microbes require high temperature for growth that would promote elevated rates of LB feedstock (e.g., cellobiose and xylose) biotransformation toward bioethanol productivity and recovery. It sometimes reduces the processing cooling costs. Moreover, bioethanol generation at high temperature would help to avoid contamination issues, which often appear during mesophilic bioethanol generation (Taylor et al., 2009). *C. thermocellum* has been metabolically engineered deleting two genes orotidine 5'-phosphate decarboxylase (*pyrF*) and phosphotransacetylase (*pta*) to efficient utilization of cellobiose where bioethanol yield is 0.71 mol/mol (Tripathi et al., 2010). *T. saccharolyticum* has also been metabolically engineered through knocking out a few genes such as hypoxanthine phosphoribosyltransferase (*hpt*), phosphotransacetylase (pta), acetate kinase (*ack*), lactate dehydrogenase (*ldh*), etc. Metabolically engineering *T. saccharolyticum* has improved to product bioethanol with molar yield of 1.26 mol/mol from cellobiose (Avicel) (Biswas et al., 2014). In a few studies, *T. saccharolyticium* HK07, *T. saccharolyticium* M0355, and *T. saccharolyticum* M1051 have also been engineered and the achieved bioethanol yield from cellobiose are 0.86, 1.73, and 1.73 mol/mol, respectively. The major gene deletions involved in these studies are Fe-Fe hydrogenase (*hfs*), acetate kinase (*ack*), phosphotransacetylase (*pta*), and lactate dehydrogenase (*ldh*) (Shaw et al., 2008; Shaw and Lynd, 2009; Argyros et al., 2011). Metabolically engineered *G. thermoglucosidasius* (TM242), with a deletion of lactate dehydrogenase (*ldh*), pyruvate formate lyase (*pfl*), and upregulated expression of pyruvate dehydrogenase, has been developed for ethanol synthesis from cellobiose and xylose and has produced yields of 0.35 g ethanol/g xylose and 0.47 g ethanol/g cellobiose (Cripps et al., 2009). However, metabolically engineered *T. mathranii* BG1L1, bearing a mutation in *ldh*, gave an ethanol yield of 0.42 g ethanol/g xylose (Georgieva and Ahring, 2008). Another metabolically engineered thermophilic strain, *T. saccharolyticum* JW/SH-YS485, bearing mutations in *ldh*, *pfl*, and *pta* to redirect the metabolic flux toward bioethanol, gave an ethanol yield of 0.38 g ethanol/g of mixed sugars, mainly glucose, xylose, galactose, and mannose. *G. thermoglucosidasius* TM242 bearing deletions in lactate dehydrogenase, pyruvate formate lyase, and with an upregulated pyruvate dehydrogenase (*pdh*), produced 0.41–0.44 g ethanol/g substrate, which consisted of mostly D-xylose (Shaw et al., 2008).

Bioisobutanol is another high value-added liquid biofuel. Bioisobutanol can be a better liquid biofuel than ethanol due to its higher energy density and lower hygroscopicity. Furthermore, the branched-chain structure of isobutanol gives a higher octane number than the isomeric n-butanol. *C. cellulolyticum* ATCC 35319 has been metabolically engineered to generate isobutanol from cellulose through incorporation of synthetic operon containing following genes such as *B. subtilis* acetolactate synthase (*Bs_alsS*), *E. coli* acetohydroxyacid isomeroreductase, *E. coli* dihydroxy acid dehydratase (Ec_*yqhD*), *L. lactis* ketoacid decarboxylase (Ll_*kivD*), and *E. coli* and *L. lactis* alcohol dehydrogenases (Ec_*adh* and Ll_*adh*) under the control of constitutive ferredoxin (Fd) promoter from *C. pasteurianum*. Metabolically engineered *C. cellulolyticum* has achieved bioisobutanol titer of 0.66 g/L using cellulose as sole carbon source (Higashide et al., 2011). *E. coli* ATCC 11303 has been metabolically engineered to produce bioisopropanol directly from cellobiose through the cellobiose catabolism involving beta-glucosidase (Tfu_0937) from *Thermobifida fusca* YX fused to the anchor protein Blc (Tfu0937/Blc) on the cell surface. Afterwards, a synthetic operon has been incorporated for isopropanol biosynthesis including *C. acetobutylicum* genes such as thiolase, Acyl-CoA: acetate/3-ketoacid CoA transferase, acetoacetate decarboxylase, and alcohol dehydrogenase (*Ca_thlA*, *Ca_atoDA*, *Ca_adc*, and *Ca_adhB-593*). Metabolically engineered *E. coli* strain has achieved bioisopropanol titer of about 4.1 g/L (Soma et al., 2012). To this end, metabolic engineering and synthetic biology approach helps to improve the production of liquid biofuel production from LB through strengthening microbial bug's biocatalytic efficiencies.

10.6 CONCLUDING REMARKS

Metabolic engineering and synthetic biology combinatorial approach provides an efficient way out for liquid biofuel production from lignocellulosic feedstock by transforming diverse group of native microbial regimes including mesophilic, thermophilic bacteria, and yeast toward efficient cell factories (Nakanishi et al., 2012; Hasheminejad et al., 2011; Sheng et al., 2017). In comparison to traditional error-prone random mutagenesis and genetic engineering approach for microbial strain improvement, metabolic synthetic biology approach is quiet rational, faster, organized, and extensively useful tool. More precisely, metabolic synthetic biology approach accelerates CBP through identification of potential novel enzymatic networks, designing novel synthetic metabolic pathways (natural or non natural), knocking out competing pathways to improve flux toward desired liquid biofuel generation. However, screening and identification of suitable novel enzyme targets and metabolic pathways are not so easy tasks. There are several factors involved which determines the efficacy of a new designed metabolic pathway including m-RNA abundance, transcriptional regulations, enzyme activities, enzyme inhibitions, translational regulations, cofactor abundance, growth factor availability, redox balance, substrate inhibition, product feedback inhibitions, intermediate precursor accumulative toxicity, transport of product and substrate, product tolerance, process parameter optimization, and finally scale up (Fig. 10.3). These key determinants actually directly or indirectly influence the LB assimilation toward improving liquid biofuel production, that is, titer, yield, and productivity. Metabolic engineers and synthetic biologist trust on different kinds of advanced methodologies to sort out the limiting

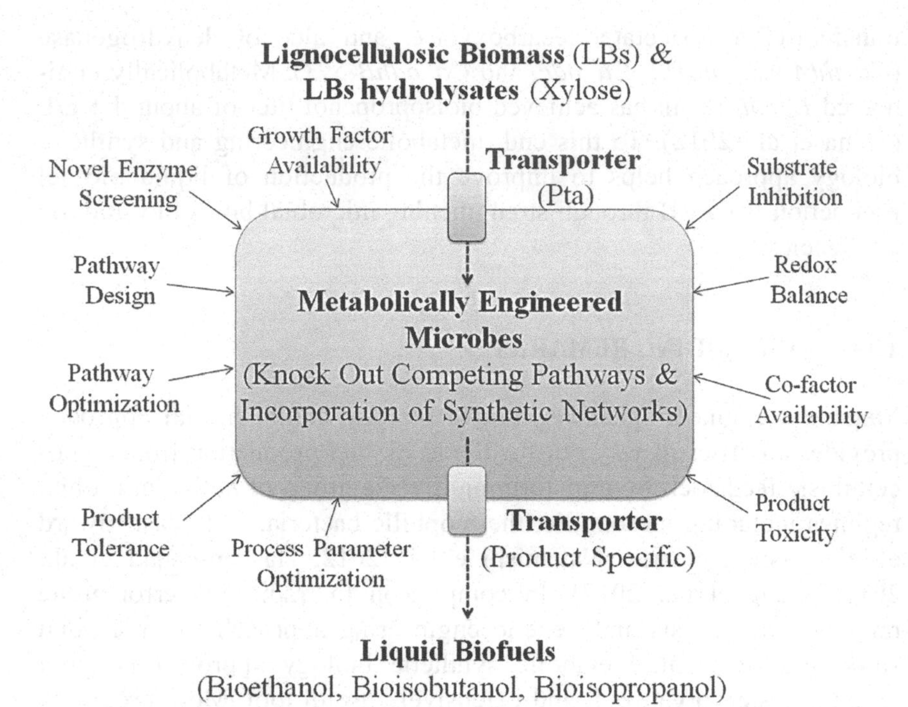

FIGURE 10.3 Metabolic engineering: a synergistic approach for liquid biofuels productions from LB.

segments of metabolic pathways (Long and Reed, 2015; Shabestary and Hudson, 2016; Roointan and Morowvat, 2017; Lee et al., 2011; Park and Lee, 2008; Lee et al., 2012). Emergence of high-throughput research platforms such as genome sequencing, transcriptomics, proteomics, metabolomics methods, influxomics, and computational tools (Weber et al., 2015; Medema et al., 2014 George et al., 2014; Liebeke et al., 2011; Wu et al., 2016) have been a nurturing field of metabolic engineering to improve liquid biofuel production from LB.

KEYWORDS

- **metabolic engineering**
- **synthetic biology**
- **liquid biofuels**
- **lignocellulosic biomass**
- **consolidated bioprocessing**
- **process parameter optimization**
- **microbial cell factories**

REFERENCES

Ajit, A.; Sulaiman, A. Z.; Chisti, Y. Production of Bioethanol by *Zymomonas mobilis* in High-gravity Extractive Fermentations. *Food Bioprod. Proc.* **2017,** *102*, 123–135.

Akinosho, H.; et al. The Emergence of *Clostridium thermocellum as* a High Utility Candidate for Consolidated Bioprocessing Applications. *Front. Chem.* **2014,** *2*, 66.

Akponah, E.; Akpomie, O. O.; Ubogu, M. Bio-ethanol Production from Cassava Effluent Using *Zymomonas mobilis* and *Saccharomyces cerevisiae* Isolated from Raffia Palm (*Elaesis guineesi*) sap. *Eur. J. Exp. Biol.* **2013,** *3*, 247–253.

Altaf, M.; et al. Influence of Micronutrients on Yeast Growth and β-D-Fructofuranosidase Production. *J. Appl. Microbiol.* **2007,** *103*, 372–380.

Anandarajah, K.; Mahendraperumal, G.; Sommerfeld, M.; Hu, Q. Characterization of Microalga *Nannochloropsis* sp. Mutants for Improved Production of Biofuels. *Appl. Energy* **2012,** *96*, 371–377.

Anderson, T. D.; et al. Assembly of Minicellulosomes on the Surface of *Bacillus subtilis*. *Appl. Environ. Microbiol.* **2011,** *77*, 4849–4858.

Ansari, S. A.; Husain, Q. Potential Applications of Enzymes Immobilized on/in Nano Materials: a Review. *Biotechnol. Adv.* **2012,** *30* (3), 512–523.

Argyros, D. A.; et al. High Ethanol Titers from Cellulose by Using Metabolically Engineered Thermophilic, Anaerobic Microbes. *Appl. Environ. Microbiol.* 2011, *77*, 8288–8294.

Badger, P. C. Biomass Transport Systems. *Encyclop. Agric. Food Biol. Eng.* **2003,** *1*, 94–98.

Balan, V.; et al. A Short Review on Ammonia Based Lignocelluloses' Biomass Pretreatment, in Chemical and Biochemical Catalysis for Next Generation Biofuel. *RSC Energy Environ.* **2011,** *4* (5), 89–114.

Banerjee, G.; et al. Synthetic Multi-component Enzyme Mixtures for Deconstruction of Lignocellulosic Biomass. *Biores. Technol.* **2010,** *101* (23), 9097–9105.

Bayer, E. A.; Shoham, Y.; Lamed, R.; Lignocellulose Decomposing Bacteria and Their Enzyme Systems in the Prokaryotes. *Prok. Physiol. Biochem.* **2013,** *1*, 215–266.

Beall, D. S.; Ohta, K.; Ingram, L. O. Parametric Studies of Ethanol Production from Xylose and Other Sugars by Recombinant *Escherichia coli*. *Biotechnol. Bioeng*. **1991,** *38*, 296–303.

Bhutto, A. W.; et al. Progress in Production of Biomass-to-liquid Biofuels to Decarbonize Transport Sector Prospectus and Challenges. *RSC Adv*. **2016,** *6*, 32140–32170.

Biswas, R.; et al. Increase in Ethanol Yield via Elimination of Lactate Production in an Ethanol-tolerant Mutant of *Clostridium thermocellum*. *PLoS One*. **2014,** *9*(2), e86389

Bokinsky, G.; et al. Synthesis of Three Advanced Biofuels from Ionic Liquidpretreated Switchgrass Using Engineered *Escherichia coli*. *Proc. Natl. Acad. Sci. USA*. **2011,** *108* (50), 19949–19954.

Caputo, A. C.; et al. Economics of Biomass Energy Utilization in Combustion and Gasification Plants: Effects of Logistic Variables. *Biomass Bioenerg*. **2005,** *28* (1), 35–51.

Carolan, J. E.; Joshi, S. V.; Dale, B. E. Technical and Financial Feasibility Analysis of Distributed Bioprocessing Using Regional Biomass Pre-processing Centers. *J. Agri. Food Ind. Org*. **2007,** *5* (2), 1–29.

Casey, E.; et al. Effect of Acetic Acid and pH on the Cofermentation of Glucose and Xylose to Ethanol by a Genetically Engineered Strain of *Saccharomyces cerevisiae*. *FEMS Yeast Res*. **2010,** *2* (10), 385–393.

Chanal, A.; et al. Scaffoldin Modules Serving as "Cargo" Domains to Promote the Secretion of Heterologous Cellulosomal Cellulases by *Clostridium acetobutylicum*. *Appl. Environ. Microbiol*. **2011,** *77*, 6277–6280.

Chen, W-H.; et al. Thermochemical Conversion of Microalgal Biomass into Biofuels: a Review. *Bioresour. Technol*. **2015,** *184*, 314–327.

Cherubini, F.; Strømman, A. H. Chemicals from Lignocellulosic Biomass: Opportunities, Perspectives, and Potential of Biorefinery Systems. *Biofuels Bioprod. Bioref.* **2011,** *5*, 548–561.

Chesson, A.; Forsberg, C. W. Polysaccharide Degradation by Rumen Microorganisms. *Rumen Microbial. Ecosys*. **1998,** *1*, 251–284.

Colquhoun, T. A.; Clark, D. G. Unraveling the Regulation of Floral Fragrance Biosynthesis. *Plant Signal. Behav*. **2011,** *6*, 378–381.

Cripps, R. E.; et al. Metabolic Engineering of *Geobacillus thermoglucosidasius* for High Yield Ethanol Production. *Metab. Eng*. **2009,** *11*, 398–408.

Cruz, A. G.; et al. Impact of High Biomass Loading on Ionic Liquid Pretreatment. *Biotechnol. Biofuels*. **2013,** *6* (1), 52.

Dashtban, M.; Schraft, H.; Qin, W. Fungal Bioconversion of Lignocellulosic Residues; Opportunities & Perspectives. *Int. J. Biol. Sci*. **2009,** *5* (6), 578–595.

Datta, S. Recent Strategies to Overexpress and Engineer Cellulases for Biomass Degradation. *Curr. Metabol*. **2016,** *4*, 14–22.

David, K.; Ragauskas, A. J. Switchgrass as an Energy Crop for Biofuel Production: a Review of Its Ligno-cellulosic Chemical Properties. *Energy Environ. Sci*. **2010,** *3* (9), 1182–1190.

Dien, B. S.; et al. Development of New Ethanologenic *Escherichia coli* Strains for Fermentation of Lignocellulosic Biomass. *Appl. Biochem. Biotechnol*. **2000,** *84*, 181–196.

Dohleman, F. G.; Long, S. P. More Productive than Maize in the Midwest: How Does Miscanthus Do It? *Plant Physiol.* **2009,** *150* (4), 2104–2115.

Duffy, M. D.; Nanhou, V. Y. *Switchgrass Production in Iowa: Economic Analysis, Special Publication for Cahriton Valley Resource Conservation District.* Iowa State University Extension Publication: Iowa State University; 2002.

Egelkrout, E.; Rajan, V.; Howard, J. A. Overproduction of Recombinant Proteins in Plants. *Plant Sci.* **2012,** *184,* 83–101.

Eliasson, A.; et al. Anaerobic Xylose Fermentation by Recombinant *Saccharomyces cerevisiae* Harbouring XYL1, XYL2 and XKS1 in Mineral Media Chemostat Cultivations. *Appl. Environ. Microbiol.* **2000,** *66,* 3381–3386.

Eudes, A.; et al. Biosynthesis and Incorporation of Side-chain-truncated Lignin Monomers to Reduce Lignin Polymerization and Enhance Saccharification. *Plant Biotechnol. J.* **2012,** *10,* 609–620.

Fernando, S.; et al. Biorefineries: Current Status Challenges and Future Direction. *Energy Fuel* **2006,** 1727–1737.

Fu, C.; et al. Genetic Manipulation of Lignin Reduces Recalcitrance and Improves Ethanol Production from Switchgrass. *Proc. Natl. Acad. Sci. USA.* **2011,** *108,* 3803–3808.

Gao, D.; et al. Binding Characteristics of *Trichoderma reesei* Cellulases on Untreated, Ammonia Fiber Expansion (AFEX), and Diluteacid Pretreated Lignocellulosic Biomass. *Biotechnol. Bioeng.* **2011,** *108* (8), 1788–1800.

Gao, D.; et al. Hemicellulases and Auxiliary Enzymes for Improved Conversion of Lignocellulosic Biomass to Monosaccharides. *Biotechnol. Biofuels* **2011,** *4,* 5.

Gao, D.; et al. Mixture Optimization of Six Core Glycosyl Hydrolases for Maximizing Saccharification of Ammonia Fiber Expansion (AFEX) Pretreated Corn Stover. *Biores. Technol.* **2010,** *101* (8), 2770–2781.

George, K. W.; et al. Correlation Analysis of Targeted Proteins and Metabolites to Assess and Engineer Microbial Isopentenol Production. *Biotechnol. Bioeng.* **2014,** *111,* 1648–1658.

Georgieva, T. I.; Mikkelsen, M. J.; Ahring, B. K. Ethanol Production from Wet-exploded Wheat Straw Hydrolysate by Thermophilic Anaerobic Bacterium *Thermoanaerobacter* BG1L1 in a Continuous Immobilized Reactor. *Appl. Biochem. Biotechnol.* **2008,** *145,* 99–110.

Ghosh, D.; Hallenbeck, P. C. Advanced Bioethanol Production. *Microbial. Technol. Adv. Biofuels Prod.* **2012,** *1,* 165–181.

Golias, H.; et al. Evaluation of a Recombinant *Klebsiella oxytoca* Strains for Ethanol Production from Cellulose by Simultaneous Saccharification and Fermentation: Comparison with Native Cellobiose-utilizing Yeast Strains and Performance in Co-culture with Thermo Tolerant Yeast *Zymomonas mobilis*. *J. Biotechnol.* **2002,** *96,* 155–168.

Gray, B. F.; Griffiths, J. F.; Hasko, S. M. Spontaneousignition Hazards in Stockpiles of Cellulosic Materials: Criteria for Safe Storage. *J. Chem. Technol. Biotechnol. Chem. Technol.* **1984,** *34* (8), 453–463.

Hall, M.; et al. Cellulose Crystallinity—a Key Predictor of the Enzymatic Hydrolysis Rate. *FEBS J.* **2010,** *277* (6), 1571–1582.

Hasheminejad, M. et al. Upstream and Downstream Strategies to Economize Biodiesel Production. *Bioresour. Technol.* **2011,** *102* (2), 461–468.

Hatakka, A.; Hammel, K. E. Fungal Biodegradation of Lignocelluloses. *Ind. Appl.* **2010,** *10*, 319–340.

Hector, R. E.; et al. Expression of a Heterologous Xylose Transporter in a *Saccharomyces cerevisiae* Strain Engineered to Utilize Xylose Improves Aerobic Xylose Consumption. *Appl. Microbiol. Biotechnol.* **2008,** *80*, 675–684.

Hess, J. R.; et al. Corn Stover Availability for Biomass Conversion: Situation Analysis. *Cellulose* **2009,** *16* (4), 599–619.

Higashide, W.; et al. Metabolic Engineering of *Clostridium cellulolyticum* for Production of Isobutanol from Cellulose. *Appl. Environ. Microbiol.* **2011,** *77*, 2727–2733.

Holtzapple, M. T. Hemicellulose. In *Encyclopedia of Food Science, Food Technology and Nutrition*; Macrae, R., Robinson, R. K., Sadler, M. J.; Eds.; **1993,** Vol. 2, pp 2324–2334.

Hong, Y.; et al. Impact of Cellulase Production on Environmental and Financial Metrics for Lignocellulosic Ethanol. *Biofuels Bioprod. Bioref.* **2013,** *7* (3), 303–313.

Horn, S. J.; et al. Novel Enzymes for the Degradation of Cellulose. *Biotechnol. Biofuels.* **2012,** *5* (1), 45.

Hoskinson, R. L.; et al. Engineering, Nutrient Removal, and Feedstock Conversion Evaluations of Four Corn Stover Harvest Scenarios. *Biomass Bioenerg.* **2007,** *31* (2–3), 126–136.

Huber, G. W. S.; et al. Synthesis of Transportation Fuels from Biomass: Chemistry, Catalysts, and Engineering. *Chem. Rev.* **2006,** *106*, 4044–4098.

Huggins, D. R.; et al. Introduction: Evaluating Long-term Impacts of Harvesting Crop Residues on Soil Quality. *Agron. J.* **2011,** *103* (1), 230–233.

Hyeon, J. E.; et al. Production of Minicellulosomes for the Enhanced Hydrolysis of Cellulosic Substrates by Recombinant *Corynebacterium glutamicum*. *Enz. Microb. Technol.* **2011,** *48*, 371–377.

Ilmén, M.; et al. High Level Secretion of Cellobiohydrolases by *Saccharomyces cerevisiae*. *Biotechnol Biofuels.* **2011,** *4*, 30.

Isikgor, F. H.; Becer, C. R. Lignocellulosic Biomass: a Sustainable Platform for the Production of Bio-based Chemicals and Polymers. *Polym. Chem.* **2015,** *6*, 4497–4559.

Jeffries, T. W.; Jin, Y. S. Metabolic Engineering for Improved Fermentation of Pentoses by Yeasts. *Appl. Microbiol. Biotechnol.* **2004,** *63*, 495–509.

Jeffries, T. W.; et al. Genome Sequence of the Lignocellulogic-Bioconverting and Xylose-fermenting Yeast *Pichia stipitis*. *Nat. Biotechnol.* **2007,** *25*, 319–326.

Jeoh, T.; et al. Cellulase Digestibility of Pretreated Biomass is Limited by Cellulose Accessibility. *Biotechnol. Bioeng.* **2007,** *98* (1), 112–122.

Jin, H.; et al. Engineering Biofuel Tolerance in Non-native Producing Microorganisms. *Biotechnol. Adv.* **2014,** *32*, 541–548.

Jin, M. et al. A Novel Integrated Biological Process for Cellulosic Ethanol Production Featuring High Ethanol Productivity, Enzyme Recycling and Yeast Cells Reuse. *Energ. Environ. Sci.* **2012,** *5* (5), 7168–7175.

Johansson, J. Transport and Handling of Forest Energy Bundles—Advantages and Problems. *Biomass Bioenerg.* **2006,** *30* (4), 334–341, 2006.

Jönsson, L. J.; Alriksson, B.; Nilvebrant, N. O. Bioconversion of Lignocellulose: Inhibitors and Detoxification. *Biotechnol. Biofuels.* **2013,** *6*, 16.

Kotter, P.; Ciriacy, M.; Xylose Fermentation by *Saccharomyces cerevisiae*. *Appl. Microbiol. Biotechnol.* **1993,** *38*, 776–783.

Kuivanen, J.; et al. Metabolic Engineering of the Fungal D-galacturonate Pathway for L-ascorbic Acid Production. *Microbial Cell Factories* **2015,** *14*, 2.

Kuivanen, J.; et al. Engineering *Aspergillus niger* for Galactaric Acid Production: Elimination of Galactaric Acid Catabolism by Using RNA Sequencing and CRISPR/Cas9. *Microb. Cell Fact.* **2016,** *15*, 210.

Kumar, A.; Sokhansanj, S.; Flynn, P. C. Development of a Multicriteria Assessment Model for Ranking Biomass Feedstock Collection and Transportation Systems. *Appl. Biochem. Biotechnol.* **2006,** *129* (1–3), 71–87.

Kurian, J. K.; et al. Feedstocks, Logistics and Pre-treatment Processes for Sustainable Lignocellulosic Biorefineries: a Comprehensive Review. *Renew. Sustain. Energ. Rev.* **2013,** *25*, 205–219.

Lange, J. P. Lignocellulose Conversion: an Introduction to Chemistry, Process and Economics. *Biofuels Bioprod. Bioref.* **2007,** *1*, 39–48.

Laursen, W. Students Take a Green Initiative. *Chem. Eng.* 2006, *774,* 32–34.

Lee, J. W.; et al. Systems Metabolic Engineering for Chemicals and Materials. *Trends Biotechnol.* **2011,** *29*, 70–78.

Lee, J. W.; et al. Systems Metabolic Engineering of Microorganisms for Natural and Non-natural Chemicals. *Nat. Chem. Biol.* **2012,** *8*, 536–546.

Liebeke, M.; et al. Etabolomics and Proteomics Study of the Adaptation of *Staphylococcus aureus* to Glucose Starvation. *Mol. Biosyst.* **2011,** *7*, 1241–1253.

Long, M. R.; Ong, W. K.; Reed, J. L. Computational Methods in Metabolic Engineering for Strain Design. *Curr. Opin. Biotechnol.* **2015,** *34*, 135–141.

Lu, J.; Sheahan, C; Fu, P. Metabolic Engineering of Algae for Fourth Generation Biofuels Production. *Energy Environ. Sci.* **2011,** *4*, 2451–2466.

Lu, Y.; et al. Influence of High Solid Concentration on Enzymatic Hydrolysis and Fermentation of Steam-exploded Corn Stover Biomass. *Appl. Biochem. Biotechnol.* **2010,** *160* (2), 360–369.

Lynd, L. R.; et al. Consolidated Bioprocessing of Cellulosic Biomass: an Update. *Curr. Opin. Biotechnol.* **2005,** *16* (5), 577–583.

Matano, Y.; Hasunuma, T.; Kondo, A. Cell Recycle Batch Fermentation of High-solid Lignocellulose Using a Recombinant Cellulase-displaying Yeast Strain for High Yield Ethanol Production in Consolidated Bioprocessing. *Bioresour. Technol.* **2013,** *135*, 403–409.

Mathur, P. B.; Vadez, V.; Sharma, K. K. Transgenic Approaches for Abiotic Stress Tolerance in Plants: Retrospect and Prospects. *Plant Cell Rep.* **2008,** *27*, 411–424.

Mazzoli, R.; Lamberti, C.; Pessione E. Engineering New Metabolic Capabilities in Bacteria: Lessons from Recombinant Cellulolytic Strategies. *Trends Biotechnol.* **2012,** *30*, 111–119.

Medema, M. H.; et al. A Systematic Computational Analysis of Biosynthetic Gene Cluster Evolution: Lessons for Engineering Biosynthesis. *PLoS Comput. Biol.* **2014,** *10*, e1004016.

Miller, E. N.; et al. Genetic Changes that Increase 5-hydroxymethyl Furfural Resistance in Ethanol-producing *Escherichia coli* LY180. *Biotechnol. Lett.* **2002,** *32* (5), 661–667.

Mojzita, D.; et al. Metabolic Engineering of Fungal Strains for Conversion of D-Galacturonate to Meso-Galactarate. *Appl. Environ. Microbiol.* **2010,** *1* (76), 169–175.

Mooney, C. A.; et al. The Effect of Initial Pore Volume and Lignin Content on the Enzymatic Hydrolysis of Softwoods. *Bioresour. Technol.* **1998,** *64*, 113–119.

Mosier, N.; et al. Features of Promising Technologies for Pretreatment of Lignocellulosic Biomass. *Bioresour. Technol.* **2005,** *96*, 673–686.

Nakanishi, A.; et al. Effect of Pretreatment of Hydrothermally Processed Rice Straw with Laccase-displaying Yeast on Ethanol Fermentation. *Appl. Microbiol. Biotechnol.* **2012,** *94*, 939–948.

Nicholas, N. N.; Dien, B. S.; Bothart, R. J. Use of Catabolite Repression Mutants for Fermentation of Sugar Mixtures to Ethanol. *Appl. Microbiol. Biotechnol.* **2001,** *56*, 120–125.

Nigam, P. S.; Singh A. Production of Liquid Biofuels from Renewable Resources. *Prog. Energy Combust. Sci.* **2011,** *37*, 52–68.

Ohta, K.; et al. Genetic Improvement of *Escherichia coli* for Ethanol Production: Chromosomal Integration of *Zymomonas mobilis* Genes Encoding Pyruvate Decarboxylase and Alcohol Dehydrogenase II. *Appl. Environ. Microbiol.* **1991,** *57*, 893–900.

Ortiz-Canavate, J.; Characteristics of Different Types of Gaseous and Liquid Biofuels and Their Energy Balance. *J. Agric. Eng. Res.* **1994,** *59*, 231–238.

Osamu, K.; Carl, H. W. *Biomass Handbook*; Gordon Breach Science Publisher: *New York, USA,* 1989.

Park, J. H.; Lee, S. Y. Towards Systems Metabolic Engineering of Microorganisms for Amino Acid Production. *Curr. Opin. Biotechnol.* **2008,** *19*, 454–460.

Parker, C.; et al. Characterization of the *Zymomonas mobilis* Glucose Facilitator Gene-product (Glf) in Recombinant *Escherichia coli*: Examination of Transport Mechanism, Kinetics and the Role of Glucokinase in Glucose Transport. *Mol. Microbiol.* **1995,** *15*, 795–802.

Payne, C. M.; et al. Glycosylated Linkers in Multimodular Lignocellulose-degrading Enzymes Dynamically Bind to Cellulose. *Proc. Natl. Acad. Sci. USA.* **2013,** *110*, 14646–14651.

Qi, B.; et al. Enzyme Adsorption and Recycling During Hydrolysis of Wheat Straw Lignocellulose. *Biores. Technol.* **2011,** *102* (3), 2881–2889.

Rentizelas, A. A.; Tolis, A. J.; Tatsiopoulos, I. P. Logistics Issues of Biomass: the Storage Problem and the Multi-biomass Supply Chain. *Renew. Sust. Eneg. Rev.* **2009,** *13* (4), 887–894.

Rintala, E.; et al. Transcription of Hexose Transporters of *Saccharomyces cerevisiae* is Affected by Change in Oxygen Provision. *BMC Microbiol.* **2008,** *8*, 53–65.

Rizk, M.; Antranikian, G.; Elleuche, S. End-to-end Gene Fusions and Their Impact on the Production of Multifunctional Biomass Degrading Enzymes, *Biochem. Biophy. Res. Comm.* **2012,** *428* (1), 1–5.

Rojas, C. A.; et al. Genetically Modified Crops for Biomass Increase. Genes and Strategies. *GM Crops.* **2010,** *1* (3), 137–142.

Roointan, A.; Morowvat, M. H. Road to the Future of Systems Biotechnology: Crisprcas-mediated Metabolic Engineering for Recombinant Protein Production. *Biotechnol. Genet. Eng. Rev.* **2017,** *32* (1–2), 1–18.

Saha, B. C. Enzymes as Biocatalysts for Conversion of Lignocellulosic Biomass to Fermentable Sugars. In *Handbook of Industrial Biocatalysis*; **2005,** Vol. 1, pp 1–12.

Saini, J. K.; Saini, R.; Tewari, L. Lignocellulosic Agriculture Wastes as Biomass Feedstocks for Second-generation Bioethanol Production: Concepts and Recent Developments. *3 Biotech*. **2015,** *5*, 337–353.

Saloheimo, A.; et al. Xylose Transport Studies with Xylose-utilizing *Saccharomyces cerevisiae* Strains Expressing Heterologous and Homologous Permeases. *Appl. Microbiol. Biotechnol.* **2007,** *74*, 1041–1052.

Sannigrahi, P.; Ragauskas, A. J.; Tuskan, G. A. Poplar as a Feedstock for Biofuels: a Review of Compositional Characteristics. *Biofuels Bioprod. Bioref.* **2010,** *4* (2), 209–226.

Satake, H.; Ono, E.; Murata, J. Recent Advances in the Metabolic Engineering of Lignan Biosynthesis Pathways for the Production of Transgenic Plant-based Foods and Supplements. *J. Agric. Food. Chem.* **2013,** *61*, 11721–11729.

Schubot, F. D.; et al. Structural Basis for the Exocellulase Activity of the Cellobiohydrolase CbhA from *Clostridium thermocellum*. *Biochemistry* **2004,** *43*, 1163–1170.

Scully, S. M.; Orlygsson, J. Recent Advances in Second Generation Ethanol Production by Thermophilic Bacteria. *Energies* **2014,** *8*, 1–30.

Seo, J-S.; et al. The Genome Sequence of the Ethanologenic Bacterium *Zymomonas mobilis* ZM4. *Nat. Biotechnol.* **2005,** *23* (1), 63–68.

Shabestary, K.; Hudson, E. P. Computational Metabolic Engineering Strategies for Growth-coupled Biofuel Production by Synechocystis. *Metab. Eng. Commun.* **2016,** *3*, 216–226.

Sharma, B.; et al. Biomass Supply Chain Design and Analysis: Basis, ovErview, Modeling, Challenges, and Future. *Renew. Sustain. Energ. Rev.* **2013,** *24*, 608–627.

Shaw, A. J.; et al. Metabolic Engineering of a Thermophilic Bacterium to Produce Ethanol at High Yield. *Proc. Natl. Acad. Sci. USA*. **2008,** *105*, 13769–13774.

Shaw, A. J.; Hogsett, D. A.; Lynd, L. R. Identification of the [FeFe]-Hydrogenase Responsible for Hydrogen Generation in *Thermoanaerobacterium saccharolyticum* and Demonstration of Increased Ethanol Yield via Hydrogenase Knockout. *J. Bacteriol.* **2009,** *191*, 6457–6464.

Shaw, A. J.; et al. Metabolic Engineering of a Thermophilic Bacterium to Produce Ethanol at High Yield. *Proc. Natl. Acad. Sci. USA*. **2008,** *105*, 13769–13774.

Sheng, L.; et al. Development and Implementation of Rapid Metabolic Engineering Tools for Chemical and Fuel Production in *Geobacillus thermoglucosidasius* NCIMB 11955. *Biotechnol. Biofuels*. **2017,** *10*, 5.

Sikarwar, V. S.; et al. Progress in Biofuel Production from Gasification. *Prog. Energy Combust. Sci.* **2017,** *61*, 189–248.

Sokhansanj, S.; Hess, J. R.; Biomass Supply Logistics and Infrastructure. In *Biofuels: Methods and Protocols*; **2009,** Vol. 581, pp 1–25.

Sokhansanj, S.; et al.; Large-scale Production, Harvest and Logistics of Switchgrass (*Panicum virgatum* L.) Current Technology and Envisioning a Mature Technology. *Biofuels Bioprod. Biorefin.* **2009,** *3* (2), 124–141.

Soma, Y.; et al. Direct Isopropanol Production from Cellobiose by Engineered *Escherichia coli* Using a Synthetic Pathway and a Cell Surface Display System. *J. Biosci. Bioeng.* **2012,** *114*, 80–85.

Song, L.; et al. Engineering Better Biomass-degrading Ability into a GH11 Xylanase Using a Directed Evolution Strategy. *Biotechnol. Biofuels*. **2012,** *5* (1), 3.

Sticklen, M. Plant Genetic Engineering to Improve Biomass Characteristics for Biofuels. *Curr. Opin. Biotechnol.* **2006,** *17*, 315–319.

Sultana, A.; Kumar, A.; Harfield, D. Development of Agripellet Production Cost and Optimum Size. *Biores. Technol.* **2010,** *101* (14), 5609–5621.

Sun, Y.; Cheng, J. Hydrolysis of Lignocellulosic Materials for Ethanol Production: a Review. *Bioresour. Technol.* **2002,** *83,* 1–11.

Taherzadeh, M. J.; Karimi, K. Pretreatment of Lignocellulosic Wastes to Improve Ethanol and Biogas Production: a Review. *Int. J. Mol. Sci.* **2008,** *9,* 1621–1651.

Tao, G.; et al. Biomass Properties in Association with Plant Species and Assortments I: a Synthesis Based on Literature Data of Energy Properties. *Renew. Sustain. Energ. Rev.* **2012,** *16* (5), 3481–3506.

Taylor, M. P.; et al. Thermophilic Ethanologenesis: Future Prospects for Second-generation Bioethanol Production. *Trend. Biotechnol.* **2009,** *27* (7), 398–405.

Thorsell, S.; et al. Economics of a Coordinated Biorefinery Feedstock Harvest System: Lignocellulosic Biomass Harvest Cost. *Biomass Bioenerg.* **2004,** *27* (4), 327–337.

Tripathi, S. A.; et al. Development of pyrF Based Genetic System for Targeted Gene Deletion in *Clostridium thermocellum* and Creation of a *pta* Mutant. *Appl. Environ. Microbiol.* **2010,** *76,* 6591–6599.

Tumuluru, J. S.; et al. A Review of Biomass Densification Systems to Develop Uniform Feedstock Commodities for Bioenergy Application. *Biofuels Bioprod. Bioref.* **2011,** *5* (6), 683–707.

Underwood, S. A.; et al. Flux Through Citrate Synthase Limits the Growth of Ethanologenic *Escherichia coli* KO11 During Xylose Fermentation. *Appl. Environ. Microbiol.* **2002,** *68,* 1071–1081.

Van-Dyk, J. S.; Pletschke, B. I. A Review of Lignocellulose Bioconversion Using Enzymatic Hydrolysis and Synergistic Cooperation Between Enzymes—Factors Affecting Enzymes, Conversion and Synergy. *Biotechnol. Adv.* **2012,** *30* (6), 1458–1480.

Vanholme, R.; et al. Metabolic Engineering of Novel Lignin in Biomass Crops. *New Phytol.* **2012,** *196,* 978–1000.

Wang, H.; et al. Very High Gravity Ethanol and Fatty Acid Production of *Zymomonas mobilis* Without Amino Acid and Vitamin. *J. Ind. Microbiol. Biotechnol.* **2016,** *43,* 861–871.

Wang, Z.; et al. An Ethanol-tolerant Recombinant *Escherichia coli* Expressing *Zymomonas mobilis pdc* and *adh*B Genes for Enhanced Ethanol Production from Xylose. *Biotechnol. Lett.* **2008,** *30,* 657–663.

Weber, T.; et al. Metabolic Engineering of Antibiotic Factories: New Tools for Antibiotic Production in Actinomycetes. *Trends Biotechnol.* **2015,** *33,* 15–26.

Weiss, N.; et al. Enzymatic Lignocellulose Hydrolysis: Improved Cellulase Productivity by Insoluble Solids Recycling. *Biotechnol. Biofuels.* **2013,** *6,* 5.

Wieczorek, A. S.; Martin, V. J. Engineering the Cell Surface Display of Cohesins for Assembly of Cellulosome-inspired Enzyme Complexes on *Lactococcus lactis. Microb. Cell Fact.* **2010,** *9,* 69.

Wood, B. E.; Ingram, L. O. Ethanol Production from Cellobiose, Amorphous Cellulose and Crystalline Cellulose by Recombinant *Klebsiella oxytoca* Containing Chromosomally Integrated *Zymomonas mobilis* Genes for Ethanol Production and Plasmids Expressing Thermostable Cellulose from *Clostridium thermocellum. Appl. Environ. Microbiol.* **1992,** *58,* 2103–2110.

Wu, J. C.; et al. Recovery of Cellulases by Adsorption/Desorption Using Cation Exchange Resins. *Kor. J. Chem. Eng.* **2010,** *27* (2), 469–473.

Wu, S. G.; et al. Rapid Prediction of Bacterial Heterotrophic Fluxomics Using Machine Learning and Constraint Programming. *PLoS Comput. Biol.* **2016,** *12*, e1004838.

Xu, J.; Wang, Z.; Cheng, J. J. Bermuda Grass as Feedstock for Biofuel Production: a Review. *Biores. Technol.* **2011,** *102* (17), 7613–7620.

Yang, S.; et al. Transcriptomic and Metabolomic Profiling of *Zymomonas mobilis* During Aerobic and Anaerobic Fermentations. *BMC Genomics* **2009,** *10*, 34–50.

Zegada-Lizarazu, W.; Monti, A. Are We Ready to Cultivate Sweet Sorghum as a Bioenergy Feedstock? A Review on Field Management Practices. *Biomass Bioenerg.* **2012,** *40*, 1–12.

Zeng, Y.; et al. Lignin Plays a Negative Role in the Biochemical Process for Producing Lignocellulosic Biofuels. *Curr. Opin. Biotechnol.* **2014,** *27*, 38–45.

Zhang, K. W.; et al. An Engineered Monolignol 4-O-methyltransferase (MOMT4) Represses Lignin Polymerization and Confers Novel Metabolic Capability in *Arabidopsis*. *Plant Cell* **2012,** *24*, 3122–3139.

Zhang, M.; et al. Metabolic Engineering of a Pentose Metabolism Pathway in Ethanologenic *Zymomonas mobilis*. *Science* **1995,** *267*, 240–243.

Zhang, X. Z.; et al. One-step Production of Lactate from Cellulose as the Sole Carbon Source Without Any Other Organic Nutrient by Recombinant Cellulolytic *Bacillus subtilis*. *Metab. Eng.* **2011,** *13* (4), 364–372.

Zhang, Z.; Donaldson, A. A.; Ma, X. Advancements and Future Directions in Enzyme Technology for Biomass Conversion. *Biotechnol. Adv.* **2012,** *30* (4), 913–919.

Zhou, C. H.; Lu, G. Q. Chemoselective Catalytic Conversion of Glycerol as a Biorenewable Source to Valuable Commodity Chemicals. *Chem. Soc. Rev.* **2008,** *37*, 527–549.

Zub, H. W.; Brancourt-Hulmel M. Agronomic and Physiological Performances of Different Species of *Miscanthus*, a Major Energy Crop. *Rev. Agron. Sustain. Dev.* **2010,** *30* (2), 201–214.

CHAPTER 11

Weed Biomass as Feedstock for Bioethanol Production: A Review

VINOD SINGH GOUR*, RAVNEET CHUG, and S. L. KOTHARI

Amity Institute of Biotechnology, Amity University Rajasthan, Kant Kalwar, Jaipur, Rajasthan, India

*Corresponding author.
E-mail: vkgaur@jpr.amity.edu; vinodsingh2010@gmail.com

ABSTRACT

Weeds are an important source of biomass which could be utilized for biofuel production. These are undesired plants, which grow even on uncultivated land at a large scale; therefore, there is no constraint in the availability of such biomass. These are rapidly growing herbs which could provide an economically viable source of feedstock in relatively short period of time.

The harvested weed can be degraded chemically or enzymatically to release simple carbohydrates. The carbohydrates can be fermented to get bioethanol. *Lantana camara, Parthenium hysterophorus, Eichhornia crassipes,* and other weeds have been treated chemically and/or enzymatically to release the sugar content from lignocellulosic material. With the help of *Saccharomyces cerevisiae, Escherichia coli, Klebsiella oxytoca, Zymomonas mobilis,* and other microorganisms the sugars have been converted to bioethanol. The current status and future prospects of bioethanol production using feedstock derived from weeds have been discussed in this chapter in detail.

11.1 INTRODUCTION

The limitation of fossil fuel had directed the attention of researchers and scientist to explore alternative sources of energy such as wind, solar, thermal, and biofuel. Biofuel includes biogas, hydrogen, biodiesel, and bioethanol. Promotion of bioethanol and blending of bioethanol with gasoline may reduce depletion of sources of fossil fuel (Demirbas, 2007). Besides, it has a great potential as an eco-friendly fuel as it has less emission of greenhouse gases than gasoline-based fuel. The first pilot plant for bioethanol production was established at South Dakota University, Brookings in 1979 (Songstad et al., 2009). The United States is the largest producer of bioethanol in the world, followed by Brazil, but they are using sugarcane as feedstock (Martins et al., 2018). Other top producers of ethanol include China, India, and France (Demirbas, 2007). According to REN21 (2016), India has bioethanol blend mandate of 22.5%. Mixture of gasoline and ethanol in 9:1 ratio can fuel the car engines without any modification (Galbe and Zacchi, 2002). According to them, even the engines are available which can run on 85% mixture of ethanol content in the fuel with gasoline. In India, the alcohol demand was reported to increase from 3740 ML (2011–2012) to 4440 ML (2014–2015) but the supply was short by 2040 ML in 2014–2015 (Swain, 2014).

Ethanol can be produced using biomass; however, use of edible crops is not recommended in producing bioethanol as it may create conflicts between food and fuel. Lignocellulosic material is available in nature at large scale with production of ~200 billion tons per year (Chandel and Singh, 2011). Khatiwada and Silveira (2017) reported the ways through which conflict could be avoided between food and fuel in Indonesia to achieve the goal of blending 20% ethanol in gasoline. That is why the other sources of feedstock for ethanol production are being explored. Here, weeds provide a viable option of getting biomass at large scale around the globe to be used as feedstock in producing bioethanol. Bioethanol derived from produces which can be used as food is considered as first-generation fuel; however, bioethanol produced using lignocellulosic raw material is known as second-generation biofuel (Kang et al., 2014). The second-generation biofuel does not create a food verses fuel crisis/conflict.

Looking at growing importance and scopes of bioethanol as a green fuel and availability of large-scale weed biomass provides tremendous opportunities to use lignocellulosic components of the plants as feedstock for the production of bioethanol. The weeds can play an important role here as feedstock in bioethanol production as they are fast growing under cultivated and/or uncultivated area. These plants are noxious and not only reduce the yield of the crop by competing with the main crop but also cause loss of soil fertility.

The present chapter focuses on utilization of biomass derived from three weeds, namely, *Lantana camara*, *Parthenium hysterophorus*, and *Eichhornia crassipes* as a feedstock in production of bioethanol.

11.1.1 THE THREE IMPORTANT WEEDS

L. camara is a weed which belongs to family Verbenaceae. The presence of this weed has been recorded in Asia, Africa, South America, North America, and Australia (Taylor et al., 2012). It shows its robustness through wide range of distribution. As far as India is concerned, the plant has been found throughout the country (Kuila et al., 2011). According to Cilliers and Neser (1991), the invasion of *L. camara* causes reduced fertility and price of land along with soil erosion. Presence of lantanedene A, B, and triterpene acid in biomass of *L. camara* make it toxic to the animals (Wolfson and Solomons, 1964; Sharma et al., 1981). Finally, these factors influence the population of herbivore, and thereby whole ecosystem gets affected. The plant is a perennial shrub. It grows in wasteland including road sides, near railway tracks, and at the boundary of farms and in open land. The plant is not preferred as fodder by any cattle species. The growth of plant reduces availability of nutrients from soil such as N, P, and K. Therefore, the plant parts are used as fuel directly and are burnt to provide heat for various purposes. The fast growth and adaptation to various agro-climatic conditions provide this plant an upper edge to survive and thrive under unfavorable conditions such as draught, salinity, and temperature variations at wide range.

P. hysterophorus is a weed, native to the subtropics of North and South America, which belongs to the family Asteraceae (Kriticos et al., 2015). The weed is widely distributed under different agroclimatic zones in several countries including India, Pakistan, Bangladesh, Nepal, Bhutan, Israel, Belgium, Ethiopia, Australia, South Africa, Swaziland, Mozambique, Kenya, Eritrea, Somalia, the islands of Mauritius, Reunion, and Madagascar (Dhileepan and Wilmot Senaratne, 2009; McConnachie et al., 2011; Kriticos et al., 2015). It is a perennial shrub, which grows in wasteland including road sides, near railway tracks, and at the boundary of farms and in open land. It covers about 4.25 million ha of land in India which leads to loss of crops up to 40% (Pandiyan et al., 2014). Dogra et al, 2011 reported that it is an indigenous plant of subtropical America which is quickly spreading in the northwestern Indian Himalayas since last two decades and is severely affecting the growth of native plant species. This fast expansion of *P. hysterophorus* is attributed to production of on an average 40,000 seeds/plant (Dhileepan, 2012). Government of India is investing a huge amount to control this world's worst weed (Tavva et al., 2015). No wild or domestic animal likes to use this plant as a source of its nutrition. The presence of this plant reduces soil fertility due to its allelopathic nature. The pollen grains of this plant act as allergens which causes skin irritation and respiratory disorders. This noxious weed is a great challenge for farmers and foresters. Worldwide distribution of this plant offers a great source of lignocelluloses.

Both above weeds cause lithospheric pollution.

E. crassipes (water hyacinth) is an aquatic free-floating pest. It is native of South America, from where it has extended all over the world by human intrusion (Bolenz et al., 1990). Patel (2012) reviewed spread of this weed and its nuisance at various sites around the globe including Nile, Zambezi, Congo, Niger rivers, Lake Victoria, Winam Gulf of Kenya, Spain, Portugal, forest of Bangladesh, wetlands of Kaziranga, Mexico, China, and San Joaquin River Delta in California. In India, this aquatic weed was first observed in West Bengal in 1890 and is now widespread all over the country except in the arid western region of Rajasthan, in some parts of the north, and Kashmir (Das et al., 2016). Its fast vegetative propagation potential makes it a source of water pollutant.

Viability of the seeds of *E. crassipes* has been reported to be up to 20 years, which facilitates the propagation and hampers the control measures (Patel, 2012). It spreads over the water surface and reduces the dissolved oxygen content of water which leads to death of aquatic fauna (Nigam, 2002). Malik (2007) reviewed the challenges and opportunities of *E. crassipes*. According to him, the profuse growth and coverage of it result in blockage of rivers and canals and finally to flood. It also increases evapotranspiration from water bodies and causes threat to aquatic biodiversity. This remains a challenge throughout the tropics to control this weed from freshwater bodies. However, it has been recognized as a potential source of hemicelluloses with 35–55% of its dry weight (Kumar et al., 2009).

Table 11.1 summarizes the bioethanol production using lignocellulosic biomass derived from above described three weeds.

TABLE 11.1 Bioethanol Production Using Biomass Derived from *Eichhornia crassipes*, *Lantana camara*, and *Parthenium hysterophorus*.

S. no.	Plant name	Biomass	Pretreatment	Percentage conversion	Microbe used	References
1	*Eichhornia crassipes*	Complete plant	Sulfuric acid	73% xylose into ethanol	*Pichia stipitis*	Kumar et al., 2009
2	*E. crassipes*	Complete plant	Sulfuric acid Followed by cellulose	Not mentioned	*Aspergillus fumigatus* ABK9 *Saccharomyces cerevisiae* MTCC 173 *Zymomonas mobilis* MTCC 2428	Das et al., 2016a
3	*E. crassipes*	Complete plant	Cellulase (GH5 and GH43)	46.2%	*S. cerevisiae* + *Candida shehatae*	Das et al., 2016b
4	*E. crassipes*	Leaves	Alkaline/oxidative pretreatment	14%	*S. cerevisiae* NBRC 2346	Mishima et al., 2008
5	*E. crassipes*	Leaves	Alkaline/oxidative pretreatment	17%	*Escherichia coli* KO11	Mishima et al., 2008
6	*E. crassipes*	Complete plant	Sulfuric acid followed by autoclave	1%	*C. shehatae*	Ayudhya et al., 2007
7	*E. crassipes*	Complete plant without roots	Acid–alkali pretreatment	0.21%	*S. cerevisiae* and *Pachysolen tannophilus*	Mukhopadhyay and Chatterjee, 2010
8	*Lantana camara*	Complete plant	Sulfuric acid and high temperature	0.51%	*Pichia stipitis*	Kuhad et al., 2010
9	*L. camara*	Complete plant	Sulfuric acid and high temperature	1.48%	*S. cerevisiae*	Kuhad et al., 2010
10	*L. camara*	Stem	Sulfuric acid and alkali, high temperature, removal of fermentation inhibitors	~0.21%	*Aspergillus niger* *S. cerevisiae VS3*	Pasha et al., 2007

TABLE 11.1 Bioethanol Production Using Biomass Derived from *Eichhornia crassipes*, *Lantana camara,* and *Parthenium hysterophorus.*

S. no.	Plant name	Biomass	Pretreatment	Percentage conversion	Microbe used	References
11	*L. camara*	Complete plant	Laccase followed by simultaneous saccharification	6.01% (v/v)	*S. cerevisiae* CM5	Kuila and Banerjee, 2014
12	*Parthenium hysterophorus*	Complete plant	Sulfuric acid Autoclave and cellulase	0.27%	*Saccharomyces pombe* R3DOM3 and *S. cerevisiae* R3DIM4	Tavva et al., 2016
13	*P. hysterophorus*	Complete plant	Sulfuric acid and autoclave	0.149%	*Saccharomyces pombe* MTCC 170	Singh et al., 2015
14	*P. hysterophorus*	Complete plant	Sulfuric acid Autoclave and ultrasound waves	0.166%	*S. cerevisiae* MTCC 170	Singh et al., 2015
15	*P. hysterophorus*	Complete plant	Alkali treatment and β-glucosidase supplementation	Not available	*S. cerevisiae*	Mahajan et al., 2014
16	*P. hysterophorus*	Complete plant	sulfuric acid Autoclave, detoxification, and β-glucosidase	25.7%	*S. cerevisiae* MTCC 170	Bharadwaja et al., 2015

11.2 STEPS INVOLVED IN PRODUCING BIOETHANOL FROM WEED BIOMASS

1. Collection of weed biomass
2. Sun drying the biomass
3. To increase the surface area, convert biomass in powder by grinding mills
4. Estimate simple sugars
5. Convert the biomass in simple sugars by chemical and/or physical and/or biological treatments
6. Detoxicate this material (as above treatment may lead to production of substances which may inhibit the fermentation process by either inhibiting growth of organism responsible for fermentation and/or reduces efficiency of the enzymes)
7. Estimate again the simple sugars present in feedstock after saccharification (simple sugars will include 5C and 6C sugars)
8. Ferment the simple sugars (product of saccharification) using suitable microbe
9. Separation of bioethanol
10. Purification of bioethanol
11. Estimate the bioethanol produced
12. Find out the economic viability of the bioethanol.

11.3 ADVANTAGES OF USE OF WEED BIOMASS IN BIOETHANOL PRODUCTION

- Availability of low-cost lignocellulosic biomass provides large scale feedstock to produce bioethanol. Ethanol is the most suitable alcohol as fuel for spark-ignition engines (Najafi et al., 2009).
- The physicochemical properties of ethanol make it more suitable for enhancement of performance of engine.
- Ethanol has high calorific value ranging from 21–22 MJ/L.
- Ethanol has high octane number (96–113). Mixing bioethanol with gasoline enhances the octane number of the fuel and reduces the requirement of toxic additives used to increase the octane number (Galbe and Zacchi, 2002).
- The blending of ethanol provides additional oxygen which facilitates the process of combustion and thereby it reduces emission of

CO and hydrocarbons (Najafi et al., 2009). It also reduces emission of particulate matter.

- According to Bailey (1996), ethanol has same compression engine efficiency as that of diesel and it has got better efficiency than gasoline.
- According to Al-Hasan (2002), blending of gasoline with ethanol increased the brake power, torque, volumetric and brake thermal efficiencies, and fuel consumption.
- Blend of ethanol with gasoline enhanced knocking resistance (Yacoub et al., 1998).
- Low cetane number and high heat of vaporization of ethanol facilitates fuel ignition process of the engine (Balat et al., 2008).
- It reduces the environmental pollution due to combustion of solid agro waste.
- As a by-product of ethanol production CO_2 is obtained, this can provide extra business to the industry involved in bioethanol production (Xiao-Yang, 2016).
- It will reduce soil deterioration, avoid allergens, and improve crop yield due to reduced competition with weed.
- Use of waste to get eco-friendly product.
- It will provide economic benefits to farmers/end users.
- Storage and dispensing is similar to petrol.
- Intermolecular H-bonds in ethanol facilitate storage of the liquid fuel at room temperature.

11.4 LIMITATIONS OF ETHANOL AS FUEL

Balat et al. (2008) have summarized the disadvantages of bioethanol as a fuel which are as follows:

- Energy density of ethanol is less than gasoline so more volume of ethanol will be required than gasoline in order to get the same amount of energy output.
- Ethanol is corrosive in nature as it can absorb water and dissolve acids and ionic species (Eppel et al., 2008; Matějovský et al., 2017).
- The Reid vapor pressure of ethanol is (17 kPa) approximately one-third of gasoline, therefore blending ethanol at more than 10% can impede cold start (Chen et al., 2011).

- The process of producing bioethanol from lignocellulosic raw material is still an energy consuming and difficult process. Very high conversion ratio of this material to bioethanol has yet to be achieved.
- Bioethanol can be produced from cellulosic feedstocks. One major problem with bioethanol production is the availability of raw materials for the production. The availability of feedstocks for bioethanol can vary considerably from season to season and depends on geographic locations.
- Lignocellulosic biomass is the most promising feedstock considering its great availability and low cost.
- Conversion technologies for producing higher yield of bioethanol are yet to be developed so that bioethanol can be produced in a cost-effective and eco-friendly manner at industrial scale for commercial production.

11.5 CONCLUSION

Weeds cover a huge land area and grow with faster pace and provide large scale biomass. This biomass, a source of lignocellulosic material, is available at low cost. The lignocellulosic material can be hydrolyzed to fermentable sugars by various processes including physical, chemical, and biological interventions. Fermentation converts sugars into ethanol. This bioethanol can serve as a renewable fuel. It can also be blended with fossil fuel. The demand of bioethanol is increasing and will increase in future also. People are working on the use of weed-derived biomass to convert it into bioethanol. In this race of use of biomass, *L. camara*, *P. hysterophorus*, and *E. crassipes* have remained among the most preferred biomass due to their high yield and easy availability.

KEYWORDS

- **weed**
- **biomass**
- **lignocellulosic material**
- **physical and chemical treatments**
- **saccharification**
- **fermentation**
- **ethanol production**

REFERENCES

Al-Hasan, M. Effect of Ethanol–Unleaded Gasoline Blends on Engine Performance and Exhaust Emission. *Energ. Convers. Manag.* **2002,** *44*, 1547–1561.

Bailey, B. K. Performance of Ethanol as a Transportation Fuel. In *Handbook on Bioethanol: Production and Utilization*; Taylor & Francis: Washington, DC, **1996,** pp 37–60.

Balat, M.; Balat, H.; Öz, C. Progress in Bio-ethanol Processing. *Prog. Energy Combust. Sci.* **2008,** *34*, 551–573.

Bharadwaja, S. T.; Singh, S.; Moholkar, V. S. Design and Optimization of a Sono-hybrid Process for Bio-ethanol Production from *Parthenium hysterophorus*. *J. Taiwan. Inst. Chem. E* **2015,** *51*, 71–78

Bolenz, S.; Omran, H.; Gierschner, K. Treatments of Water Hyacinth Tissue to Obtain Useful Products. *Biol. Wastes* **1990,** *33*, 263–274.

Chandel, A. K.; Singh, O. V. Weedy Lignocellulosic Feedstock and Microbial Metabolic Engineering: Advancing the Generation of 'Biofuel.' *Appl. Microbiol. Biotechnol.* **2011,** *89*, 1289–1303

Chen, R. H.; Chiang, L. B.; Chen, C. N.; Lin, T. H. Cold-start Emissions of an SI Engine Using Ethanol–Gasoline Blended Fuel. *Appl. Therm. Eng.* **2011,** *31*, 1463–1467.

Cilliers, C. J.; Neser, S. Biological Control of *Lantana camara* (Verbenaceae) in South Africa. *Agric. Ecosys. Environ.* **1991,** *37*, 57–75.

Das, A.; Ghosh, P.; Paul, T.; Ghosh, U.; Pati, B. R.; Mondal, K. C. Production of Bio-ethanol as Useful Biofuel Through the Bioconversion of Water Hyacinth (*Eichhornia crassipes*). *3 Biotech.* **2016a,** *6*, 70.

Das, S. P; Gupta, A; Das, D; Goyal, A. Enhanced Bioethanol Production from Water Hyacinth (*Eichhornia crassipes*) By Statistical Optimization of Fermentation Process Parameters Using Taguchi Orthogonal Array Design. *Int. Biodeterior. Biodegrad* **2016b,** *109*, 174–184.

Demirbas, A. Producing and Using Bio-ethanol as an Automotive Fuel. *Energ. Source.* **2007,** *2*, 391–401.

Dhileepan, K. Reproductive Variation in Naturally Occurring Populations of the Weed *Parthenium hysterophorus* (Asteraceae) in Australia. *Weed Sci.* **2012,** *60*, 571–576.

Dhileepan, K.; Wilmot Senaratne, K. A. How Widespread is *Parthenium hysterophorus* and Its Biological Control Agent *Zygogramma bicolorata* in South Asia? *Weed Res.* **2009,** *49*, 557–562.

Dogra, K. S.; Sood, S. K.; Sharma, R. Distribution, Biology and Ecology of *Parthenium hysterophorus* l. (Congress Grass) an Invasive Species in the North-Western Indian Himalaya (Himachal Pradesh). *Afr. J. Plant. Sci.* **2011,** *5*, 682–687.

Elder, M.; Prabhakar, S. V. R. K.; Romero, J.; Matsumoto, N. Prospects and Challenges of Biofuels in Asia: Policy Implications. In *Climate Change Policies in the Asia-Pacific: Re-uniting Climate Change and Sustainable Development*; Hamanaka, H., Morishima, A., Mori, H., King, P., Eds.; Institute for Global Environmental Strategies: Hayama, Japan, **2008;** pp 105–131.

Eppel, K.; Scholz, M.; Troßmann, T.; Berger, C. Corrosion of Metals for Automotive Applications in Ethanol Blended Biofuels. *Energy Mater.* **2008,** *3*, 227–231.

Galbe, M.; Zacchi, G. A Review of the Production of Ethanol from Softwood. *Appl. Microbiol. Biotechnol.* **2002,** *59,* 618–628.

Isarankura-Na-Ayudhya, C.; Tantimongcolwat, T.; Kongpanpee, T.; Prabkate, P.; Prachayasittikul, V. Appropriate Technology for the Bioconversion of Water Hyacinth (*Eichhornia crassipes*) to Liquid Ethanol. *EXCLI. J.* **2007,** *6,* 167–176.

Kang, Q.; Appels, L.; Tan, T.; Dewil, R. Bio-ethanol from Lignocellulosic Biomass: Current Findings Determine Research Priorities. *Sci. World J.* **2014,** *2014.*

Khatiwada, D.; Silveira, S. Scenarios for Bio-ethanol Production in Indonesia: How can We Meet Mandatory Blending Targets? *Energy* **2017,** *119,* 351–361.

Kriticos, D. J.; Brunel, S.; Ota, N.; Fried, G.; Oude Lansink, A. G. J. M.; Panetta, F. D.; et al. Downscaling Pest Risk Analyses: Identifying Current and Future Potentially Suitable Habitats for *Parthenium hysterophorus* with Particular Reference to Europe and North Africa. *PLoS One* 2015, *10* (9), e0132807. DOI: 10.1371/journal.pone.0132807.

Kuhad, R. C.; Gupta, R.; Khasa, Y. P.; Singh, A. Bioethanol Production from *Lantana camara* (Red Sage): Pretreatment, Saccharification and Fermentation. *Bioresour. Technol.* **2010,** *101,* 8348–8354.

Kuila, A.; Banerjee, R. Simultaneous Saccharification and Fermentation of Enzyme Pretreated *Lantana camara* Using *S. cerevisiae*. *Bioprocess Biosyst. Eng.* **2014,** *37,* 1963–1969.

Kuila, A.; Mukhopadhyay, M.; Tuli, D. K.; Banerjee, R. Production of Ethanol from Lignocellulosics: An Enzymatic Venture. *EXCLI J.* **2011,** *10,* 85.

Kumar, A.; Singh, L. K.; Ghosh, S. Bioconversion of Lignocellulosic Fraction of Water-hyacinth (*Eichhornia crassipes*) Hemicellulose Acid Hydrolysate to Ethanol by *Pichia stipitis*. *Bioresour. Technol.* **2009,** *100,* 3293–3297.

Mahajan, C.; Chadha, B. S.; Nain, L.; Kaur, A. Evaluation of Glycosyl Hydrolases from Thermophilic Fungi for Their Potential in Bioconversion of Alkali and Biologically Treated *Parthenium hysterophorus* Weed and Rice Straw into Ethanol. *Biores. Tech.* **2014,** *163,* 300–307.

Malik, A. Environmental Challenge Vis a Vis Opportunity: The Case of Water Hyacinth. *Environ. Int.* **2007,** *33,* 122–138.

Martins, A. L.; Wanke, P.; Chen, Z.; Zhang, N. Ethanol Production in Brazil: An Assessment of Main Drivers with MCMC Generalized Linear Mixed Models. *Resour. Conserv. Recycl.* **2018,** *132,* 16–27.

Matějovský, L.; Macák, J.; Pospíšil, M.; Baroš, P.; Staš, M.; Krausová A. Study of Corrosion of Metallic Materials in Ethanol–Gasoline Blends: Application of Electrochemical Methods. *Energy Fuels* **2017,** *31,* 10880–10889.

McConnachie, A. J.; Strathie, L. W.; Mersie, W.; Gebrehiwot, L.; Zewdie, K.; Abdurehim, A.; Abrha, B.; Araya.; Asaregew, F.; Assefa, F.; Gebre-Tsadik, R. Current and Potential Geographical Distribution of the Invasive Plant *Parthenium hysterophorus* (Asteraceae) in Eastern and Southern Africa. *Weed Res.* **2011,** *51,* 71–84.

Mishima, D.; Kuniki, M.; Sei, K.; Soda, S.; Ike, M.; Fujita, M. Ethanol Production from Candidate Energy Crops: Water Hyacinth (*Eichhornia crassipes*) and Water Lettuce (*Pistia stratiotes* L.). *Bioresour. Technol.* **2008,** *99,* 2495–2500.

Mukhopadhyay, S. B.; Chatterjee, N. C. Bioconversion of Water Hyacinth Hydrolysate into Ethanol. *Bioresources* **2010,** *5,* 1301–1310.

Najafi, G.; Ghobadian, B.; Tavakoli, T.; Buttsworth, D. R.; Yusaf, T. F.; Faizollahnejad, M. Performance and Exhaust emissions of a Gasoline Engine with Ethanol Blended Gasoline Fuels Using Artificial Neural Network. *Appl. Energy* **2009,** *86*, 630–639.

Nigam, J. N. Bioconversion of Water-hyacinth (*Eichhornia crassipes*) Hemicelluloses Acid Hydrolysate to Motor Fuel Ethanol by Xylose-fermenting Yeast. *J. Biotechnol.* **2002,** *97*, 107–116.

Pandiyan, K.; Tiwari, R.; Rana, S.; Arora, A.; Singh, S.; Saxena, A. K.; Nain, L. Comparative Efficiency of Different Pretreatment Methods on Enzymatic Digestibility of *Parthenium* sp. *World J. Microbiol. Biotechnol.* **2016,** *30*, 55–64.

Pasha, C.; Nagavalli, M.; Venkateswar Rao, L. *Lantana camara* for Fuel Ethanol Production Using Thermotolerant Yeast. *Lett. Appl. Microbiol.* **2007,** *44*, 666–672

Patel, S. Threats, Management and Envisaged Utilizations of Aquatic Weed *Eichhornia crassipes*: An Overview. *Rev. Environ. Sci. Bio.* **2012,** *11*, 249–259.

REN21. *Renewables 2017*, Global Status Report; REN21 Secretariat: Paris, 2017. ISBN 978-3-9818107-6-9.

Sharma, O. P.; Makkar, H. P. S.; Dawra, R. K.; Negi, S. S. A Review of the Toxicity of *Lantana camara* (Linn.) in Animals. *Clin. Toxicol.* **1981,** *18*, 1077–1094.

Singh, S.; Sarma, S.; Agarwal, M.; Goyal, A.; Moholkar, V. S. Ultrasound Enhanced Ethanol Production from *Parthenium hysterophorus*: A Mechanistic Investigation. *Biores. Tech.* **2015,** *188*, 287–294.

Songstad, D. D.; Lakshmanan, P.; Chen, J.; Gibbons, W.; Hughes, S.; Nelson, R. Historical Perspective of Biofuels: Learning from the Past to Rediscover the Future. *In Vitro Cell Dev. Biol. Plant* **2009,** *45*, 189–192.

Swain, K. C. Biofuel Production in India: Potential, Prospectus and Technology. *J. Fundam. Renew. Energy Appl.* 2014, *4*, 129. DOI: 10.4172/2090-4541.1000129.

Tavva, S. M. D.; Deshpande, A.; Durbha, S. R.; Palakollu, V. A. R.; Goparaju, A. U.; Yechuri, V. R.; Bandaru, V. R.; Muktinutalapati, V. S. R. Bioethanol Production Through Separate Hydrolysis and Fermentation of *Parthenium hysterophorus* Biomass. *Renew. Energy* **2016,** *86*, 1317–1323.

Taylor, S.; Kumar, L.; Reid, N.; Kriticos, D. J. Climate Change and the Potential Distribution of an Invasive Shrub, *Lantana camara* L. *PLoS One* 2012, *7*, e35565. DOI: 10.1371/journal.pone.0035565.

Wolfson, S. L.; Solomons, T. W. G. Poisoning by Fruit of *Lantana camara*. An Acute Syndrome Observed in Children Following Ingestion of the Green Fruit. *Am. J. Dis. Child.* **1964,** *107*, 173–176.

Yacoub, Y.; Bata, R.; Gautam, M. The Performance and Emission Characteristics of C1–C5 Alcohol–Gasoline Blends with Matched Oxygen Content in a Single-cylinder Spark Ignition Engine. *Proc. Inst. Mech. Eng. A* **1998,** *212*, 363–379.

Zhang, X. Y. Developing Bioenergy to Tackle Climate Change: Bioenergy Path and Practice of Tianguan Group. *Adv. Clim. Chan. Res.* **2016,** *7*, 17–25.

CHAPTER 12

Implication of Anaerobic Digestion for Large-Scale Implementations

SAMUEL JACOB[1,*] AND RINTU BANERJEE[2]

[1]*Department of Biotechnology, School of Bioengineering, SRM Institute of Science and Technology, Kattankulathur 603203, Tamil Nadu, India*

[2]*Department of Agricultural and Food Engineering, Indian Institute of Technology, Kharagpur 721302, West Bengal, India*

[*]*Corresponding author. E-mail: samueljacob.b@ktr.srmuniv.ac.in*

ABSTRACT

In a world of diminishing resources and increasing needs, each opportunity for the sustainable reuse of waste materials must be examined. Organic refuse from household, agriculture, and agro-industrial processing can help to fulfill the requirements for fuel and fertilizer. Biomethane production from organic wastes through anaerobic digestion is perceived as one of the viable options for the safe disposal of wastes without hindering the natural ecosystem. Despite its numerous advantages, anaerobic digestion at times tends to be ineffective due to certain bottlenecks associated with the inoculum and process instability owing to inhibitory products. This chapter envisages different parametric controls of anaerobic digestion that impede adoption of this technology in large scale and the mitigation strategies to overcome.

12.1 INTRODUCTION

Anaerobic digestion (AD) is a chain of natural biological process that takes place in the absence of oxygen where the organic materials are broken down to produce mainly methane and carbon dioxide. This is achieved due to the interaction of a wide range of microorganisms (hydrolysers, acidogens, acetogens, and methanogens) with the substrate fed in the reactor. It involves four stages, that is, hydrolysis, acidogenesis, acetogenesis, and methanogenesis (Fig. 12.1).

a) Hydrolysis is a primary stage where the complex macromolecules mainly carbohydrates, proteins, and lipids are broken down into sugars, amino acids, and fatty acids, respectively, with the help of hydrolytic bacteria.
b) Acidogenesis is performed by acidogenic bacteria. Here, sugars, amino acids, and fatty acids produced in the previous step act as substrates which are converted to CO_2, ammonia, organic acids, and hydrogen.

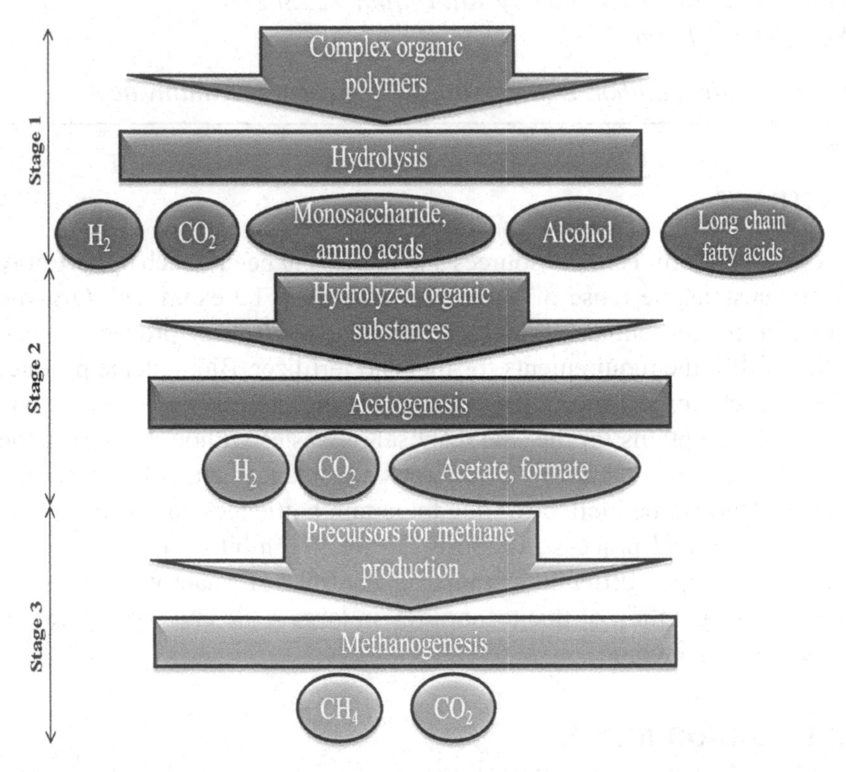

FIGURE 12.1 Schematic illustrations of the anaerobic digestion process for biogas production.

c) Acetogenesis is the process of generation of acetic acid along with other components such as CO_2 and hydrogen using the previously produced organic acids by acetogenic bacteria.
d) Methanogenesis is the final stage which includes two different pathways, that is, acetoclastic methanogenesis and hydrogenotrophic methanogenesis. The end products for both are methane and CO_2 using acetic acids for the former and CO_2 and hydrogen for the latter.

12.2 BIOMETHANE

Biogas is principally a mixture of methane (CH_4), carbon dioxide (CO_2), and minute traces of hydrogen sulfide (H_2S), nitrogen (N_2), ammonia (NH_3), and moisture (Fig. 12.2). Methane is the only constituent of biogas with significant fuel value. The energy released during combustion of methane allows biogas to be used as a fuel for any heating purpose such as cooking and electricity generation. Biomethane is one of the most

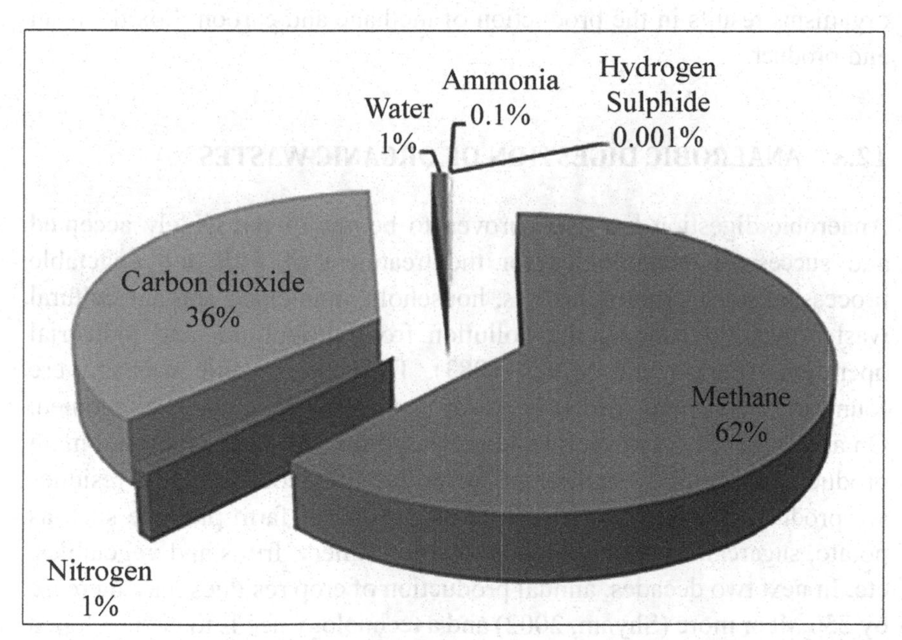

FIGURE 12.2 A typical biogas composition (values in approximate figures).

efficient and effective options among the various other alternative sources of renewable energy currently available.

Biogas is conventionally produced through anaerobic digestion processes where the microorganisms convert complex organic matter into methane and carbon dioxide. Biodegradable organic wastes such as animal manure, municipal waste, and plant biomass such as crops, fruit, and vegetables wastes have been traditionally used as the feedstock for biogas production (Federal Energy Management Program, 2004; Li et al., 2011). It is a sustainable solution to treat waste as the cost of the waste treatment is low with additional energy output (Verstraete et al., 2005). In addition, the effluent or residual slurry from the biogas production process supplies essential nutrients which can also be utilized as fertilizer (Vasudeo, 2005). The anaerobic digestion of biomass for biomethane generation requires less capital investment per unit production cost compared to other renewable energy sources, such as hydro-, solar, and wind energy (Rao et al., 2010). Also, there is limited competition with food by using agro and industrial wastewater and residues to produce biogas.

Anaerobic digestion of wastes is a complex multistep process where a concerted effect of aerobic, facultative, and obligate anaerobic microorganisms results in the production of methane and carbon dioxide as an end product.

12.3 ANAEROBIC DIGESTION OF ORGANIC WASTES

Anaerobic digestion has been proven to be one of the widely accepted and successful technologies for the treatment of fruit and vegetable processing waste from industries, household, municipal, and agricultural wastes thereby reducing the pollution from agricultural and industrial operations (Parkin and Miller, 1983). Thus, the organic wastes were found to be potential feedstock as far as the energy sector is concerned. On an average, 1.5 t of crop residue is used for 1 t of the processed main product. In addition, significant quantities of secondary crop residues are produced during agro-industrial processing of farm produce such as potato, sugarcane, coconut, paddy, dairy, distillery, fruits and vegetables, etc. In next two decades, annual production of crop residues may increase by 250 Mt or more (Shyam, 2002) and a technology needs to be developed to handle such quantum of wastes in a sustainable manner. In spite of the beneficial aspects of anaerobic digestion, process instabilities are often encountered which leads to low methane yield and thus averting it from being widely applied. The full-fledged utilization of this technology in the agro-industrial sector is impeded due to various factors. Some of these limitations that are seemingly inevitable include:

- Organic compounds present in the waste undergo only partial degradation.
- Rate of the reaction is slow and requires digester of high capacity which incurs more cost.
- Inhibitors released during the process are more vulnerable.
- Presence of noncombustible components such as CO_2, hydrogen sulfide, and moisture in biogas makes it less economical to be used as fuel gas.
- Presence of hazardous undegraded volatile and nonvolatile matter (heavy metals) in the resulting sludge causes secondary pollution.

In order to strike an ecological and economical balance, formidable technological and dissemination of challenges need to be addressed (Venkataraman et al., 2010; Smith, 2010). In India, it has been reported that the percentage of functional plants ranges between 40% and 81% (Dutta et al., 1997; Bhat et al., 2001). The nonfunctional plants were due to failures caused by design, feedstock, inoculum, and maintenance problems. Thus, with biogas being recognized as a potential energy source, currently technologies are developed to alleviate the challenges, upgrade performance quality, and to improve energy usage.

12.4 BARRIERS IN RURAL FAMILY-SIZE BIOGAS PLANTS

The statistics show that at present in India the total biogas yield is about 2.07 billion m^3/year. Compared to the estimated range of biogas production, that is, 29–48 billion m^3/year, the current production of biogas is low. The benefits of biogas production are highly related to the environment and health, but there are various obstacles for the utilization of biogas technologies which has to be overcome. In rural areas, there exist many family-size biogas plants with capacities ranging from 1 to 10 m^3 biogas per day. The primarily used feed stocks in rural areas include animal manure and agricultural wastes which will lead to the production of biogas and slurry. These biogas plants are mainly used to produce energy that can be used for self-consumption in each household (Mittal et al., 2018). Some of the barriers observed for the adoption of biogas plant in rural areas include:

- High investment cost: This comprises the capital cost required for the installation of biogas plant and also the transaction cost. A high amount of transaction cost is required because of the procedural delay in getting the subsidies and also the amount paid to the bureaucracy (Rao and Ravindranath, 2002).
- Technical knowledge gap: Another important limitation observed is the unawareness about the technical aspect of biogas reactor. There is a lack of skilled technical laborer in rural areas for the services in case if the bioreactor stopped working. Another problem seen is on the amount of feed (substrate) and inoculum fed to the reactor (Bansal et al., 2013). There is a possibility of shortage of the resources to be fed into the reactor. Also, many do not have the ownership of cattle so as to put the cattle dung as an inoculum or substrate to the reactor (Mittal et al., 2018).
- Market barriers: In rural areas, biogas mainly strives with cheaper alternatives such as solid biomass, firewood, and cow dung which are generally used for the cooking purposes and which can be obtained from the local area very quickly.

But people are less aware about the negative impact caused by this traditional biomass such as the interior air pollution, damage to forest resources, and also time consumed in firewood collection (Rao and Ravindranath, 2002).

12.5 PROCESS PARAMETERS THAT INFLUENCE THE EFFICIENCY OF ANAEROBIC DIGESTION

12.5.1 MIXING IN ANAEROBIC DIGESTER

Biogas yield is affected by several factors such as total solids (TS), inoculum, stirring effect, temperature, and volatile solids (VS). Amongst these TS and agitation time are the crucial factors that affect the biogas production. Agitation helps in making the reactor stable.

Numerous studies specified that improper mixing resulted in floating layers of solids when the reactor was fed with municipal waste (Stenstrom et al., 1983). To achieve efficient substrate conversion many researchers have worked on the importance of mixing. By mixing the digester properly the organisms are distributed uniformly throughout and also it transfers heat to different portions of the reactor. This leads to the collection of gas from the reactor. Several methods can be used for mixing purposes which include mechanical mixers, recirculating either the produced biogas using pumps or the digester materials. The two important factors of mixing involve the time of mixing and its intensity (Karim et al., 2005). The literature on anaerobic digestion highlights the significance of proper mixing so as to improve uniform distribution of the microorganisms, substrates, and enzymes within the reactor (Parkin and Owen, 1986; Lema et al., 1991).

12.5.2 ORGANIC LOADING RATE

One of the important parameters of the anaerobic digestion process is the organic loading rate (OLR). So for the system stability higher OLRs are suggested which could provide high biogas yield and also high waste treatment capacity (Liu et al., 2012). An advantage of anaerobic digestion process over other biological operations is operation under high OLR and low production of sludge (Bouallagui et al., 2004). OLRs in the range of 1–4 g of VS/L/day were used generally in conventional anaerobic digester for treatment of biological wastes (Zhang et al., 2007). Higher range of OLR helps in enhancing the growth of bacterial community; requires less energy for heating, and it moderates the reactor requirements (Nagao et al., 2012). Sometimes, providing a high OLR range can ultimately lead to the failure of the reactor. This happens due to the accumulation of ethanol and volatile fatty acid (VFA), less transfer of heat, and uneven distribution of materials during mixing. The substrates and the operational conditions play an important role in finding out the optimum OLR (Liu et al., 2007).

12.5.3 MICROBIAL DIVERSITY

The microbial communities present in the anaerobic digestion process are classified mainly into three groups, that is, primary fermenting bacteria, anaerobic oxidizing bacteria, and methanogenic archaea (Angelidaki et al., 2009). As the first step of AD process, primary fermenting bacteria hydrolyze polymers to monomers. This step is characterized by the act of some hydrolytic enzymes which are secreted by this group of organisms.

These monomers are further reduced to organic acids, fatty acids, alcohols, hydrogen, and CO_2. The products formed are further reduced to acetate, hydrogen, formate, and carbon dioxide by the oxidizing bacteria (Angelidaki et al., 2009). Now the degradation of acetate to methane is done by acetoclastic methanogens and hydrogenotrophic methanogens convert hydrogen and carbon dioxide to methane (Francisci et al., 2015). For a good yield of methane gas the balance within the microbial community is very essential, where this is directly related to the overall constancy of the process.

The laboratory-scale bioreactor has shown that Bacteroidetes and Firmicutes phyla are the dominant bacterial groups (Bengelsdorf et al., 2013). The prevalence of the Firmicutes phylum was explained by its capability of producing many enzymes which performs hydrolysis, acidogenesis and acetogenesis. In case of methanogenic archaea, the acetoclastic *Methanosaeta* genus was prevalent at the low level of acetate in the digester. Later on, two hydrogenotrophic species were also observed, *Methanospirillum hungatei* and *Methanoculleus receptaculi* (Lerm et al., 2012). The microbial communities present in the mesophilic reactors include Firmicutes, Chloroflexi, Bacteroidetes and Actinobacteria. Similarly in the thermophilic reactors, the microbial communities involved are Thermotogae, Firmicutes, Synergistetes, and Bacteroidetes (Zamanzadeh et al., 2017).

12.5.4 VOLATILE FATTY ACIDS

VFAs are one of the main intermediate products for anaerobic digestion process. Some examples of VFA include acetic acid, butyric acid, propionic acid, and valeric acid. The VFAs formed in the process are transformed to methane and carbon dioxide by acetogens and methanogenic bacteria. On the other hand, VFAs can lead to the failure of anaerobic digestion due to their accumulation at high OLR. Among these VFAs, propionic acid and acetic acid play the important role in the production of biogas and the concentration of the same is used as the indicator for the AD performance (Buyukkamaci et al., 2004). VFAs help in determining the pH which acts as one of the important factors affecting AD (Zhang et al., 2015).

12.5.5 pH

pH value is a function of bicarbonate and VFA concentration, alkalinity, and also the CO_2 level in the biogas. It is important to adjust the relationship between the VFA and bicarbonate concentrations for attaining constant pH (Liu et al., 2007). pH of the substrate and medium is an important factor in the anaerobic digestion of the waste as the growth and proliferation of methanogenic bacteria is inhibited under acidic conditions leading to cessation of biogas production. Factors such as retention time and loading rate in batch reactors determine the pH of the reaction medium. Steps involved in anaerobic digestion process have different optimal pH; fluctuations may occur due to the biological transformations and release of toxic intermediates higher than their critical limit. During the acido/acetogenesis step, pH would fall below 5, which is detrimental for methanogens. This would lead to acid accumulation and digester failure. On the other hand, an excessive rise in population of methanogens can lead to rapid utilization of VFAs and causes accumulation of ammonia, thereby increasing the pH above 8 which is inhibitory for acidogenesis (Lusk, 1999). Optimal pH range 6.5–7.5 is required for attaining maximum biogas yield in anaerobic digestion which then depends on type of substrate used and digestion technique (batch or continuous) (Liu et al., 2007). During start-up phase maintenance of constant pH is essential as freshly added substrate has to undergo the stage of hydrolysis and acidogenesis, which eventually reduces the pH. In order to maintain equilibrium pH above the acidic level, neutralizing agents such as calcium carbonate or lime has to be added into the system which might incur additional process cost. Modulation of reaction pH with the help of additional substrates which provide buffering effect could be an economically viable option.

12.5.6 INHIBITORY ACTION OF VFA

VFAs are the rate-limiting intermediate in the anaerobic digestion process. When the concentration of VFA is at 6.7–9.0 mol/m^3, it can inhibit the growth of microorganisms, especially methanogens (Batstone et al.,

2000). Increased concentrations of toxic compounds, variation in temperature, and organic overloading may cause accumulation of VFA. Under such circumstances, the methanogens were unable to utilize the VFA and hydrogen rapidly. This leads to acid accumulation followed by decrease in the pH causing the inhibition of hydrolysis/acetogenesis. Undissociated VFAs can be toxic to microorganisms in which they were transported through the cell membrane and dissociate. This causes a reduction in pH followed by disruption of homeostasis.

In a batch anaerobic reactor, presence of increasing concentration of VFA has differential effect on the hydrolytic, acidogenic, and methanogenic phases. Increased concentrations of acetate and propionate cause substrate inhibition to methanogenesis. Concomitantly, acetate leads to noncompetitive inhibition of propionate degradation and uncompetitively inhibits degradation of benzoate. This leads to enhancement of inhibitory effect of pH on methanogenesis.

A cost-effective approach to deal with the pH and VFA accumulation (souring) problem is adoption of codigestion process where the intermediary degradatory products such as acids and ammonia produced from the substrate degradation provide buffering effect which then stabilizes the pH.

12.5.7 VARIATION IN CARBON TO NITROGEN RATIO

During anaerobic digestion, the carbon to nitrogen ratio (C:N) of the substrate added to a reactor has an important role on the growth rate of the microbial population and biogas production. High C:N reduces the efficiency of the process due to limited availability of nitrogen and leads to accumulation of VFA. Whereas, low C:N results in excess formation of ammonia which is toxic to bacterial population. Microorganisms require a suitable C:N of 20:1–30:1 for their metabolic processes (Bhattacharyya and Banerjee, 2007). In thermophilic anaerobic codigestion of poultry, cow, and mixtures of manures with paper or cellulosic materials, it has been seen that C:N higher than 23:1 were found to be unsuitable for optimal digestion, and ratios lower than 10:1 were inhibitory (Kimchie, 1984). Dairy and swine manure typically contain C:N of approximately 9:1 and 6:1, respectively. The addition of codigestion materials with higher carbon contents to manure feedstock can improve the C:N, thereby increasing methane production. For example, the C:N of the dairy and swine manures may be enhanced by adding food processing residues such as potato waste with a C:N of 28:1 or crop residues, such as oat straw with C:N of 48:1.

12.6 MITIGATION STRATEGY FOR LIMITATIONS IN ANAEROBIC DIGESTION

12.6.1 ANAEROBIC CODIGESTION OF AGROWASTES FOR BIOMETHANE PRODUCTION

During codigestion process, two or more wastes were mixed together to form a homogenous mixture and fed into a reactor for their simultaneous treatment. In recent times, anaerobic codigestion became more prominent area of research since digestion of more than two wastes provides more stability as compared to monosubstrate digestion (Lansing et al., 2010). Moreover, codigestion of waste provides the missing nutrient in the substrate thereby it stabilizes the digestion process. In addition to it, the quantum of one type of organic waste generated at a particular site during certain time may not be adequate to make anaerobic digestion cost effective throughout the year.

Anaerobic codigestion between sewage sludge and organic fraction of municipal solid waste is the most reported codigestion research. However, between 2010 and 2013 several combinations of substrate mixture starting from animal manure to agro-industrial processing wastes have been widely reported (Mata-Alvarez et al., 2014).

Most industrial codigestion plants treat municipal solid waste along with sewage. Lehtomäki et al. (2007) reported that specific methane production potential of cattle manure was found to improve when codigested with other substrates such as sewage sludge, fruit and vegetable waste (FVW), energy crops, and municipal solid waste.

FVWs have methane production potential of about 0.37 m^3/kg VS when used as a digester feedstock. Whereas, under continuous codigestion with cattle slurry and chicken manure, there was enhancement of methane yield from 0.23 to 0.45 $m^3/kg\ VS_{added}$ when the proportion of FVW was increased from 20% to 50% with a retention time of 21 days (Callaghan et al., 2002). Usually, FVW contains low quantity of nitrogen (~0.8–1.2%). This waste can be codigested with other nitrogen-rich wastes in order to balance the C:N for efficient digestion. The aim of this work is to explore some new combination of substrates as cosubstrate for effective utilization of vegetable wastes. Table 12.1 represents some of the successful codigestion studies reported in the literature.

12.6.2 KINETIC MODELING OF ANAEROBIC DIGESTION PROCESS

TABLE 12.1 Studies on Anaerobic Codigestion of Different Agrowastes.

Substrate combinations	Inoculum used	Reactor type	Biogas/methane yield	References
Cow manure + grass			268 L/kg VS	
Cow manure + sugar beet top	Digested sludge	CSTR	229 L/kg VS	Lehtomaki et al. (2007)
Cow manure + straw			213 L/kg VS	
VPW + swine manure	Anaerobic sludge	CSTR	227 L/kg VS	Molinuevo-Salces et al. (2012)
		Batch	257–286 L/kg VS	
VPW + poultry litter		Batch	426 L/kg VS	Molinuevo-Salces et al. (2012)
Potato tuber + industrial by-products + swine manure	Municipal sewage sludge	CSTR	280–300 L/kg VS	Kaparaju and Rintala (2005)
FVW + chicken manure + cattle slurry	NA	CSTR	230–400 L/kg VS	Callaghan et al. (2002)
Potato waste + sugar beet leaves	Anaerobic digested sewage sludge	Batch	320 L/kg VS	Parawira et al. (2004)
Potato processing waste (pulp, peel, and fluid waste) + by-products of sugar beet leaves (pulp and tail silage)	Active sludge from biogas plants	CSTR	320–330 L/kg VS	Kryvoruchko et al. (2009)
Grass + FVW + cow manure	Sludge from biogas plant	CSTR	260–300 L/kg VS	Ganesh et al., (2013)
Crop silage + cow manure	Bacterial inoculum from anaerobic digester	CSTR	169–237 L/kg VS	Comino et al., (2010)
Cattle manure + maize silage + triticale + potato and onion	Sludge	CSTR	470–480 L/kg VS	Giuliano et al., (2013)

CSTR, continuous stirred tank reactor; FVW, fruit and vegetable waste; NA, not available; VPW, vegetable processing waste.

Anaerobic codigestion of organic wastes from different origin need precise management since random or heuristic decisions on the proportion of waste or feedstock added to large-scale plants may lead to process instabilities and significant reduction in methane yield (Zaher et al., 2009). Consequently, there is a requirement for accurate modeling of the anaerobic degradation of waste. The advantage of employing mathematical models lies in their ability to reproduce dynamic process behavior on a computer in a precise and quantifiable manner. The mathematical equations of the model are used for simulating the physical, chemical, and biological processes (Esposito et al., 2008; Galí et al., 2009).

The first codigestion modeling study was performed by Bozinis et al. (1996), using an operation model based on a simple uninhibited Monod kinetics depending on the composition of the waste (lipids, proteins, and carbohydrates). Another mathematical model has been reported by Gavala et al. (1996) for the codigestion of olive mill wastes, pig sludge, and dairy wastewater. This model considered a four-step pathway (hydrolysis, acidogenesis, acetogenesis, and methanogenesis) in which three bacterial groups were involved in the anaerobic digestion process. The wastes were defined by a simplified composition: carbohydrates (soluble and insoluble), proteins (soluble and insoluble), and VFA. However, the model neither could predict pH and biogas composition nor did it take into account the inhibitory effect of low pH values, high VFA concentration, or shortage of ammonium nitrogen (Fezzani and Cheikh, 2009).

The main aspects of a biological system can be represented in terms of mathematical models. Examples of such descriptive models include bacterial growth representation through Monod equation, enzyme–substrate reaction behavior through Michaelis–Menten equation, etc. These mathematical models based on biological system improve the basic understanding of the system and their underlying mechanism, which helps in formulation and validation of hypothesis, prediction of behavior of the system under different situations, and environmental conditions. Consequently, it reduces the risk factors, manual errors, time, and requirement of experimental information. A suitable model for any process should be chosen based on the four criteria as described below:

Simplicity: the model should be as simple as possible.

Causality: the relevant cause–effect relationship should be represented by the model.

Identifiability: from the available experimental data, the unknown parameters should be identifiable.

Predictability: the model should be capable of predicting the ultimate goal at alternative reasonable conditions.

Some of the important kinetic models used for anaerobic digestion of wastes have been represented in Table 12.2.

12.7 DIGESTED SLURRY FOR AGRICULTURE APPLICATION

12.7.1 AS BIOFERTILIZER

The residual slurry obtained after anaerobic digestion is termed as digestate which contains partially degraded organic materials along with sludge and inorganic matter. Recycling of this digestate by applying back to the soil as an organic fertilizer could be the best option for its safe disposal. In some cases, it can be used as a composting material with further addition of cyanobacterial species that could enrich the nutrient contents, namely, N, P, and K which are considered to be an important component for soil fertility. Thus, additional income can be obtained by sale of digestate as an organic fertilizer.

Application of digestate onto the farm soil is dependent on three parameters such as:

a) chemical properties,
b) stability, and
c) hygienization.

Application of biogas residual slurry leads to improvement in soil quality parameters such as bulk density, porosity, soil organic matter, pH, and total carbon and nitrogen.

12.7.2 AS PLANT GROWTH PROMOTER

Liu et al. (2007) reported that the residual digested slurry from anaerobic digestion process contained organic nitrogen, micronutrients as mineral elements, and bioactive components such as humic acid, vitamin,

TABLE 12.2 Different Kinetic Models Applied in Anaerobic Digestion Process.

Kinetic model	Model equation	References
First order	$\ln\frac{M_u}{(M_u-M)}=kt$	Lo et al. (2012)
Exponential rise to maximum	$M=M_u\left(1-e^{-kt}\right)$	Bilgili et al. (2009); De Gioannis et al. (2009)
Chen–Hashimoto	$B=B_O\left(1-\frac{\mathrm{K}}{\mu_m\theta-1+\mathrm{K}}\right)$ $\mu=\frac{\mu_m S}{\mathrm{K}S_O+(1+\mathrm{K})S}-b$	Chen and Hashimoto (1980)
Grau model	$\mu=\frac{\mu_m S}{S_O}$ $-\frac{ds}{dt}=\frac{\mu_m\times S}{YS_O}$ $S=\frac{S_O(1+bt)}{\mu_m\times t}$	Grau et al. (1975)
Contois model	$\mu=\frac{\mu_m S}{\mathrm{K_x}S+S}-b$	Contois (1959)
Haldane model	$\frac{dS}{dt}=-\frac{\mu_m S\times B}{Y\times \mathrm{K_S}+S+S\left(\frac{S}{\mathrm{K_I}}\right)}n$	Lokshina et al. (2001)
Gompertz model	$P=P_0\times\exp\left\{-\exp\left[\frac{R_m\times e}{P_0}(\lambda-t)+1\right]\right\}$	Zwietering et al. (1990)

μ, specific growth rate; μ_m, maximum specific growth rate; n, Haldane index (n = 1 or 2); S_O, influent substrate concentration; S, substrate concentration; X, microbial concentration; K_s,m, maximum specific substrate utilization rate; b, specific microorganism decay rate; t, retention time; Y, yield coefficient; K_x, Contois kinetic constant; K, Chen and Hashimoto dimensionless kinetic constant; K_s, half saturation constant; K_I, inhibition constant.

hormones, etc. This slurry can be directly applied to the plants either in liquid or dried form as basal supplement or top dressing (Mikled et al., 1994). It has been reported that application of residual slurry increases the yield of many crops and produced superior quality products as compared to chemical fertilizer (Gurung, 1997).

Alburquerque et al. (2012) evaluated the plant growth of two crops, namely, watermelon and cauliflower by applying digestate obtained from the full-scale anaerobic codigestion plant treating pig manure and biodiesel wastewater. They reported that application of digestate had a positive effect on watermelon in terms of yield.

John Walsh undertook a 3 year real-time trial with clover grass using digestate from a North Wales anaerobic digestion plant to analyze the application of nitrogen, its plant uptake, and effect on yield (http://www.streetkleen.co.uk/case-studies.html). The outcome of that study suggests that crop yield obtained from digestate application was similar to synthetic fertilizer at an application rate of 150 kg ha^{-1}. For every 150 kg of digestate applied per hectare will replace 850 kg CO_2e (equivalent) in the production of mineral fertilizer, that is, 744 kg from N and 106 kg from P. In addition, the application of digestate as biofertilizer had no deleterious effect on the decomposer microbial community and thus it can be treated same as chemical fertilizer. The added value of the pollution abatement properties and the recycling of nutrients economically support sustainable fertilizer management.

12.8 CONCLUSION

It has been observed that anaerobic digestion process still requires technological improvements to mitigate the bottlenecks such as souring of the reactor, nonavailability of sufficient inoculum for large-scale implementation, and lack of technical validation at small scale that leads to process disruptions in high volumetric productivity plants. It has been reported that approximately 25% of total installed biogas plants in India would remain nonfunctional due to the aforementioned constraints. Thus, there is a need to overcome the operational difficulties of anaerobic digestion of organic wastes by adoption of anaerobic codigestion that allows treating different streams of wastes in a single facility, thereby reducing constraint caused by the feestock availability limitations.

KEYWORDS

- anaerobic digestion
- biomethane
- organic wastes
- fertilizer
- methanogens

REFERENCES

Alburquerque, J. A.; et al. Agricultural Use of Digestate for Horticultural Crop Production and Improvement of Soil Properties. *Eur. J. Agron.* **2012,** *43*, 119–128.

Angelidaki, I.; et al. Defining the Biomethane Potential (BMP) of Solid Organic Wastes and Energy Crops: A Proposed Protocol for Batch Assays. *Water Sci. Technol.* **2009,** *59*, 927–934.

Bansal, M.: et al.; Development of Cooking Sector in Rural Areas in India: A Review. *Renew. Sustain. Energy Rev.* **2013,** *17*, 44–53.

Batstone, D. J.; et al.; Modeling Anaerobic Degradation of Complex Wastewater I: Model Development. *Bioresour. Technol.* **2000,** *75*, 67–74.

Bengelsdorf, F. R.; et al. Stability of a Biogas-producing Bacterial, Archaeal and Fungal Community Degrading Food Residues. *FEMS Microbiol. Ecol.* **2013,** *84*, 201–212.

Bhat, P. R.; et al. Biogas Plant Dissemination: Success Story of Sirsi, India. *Energy Sustain. Dev.* **2001,** *5*, 39–46.

Bhattacharyya, B. C.; Banerjee, R. *Environmental Biotechnology*, 1st ed.; Oxford University Press: New Delhi, India, 2007.

Bilgili, M. S.; et al. Evaluation and Modeling of Biochemical Methane Potential (BMP) of Land Filled Solid Waste: A Pilot Scale Study. *Bioresour. Technol.* **2009,** *100*, 4976–4980.

Bozinis, N. A. A Mathematical Model for the Optimal Design and Operation of an Anaerobic Co-digestion Plant. *Water Sci. Technol.* **1996,** *34*, 383–391.

Buyukkamaci, N.; Filibeli, A. Volatile Fatty Acid Formation in Anaerobic Hybrid Reactor. *Process Biochem.* **2004,** *39*, 1491–1494.

Callaghan, F. J.; et al. Continuous Codigestion of Cattle Slurry with Fruit and Vegetable Wastes and Chicken Manure. *Biomass Bioenerg.* **2002,** *22*, 71–77.

Chen, Y. R.; Hashimoto, A. G. Kinetics of Methane Fermentation. *Biotechnol. Bioeng. Symp.* **1980,** *8*, 269–282.

Comino, E.; et al. Investigation of Increasing Organic Loading Rate in the Codigestion of Energy Crops and Cow Manure Mix. *Bioresour. Technol.* **2010,** *101*, 3013–3019.

Contois, D. Kinetics of Bacterial Growth: Relationship Between Population Density and Specific Growth Rate of Continuous Cultures. *J. Gen. Microbiol.* **1959,** *21*, 40–50.

De Gioannis, G. Landfill Gas Generation After Mechanical Biological Treatment of Municipal Solid Waste: Estimation of Gas Generation Rate Constants. *Waste Manag.* **2009,** *29*, 1026–1034

Dutta, S.; et al. *Biogas: The Indian NGO Experience*; Tata Energy Research Institute: New Delhi, India, 1997.

Esposito, G.; et al. Mathematical Modeling of Disintegration-limited Codigestion of OFMSW and Sewage Sludge. *Water Sci. Technol.* **2008,** *58*, 1513–1519.

Federal Energy Management Program. Biomass Energy-Focus on wood. In *Biomass and Alternative Methane Fuels*; Waste BAMF Fact Sheet: Oak Ridge, USA, 2004.

Fezzani, B.; Cheikh, R. B. Extension of the Anaerobic Digestion Model No. 1 (ADM1) to Include Phenolic Compounds Biodegradation Processes for Simulating of Anaerobic Co-digestion of Olive Mill Wastes at Mesophilic Temperature. *J. Hazard. Mater.* **2009,** *162*, 1563–1570.

Francisci, D. D.; et al. Microbial Diversity and Dynamicity of Biogas Reactors due to Radical Changes of Feedstock Composition. *Bioresour. Technol.* **2014,** *176*, 56–64.

Galí, A.; et al. Modified Version of ADM1 Model for Agro-waste Application. *Bioresour. Technol.* **2009,** *100*, 2783–2790.

Ganesh, R.; et al. Anaerobic Codigestion of Solid Waste: Effect of Increasing Organic Loading Rates and Characterization of the Solubilized Organic Matter. *Bioresour. Technol.* **2013,** *130*, 559–569.

Gavala, H. N.; et al. Anaerobic Codigestion of Agricultural Industries Wastewaters. *Water Sci. Technol.* **1996,** *34*, 67–75.

Giuliano, A.; et al. Codigestion of Livestock Effluents, Energy Crops and Agro-waste: Feeding and Process Optimization in Mesophilic and Thermophilic Conditions. *Bioresour. Technol.* **2013,** *128*, 612–618.

Grau, P.; et al. Kinetics of Multicomponent Substrate Removal by Activated Sludge. *Water Res.* **1975,** *9*, 637–642.

Gurung, J. B. *Review of Literature on Effects of Slurry Use in Crop Production*; Final Report Submitted to the Biogas Support Program, Kathmandu, Nepal, **1997;** pp 19–54.

Kaparaju, P.; Rintala, J. Anaerobic Codigestion of Potato Tuber and Its Industrial By-products with Pig Manure. *Resour. Conserv. Recyc.* **2005,** *43*, 175–188.

Karim K.; et al. Anaerobic Digestion of Animal Waste: Waste Strength Versus Impact of mixing. *Bioresour. Technol.* **2005,** *96*, 1771–1781

Kimchie, S. High-rate Anaerobic Digestion of Agricultural Wastes. PhD Thesis, Technion, Israel, 1984.

Kryvoruchko, V.; et al. Anaerobic Digestion of By-products of Sugar Beet and Starch Potato Processing. *Biomass Bioenerg.* **2009,** *33*, 620–627.

Lansing, S.; et al. Methane Production in Low-cost, Unheated, Plug-flow Digesters Treating Swine Manure and Used Cooking Grease. *Bioresour Technol.* **2010,** *101*, 4362–4370.

Lehtomäki, A. S.; et al. Laboratory Investigations on Codigestion of Energy Crops and Crop Residues with Cow Manure for Methane Production: Effect of Crop to Manure Ratio. *Resour. Conserv. Recycl.* **2007,** *51*, 591–609.

Lema J. M.; et al. Chemical Reactor Engineering Concepts in Design and Operation of Anaerobic Treatment Processes. *Water Sci. Technol.* **1991,** *24*, 79–86.

Lerm, S.; et al. Archaeal Community Composition Affects the Function of Anaerobic Co-digesters in Response to Organic Overload. *Waste Manag.* **2012,** *32*, 389–399.

Li, Y.; et al. Solid-state Anaerobic Digestion for Methane Production from Organic Waste. *Renew. Sustain. Energy Rev.* **2011,** *15*, 821–826.

Liu, X.; et al. Pilot-scale Anaerobic Co-digestion of Municipal Biomass Waste and Waste Activated Sludge in China: Effect of Organic Loading Rate. *Waste Manag.* **2004,** *32*, 2056–2060.

Liu, C.; et al. Prediction of Methane Yield at Optimum pH for Anaerobic Digestion of Organic Fraction of Municipal Solid Waste. *Bioresour. Technol.* **2007,** *99*, 882–888.

Lo, N. L.; et al. Methane Production from Solid State Anaerobic Digestion of Lignocellulosic Biomass. *Biomass Bioenergy* **2012,** *46*, 125–132.

Lokshina, L. Y.; et al. Evaluation of Kinetic Coefficients Using Integrated Monod and Haldane Models for Low-temperature Acetoclastic Methanogenesis. *Water Res.* **2001,** *35*, 2913–2922.

Lusk, P. Latest Progress in Anaerobic Digestion. *Biocycle* **1999,** *40*, 52.

Mata-Alvarez, J.; et al. A Critical Review on Anaerobic Co-digestion Achievements Between 2010–2013. *Renew. Sustain. Energy Rev.* **2014,** *36*, 412–427.

Mikled, C.S.; Jiraporncharoen, S.; Potikanond, N.; Fermented Slurry as Fertilizer for the Production of Forage Crops. Final Report, Thai German Biogas Programme Application. Fertil. Res. **1994,** *8*, 297-306.

Mittal, S.; et al. Barriers to Biogas Dissemination in India: A Review. *Energy Policy* **2018,** *112*, 361–370.

Molinuevo-Salces, B. Vegetable Processing Wastes Addition to Improve Swine Manure Anaerobic Digestion: Evaluation in Terms of Methane Yield and SEM Characterization. *Appl. Energy* **2012,** *91*, 36–42.

Nagao, N. Maximum Organic Loading Rate for the Single-stage Wet Anaerobic Digestion of Food Waste. *Bioresour. Technol.* **2012,** *118*, 210–218.

Parawira, W. Anaerobic Batch Digestion of Solid Potato Waste Alone and in Combination with Sugar Beet Leaves. *Renew. Energy* **2004,** *29*, 1811–1823.

Parkin, G. F.; Miller, S. W. In *Response of Methane Fermentation to Continuous Addition of Selected Industrial Toxicants*, Proceedings of the 37th Purdue Industrial Waste Conference, West Lafayette, USA, 1983.

Rao, K. U.; Ravindranath N. H. Policies to Overcome Barriers to the Spread of Bioenergy Technologies in India. *Energy Sustain. Dev.* **2002,** *6*, 59–73.

Rao, M. S.; et al. Bioenergy Conservation Studies of the Organic Fraction of MSW: Assessment of Ultimate Bioenergy Production Potential of Municipal Garbage. *Appl. Energy.* **2000,** *66*, 75–87.

Shyam, M.; Agroresidue Based Renewable Energy Technologies for Rural Development. *Energy Sustain. Dev.* **2002,** *2*, 37–42.

Smith, K. R. What's Cooking? A Brief Update. *Energy Sustain. Dev.* 2010, *14*, 251–252.

Stenstrom M. K.; et al. Anaerobic Digestion of Municipal Solid Waste. *J. Environ. Eng.* **1983,** *109*, 1148–1158.

Vasudeo, G. In *Biogas Manure: A Viable Input in Sustainable Agriculture: An Integrated Approach*, International Seminar on Biogas Technology for Poverty Reduction and Sustainable Development, Beijing, China. **2005;** pp 1–5.

Venkataraman, C.; et al. The Indian National Initiative for Advanced Biomass Cookstoves: The Benefits of Clean Combustion. *Energy Sustain. Dev.* **2010,** *14*, 63–72.

Verstraete, W.; et al. Anaerobic Digestion as a Core Technology in Sustainable Management of Organic Matter. *Water Sci. Technol.* **2005,** *52*, 59–66.

Zaher, U.; et al. GISCOD: General Integrated Solid Waste Codigestion Model. *Water Res.* **2009,** *43*, 2717–2727.

Zamanzadeh, M.; et al. Biogas Production from Food Waste via Co-digestion and Digestion Effects on Performance and Microbial Ecology. *Sci. Rep.* 2017. DOI: 10.1038/s41598-017-15784-w.

Zhang, D.; et al. A Review: Factors Affecting Excess Sludge Anaerobic Digestion for Volatile Fatty Acids Production. *Water Sci. Technol.* **2015,** *72*, 678–688.

Zhang, R.; et al. Characterization of Food Waste as Feedstock for Anaerobic Digestion. *Bioresour. Technol.* **2007,** *98*, 929–935.

Zwietering, M. H.; et al. Modeling of the Bacterial Growth Curve. *Appl. Environ. Microbiol.* **1990,** *56*, 1857–1881.

CHAPTER 13

Bioenergy Production: Biomass Sources and Applications

VARTIKA VERMA, PRIYA SINGH, GAURI SINGHAL, and NIDHI SRIVASTAVA*

Department of Bioscience and Biotechnology, Banasthali Vidyapith, Rajasthan, India

**Corresponding author. E-mail: nidhiscientist@gmail.com*

ABSTRACT

Bioenergy has been proved to be effective source of biomass for the farmers and as a result agricultural market has hiked to a great extent. Bioenergy is known to be an alternative source to the petroleum-based energy sources. Some new market opportunities have been opened after the introduction of the bioenergy for the farm products. During the processing and conversion of biomass feedstock, various useful energies are produced for the farmers. The large-scale bioenergy production has been used to support the agriculture and urban areas to meet their basic demands for the life sustainability.

13.1 INTRODUCTION

Bioenergy is a broad term defined as the energy produced or generated from the organic matter or residue. The bioenergy production is done with the help of the reusable feedstock. The biomass feedstock utilizes the process of photosynthesis to produce the carbohydrate molecules and other oil complex entities. These molecules are further harvested and later used for the development of various kinds of bioenergy. Numerous technologies are available for the conversion and development of biomass into bioenergy (www.attra.ncat.org).

13.2 BIOMASS AS A SOURCE OF RENEWABLE ENERGY

Biomass is defined as any organic matter belonging to the living plant or animal origin. It exists in different forms such as agricultural, forestry, municipal, and other wastes (https://www.eia.gov/energyexplained/index.php?page=biomass_home).

13.2.1 TYPES OF BIOMASS

The main sources of biomass include animal and plant wastes, agricultural crops, wood, algae, as well as other organic residential and industrial waste. The sort (nature) of biomass has been used to determine the amount and type of bioenergy that can be used to produce by the technology. (http://enviroheat.net.au/bioenergy/what-is-bioenergy/).

Wood and its processing wastes are used in the production of heat and electricity in industries after burning. In addition with food and yard, wood waste is utilized in various power plants for the electricity and biogas after collecting in the landfills. The agricultural waste is burned for the generation of liquid biofuel. Human and animal waste is also used for the biogas production which can be further used as a fuel. These are some main examples of the biomass and their uses for the energy production.

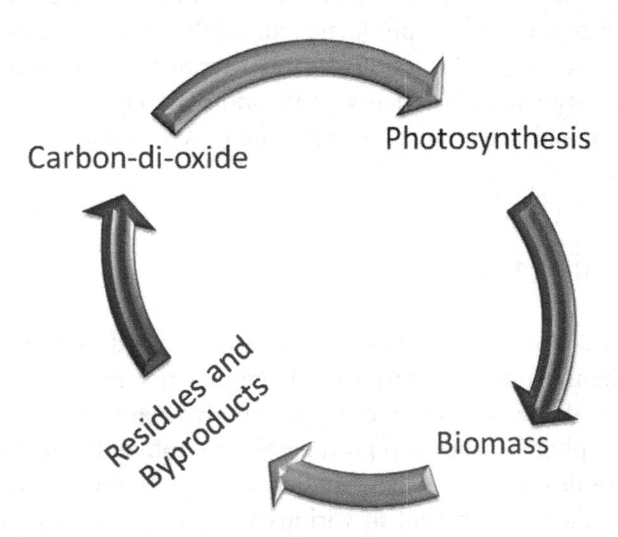

FIGURE 13.1 Pathway of sustainable bioenergy development.

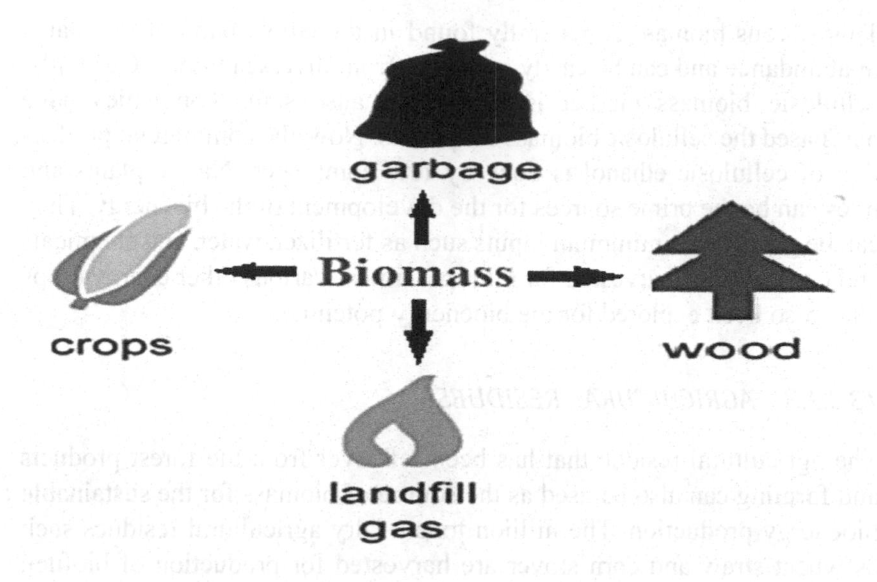

FIGURE 13.2 Various types of biomass.

13.2.1.1 OIL AND SUGAR BIOMASS

This type of biomass includes grains, seeds, and plants that are made up of starch, sugar, or oil in addition with animal fats. This type of feedstock is used for the production of biodiesel and bioethanol. In the United States, over 90% of the bioethanol is still produced from the corn. As the world's largest producer, Brazil uses sugar cane as its primary stock for the bioethanol production. Feedstocks such as wheat, barley, potatoes, milo, cheese whey, sugar beets, brewery, and beverage waste can also be easily fermented for the bioethanol production. Some feedstocks act in multiple ways, that is, either as the food or fuel sources. Plants having oil-rich seeds, also known as oilseeds, are one of the major and sustainable sources of bioenergy. Safflower, sunflower, cottonseed, soybeans, camelina, and canola are some of the common oilseed crops for the bioenergy production.

13.2.1.2 CELLULOSIC BIOMASS

Biomass that is obtained from the cellulosic plants, such as willows, poplars, and switchgrass, has gained more interest among the scientists.

This fibrous biomass is generally found in the edible parts of the plants in abundance and can be easily collected from diverse regions. Currently, cellulosic biomass market is limited because some companies have purchased the cellulosic biomass for pellets. Now, the commercial production of cellulosic ethanol is done by few companies. Native plants and trees can be the prime sources for the development of the bioenergy. They can be grown with minimum inputs such as fertilizer, water, and chemicals and can easily be harvested for multiple times. Various other energy crops have also been explored for the bioenergy potential.

13.2.1.3 AGRICULTURAL RESIDUES

The agricultural residue that has been left over from the forest products and farming can also be used as the source of biomass for the sustainable bioenergy production. The million tons of dry agricultural residues such as wheat straw and corn stover are harvested for production of biofuel. However, scientists have revealed that less than 25% of the agricultural residue has been sustainably harvested.

13.2.1.4 ANIMAL MANURE BIOMASS

The cattle manure especially from the cows and buffaloes is also used as the biomass. During the production of biogas, anaerobic digesters are primarily used to produce the biogas. This biogas was burned to generate electricity by taking these manures as the raw material. The practice of using manure digestion has increased as it is used to control a variety of environmental problems, reduce odors, generate bioenergy for the farm, and provide additional income through the sale of power (www.attra.ncat.org).

13.3 CONVERSION OF BIOMASS TO BIOENERGY

The solid biomass which includes garbage and wood can directly be burned into the landfills to generate heat. The gas that has directly been converted from the biomass is known as the biogas. Biogas is produced after the decomposition of paper, yard, and food scraps wastes in the landfills. It is also produced by the animal manure and sewage processing in the special vessels known as digesters.

Liquid biofuel can also be produced from this biogas such as biodiesel and bioethanol. These biofuels can further be used for the energy generation after burning. Ethanol is produced from the sugarcane and corns which are further fermented for the production of biofuel, and this biofuel can be used in the vehicles. Biodiesel is produced from the animal fats and vegetable oils and is also used in the vehicles as well as heating oils (https://www.eia.gov/energyexplained/index.php?page=biomass_home).

13.4 PRODUCTION OF BIOENERGY

The biomass undergoes a conversion process in order to generate bioenergy, that is, biodiesel, electricity, heat, biobutanol, bioethanol, biogas (methane), and other bioenergy products. The conversion process for different feedstocks is slightly different, but most of them are converted using similar processes and technologies.

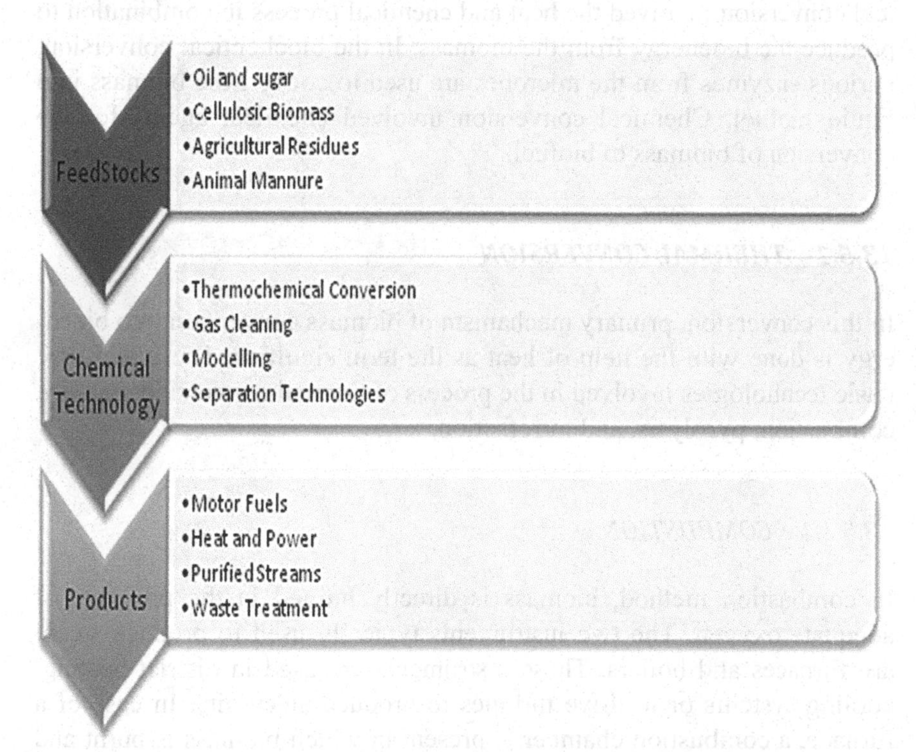

FIGURE 13.3 Chemical technologies for biomass conversion into bioenergy.

There are various pathways that result in the formation of bioenergy. The starter material has one of the most significant points that helps in the selection of method or a technology that would be used to produce bioenergy. The process that is involved in the production is relatively simple which included harvesting, growing, and burning the wood for the generation of heat. An array of conversion pathways are considered for the renovation of biomass into bioenergy in various forms of electricity, heat, or transportation fuels (https://www.energy.vic.gov.au/renewable-energy/bioenergy/bioenergy-sustainable-renewable-energy).

13.5 PATHWAYS FOR BIOENERGY PRODUCTION

Currently, four biomass conversion technologies are available that result in the production of specific bioenergy. In the thermal conversion, heat is used to convert the biomass or feedstock into the bioenergy. Thermochemical conversion involved the heat and chemical process in combination to produce the bioenergy from the biomass. In the biochemical conversion, various enzymes from the microbes are used to convert the biomass into liquid biofuel. Chemical conversion involved chemical agents for the conversion of biomass to biofuel.

13.5.1 THERMAL CONVERSION

In this conversion, primary mechanism of biomass conversion into bioenergy is done with the help of heat as the term signifies. There are three basic technologies involved in the process of thermal conversion, namely, combustion, pyrolysis, and torrefaction.

13.5.1.1 COMBUSTION

In combustion method, biomass is directly burned in the presence of adequate oxygen. The two instruments typically used to produce steam are furnaces and boilers. These instruments are used in district heating/cooling systems or to drive turbines to produce electricity. In case of a furnace, a combustion chamber is present in which biomass is burnt and converted into heat. The heat then dispersed in form of hot air or water. In

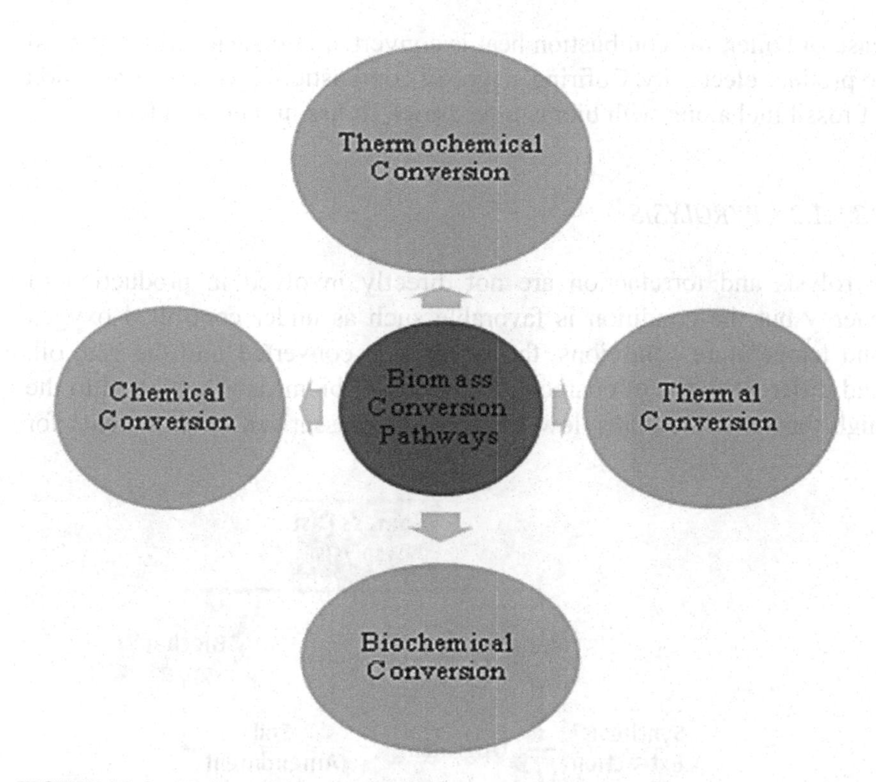

FIGURE 13.4 Various pathways of the biomass conversion into bioenergy.

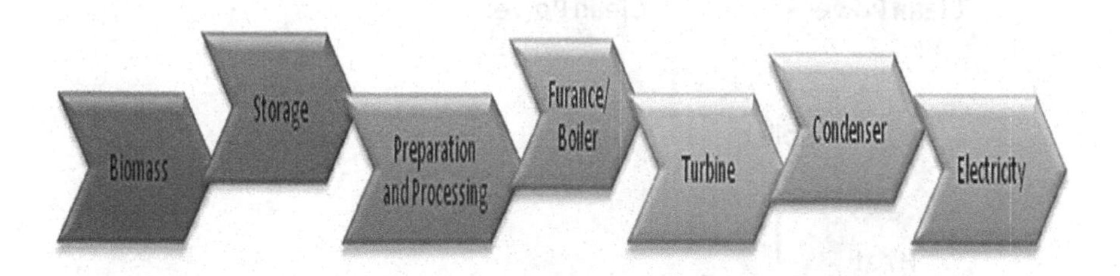

FIGURE 13.5 Flow diagram of combustion process.

case of boiler, the combustion heat is converted into steam which is used to produce electricity. Cofiring, a type of combustion, involved conversion of fossil fuel along with biomass feedstock. It has numerous advantages.

13.5.1.2 PYROLYSIS

Pyrolysis and torrefaction are not directly involved in production of energy but the condition is favorable such as under controlled oxygen and temperature conditions, the biomass is converted into the gas, oil, and different forms of charcoal. In pyrolysis, biomass is subjected to the high temperatures under low oxygen and pressurized environments for

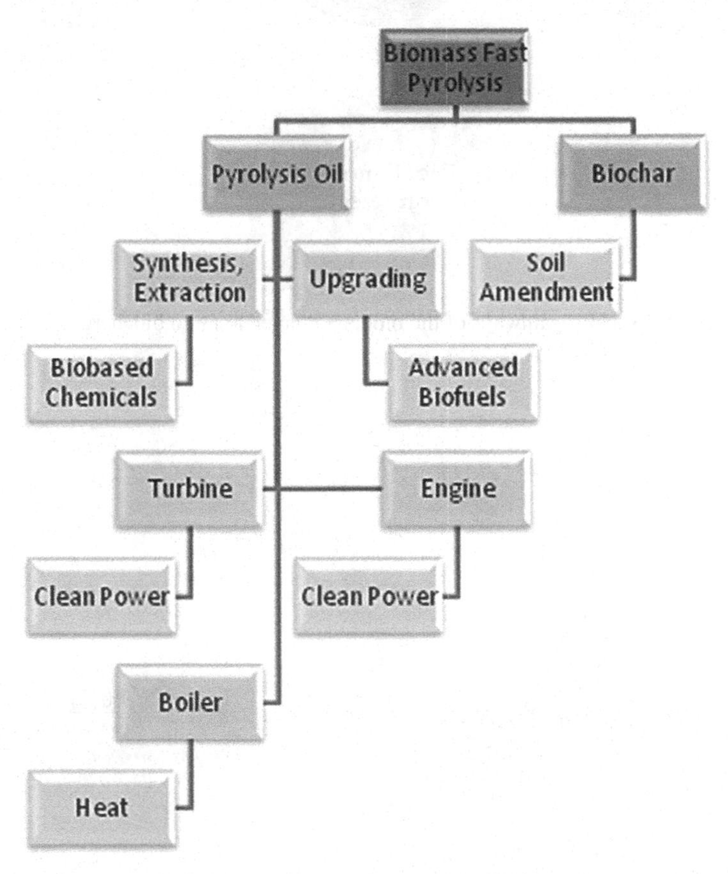

FIGURE 13.6 Flow chart of pyrolysis.

the energy production. In this process, biomass undergoes partial combustion. In turn, solid residue (char), liquid fuel, and biochar (charcoal rich in carbon) are produced.

13.5.1.3 TORREFACTION

It is a process just like the pyrolysis which involved the heating of biomass. This process is slightly different from the pyrolysis as it occurs in the absence of oxygen and at lower temperatures (200–320°C). In this process, cellulose, hemicelluloses, and lignin are partially decomposed, water is removed, and the final product biocoal (energy dense solid fuel) is produced.

13.5.2 THERMOCHEMICAL CONVERSION

In thermochemical conversion, biomass is converted into fuel gases and other chemicals. This process involved multiple stages as described in Fig. 13.7. In last stage, syngas is produced. Syngas is composed of carbon and hydrogen which can be used to produce ammonia and lubricants. When

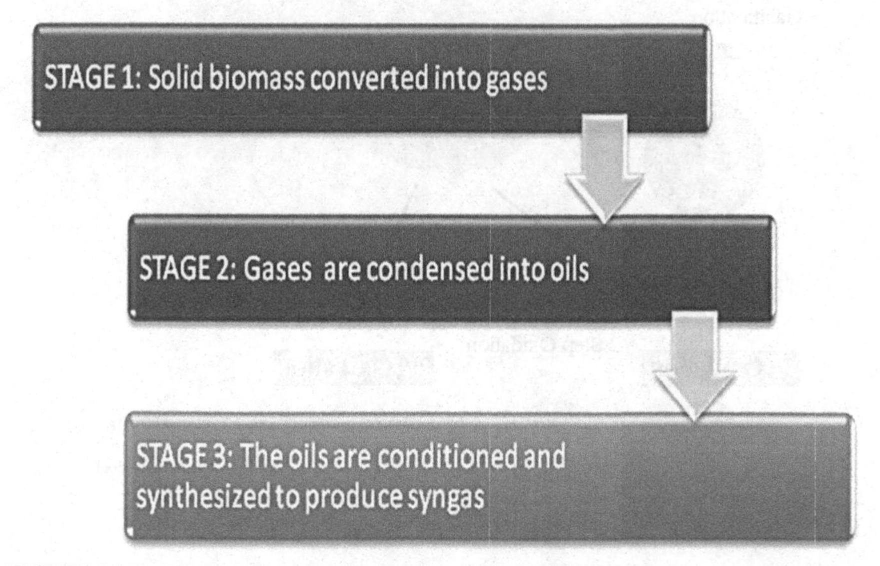

FIGURE 13.7 Various stages of thermochemical conversion.

syngas undergoes the process of Fischer–Tropsch, it produces biodiesel. Thermochemical conversion involved the method of gasification.

13.5.2.1 GASIFICATION

This technology has considered thermochemical conversions of the biomass. In this process, biomass is converted into the gas under high temperatures and a controlled environment. It required the 800°C temperature for gasification process. This technology involved two stages, that is, partial combustion to form producer gas and charcoal and chemical reduction.

13.5.3 BIOCHEMICAL CONVERSION

Since ages, microbes are used for the ethanol production. However, those microbes are considered as biochemical factories for the conversion of

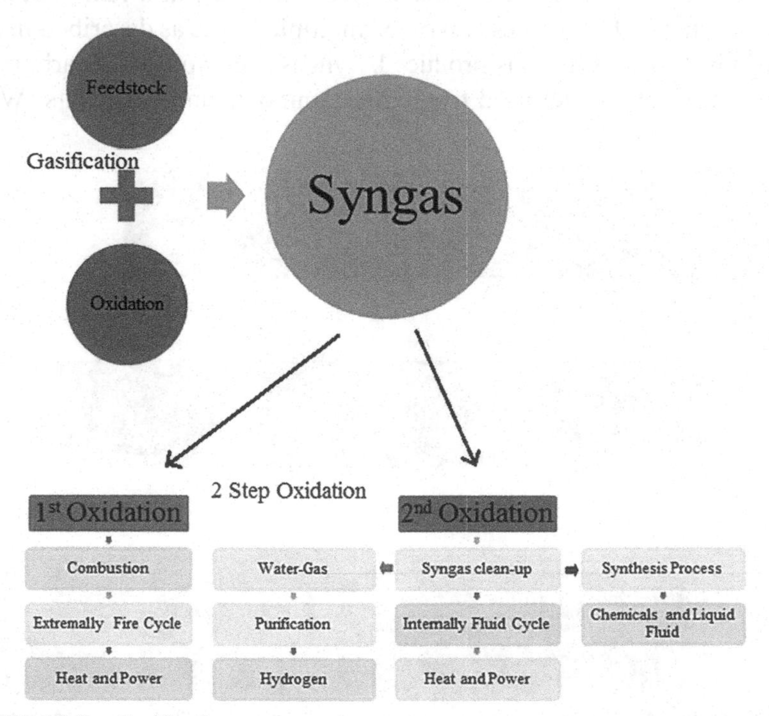

FIGURE 13.8 Gasification process.

biological materials to various forms of energy. The key technologies used for the biochemical conversion are anaerobic digestion and fermentation.

13.5.3.1 ANAEROBIC DIGESTION

In anaerobic digestion, various microbes are used for the organic material breakdown in oxygen-free environments. This technology is mostly used in the methane and carbon-rich biogas production from the agricultural residues, manure, and food scraps. This process is composed of three processes as described in Figure 13.9.

The optimization of temperature for growth of mixed bacterial cultures is done and then digesters are allowed to work for wide temperature range. When conditions get optimized, bacteria converted about 90% of the biomass into readily useable energy source, that is, biogas.

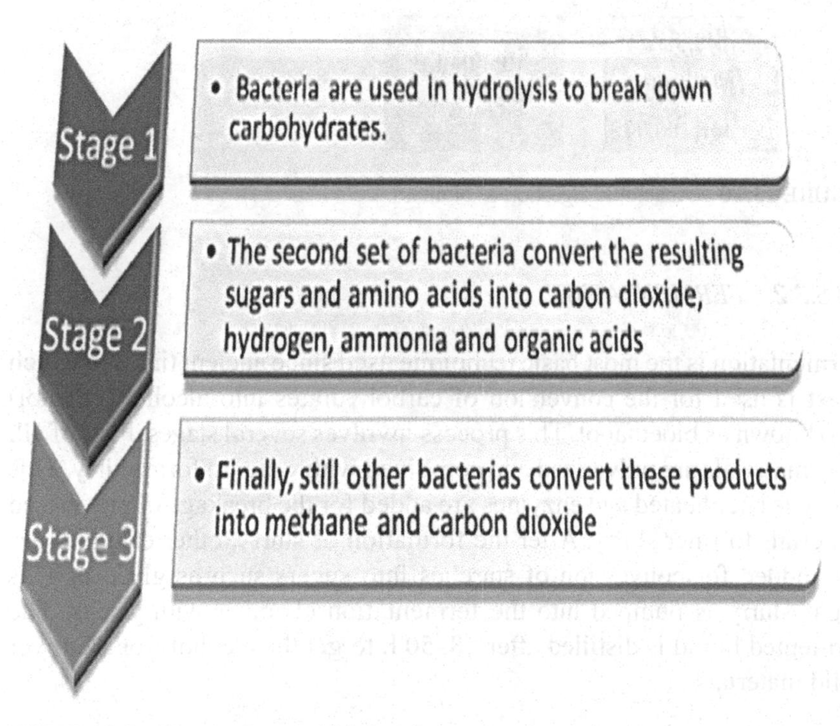

Stage 1
• Bacteria are used in hydrolysis to break down carbohydrates.
Stage 2
• The second set of bacteria convert the resulting sugars and amino acids into carbon dioxide, hydrogen, ammonia and organic acids
Stage 3
• Finally, still other bacterias convert these products into methane and carbon dioxide

FIGURE 13.9 Stages of anaerobic digestion.

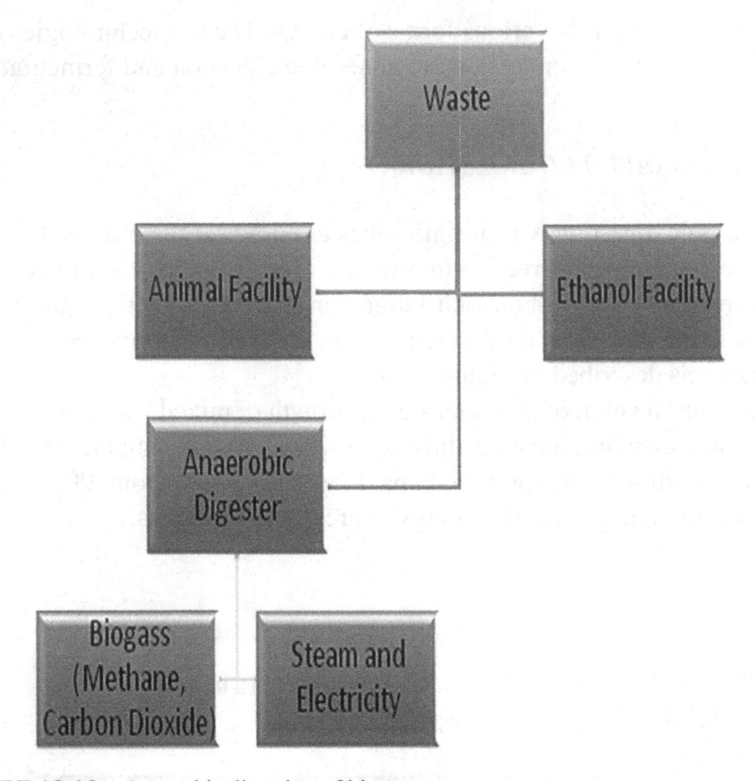

FIGURE 13.10 Anaerobic digestion of biomass.

13.5.3.2 FERMENTATION

Fermentation is the most basic technique used since ancient times in which yeast is used for the conversion of carbohydrates into alcohol (ethanol) also known as bioethanol. This process involves several stages. First of all, crop materials are pulverized and combined with water to form slurry. This slurry is later heated and enzymes are added for the breakage of pulverized materials to finer slurry. After the formation of slurry, other enzymes are also added for conversion of starches into sugars such as glucose. This sugar slurry is pumped into the fermentation chamber with yeasts. The fermented liquid is distilled after 48–50 h to get the alcohol from leftover solid materials.

13.5.4 CHEMICAL CONVERSION

Chemical conversion of biomass involves the usage of chemical agents which interact with the biomass and transform them into other useable energy. Transesterification is one of the methods used under this conversion.

13.5.4.1 TRANSESTERIFICATION

This method is the most common type of chemical conversion. In this process, fatty acids of greases and oils are converted into alcohol. In this process, the viscosity of fatty acids lowered down and made them combustible. Biodiesel is the most common product of this process. Almost any animal fat, bio-oil (soybean oil), or tree oil can be transformed to biodiesel through transesterification (http://www.wgbn.wisc.edu/conversion/bioenergy-conversion-technologies).

13.6 MAJOR BARRIERS IN THE BIOENERGY PRODUCTION

Generally, financial, economic, and technical factors are the major barriers to affect the bioenergy production. Nowadays, biomass has become limited due to the increased need of bioenergy at the industrial and domestic level and has been subjected to the market cycles. Some technical issues such as power purchase agreements and tax policies are the major constrains due to which biomass energy projects are suffering. Economically, some of the biomass get limited due to the competition with their other applications, for example, wood is a good source of biomass but is also used in the furniture industry. Available biomass energy technologies may not be sufficiently matured to represent an acceptable risk to private sector investors and have not offered adequately high returns. Beside these barriers, institutional constrains are also there as current energy policies are often biased against renewable energy sources. The researcher, developer, manufacturers, and users' cooperation has not been coordinated well (Koopmans and Koppejan, 1998).

13.7 APPLICATIONS OF BIOENERGY

Despite having some barriers, bioenergy is a sustainable source of energy. There is a demand for new sources for the bioenergy production due to increasing world population. Bioenergy is a versatile source of energy and in contrast to other energy it can be produced in the form of solid, liquid, and gas fuel. It is a chief renewable energy source by participating 14% of the renewable energy out of 18% and supplies 10% of the global energy. Initially, it was a traditional and indigenous energy source but now it has shifted toward the modern and globally traded commodity. In the process of bioenergy production, thousands of people have been globally employed. Biofuel is the most sustainable and viable source of bioenergy and has almost decreased the dependency on oil. It has more importance as it can replace the petroleum fuels. Biofuel can also be utilized as a substitute for power, heat, and chemicals production from the fossil fuel. It is the most important source of bioenergy as it has many advantages, for example, it has reduced the greenhouse gas emission, sustainable development, social arrangement, and regional advancement. Around 27 million tons of the international trade is determined by the liquid biofuel (Demirbas and Demirbas, 2007). The production of bioenergy from forest biomass has offered a distinctive solution for the reduction of wildfire hazard. The addition of biochar into carbonless soil has mitigated the carbon loss; thus returning the cycle nutrients into the forest. Biochar addition in soil has also improved the water holding capacity of the site (McElligott et al., 2011).

13.8 CONCLUSION

The development of modern industry has come with plenty of new challenges to find the alternative source of energy. Due to increased world population, continuously more sources of energy are in demand which can be fulfilled by the bioenergy. As we all know, an endless amount of biomass is available in our earth which is environment friendly, relatively cheap, well managed, and renewable. Bioenergy, an important fuel is used for the various purposes in different forms and can further be used for most of the applications.

KEYWORDS

- **bioenergy**
- **biomass sources**
- **production**
- **pathways applications**

REFERENCES

ATTRA, Sustainable Agriculture. www.attra.ncat.org (accessed Aug 18, 2018).

Demirbas, A. H.; Demirbas, I. Importance of Rural Bioenergy for Developing Countries *Energy Convers. Manag.* **2007,** *48*, 2386–2398.

EnviroHeat Australia. *What is Bioenergy*? 2014. http://enviroheat.net.au/bioenergy/what-is-bioenergy/ (accessed Aug 18, 2018).

Independent Statistics & Analysis. U.S. Energy Information Administration. *Biomass Explained.* https://www.eia.gov/energyexplained/index.php?page=biomass_home (accessed Aug 18, 2018).

Koopmans, A.; Koppejan, J. In *Agricultural and Forest Residues–Generation, Utilization and Availability*, Proceedings of the Regional Expert Consultation on Modern Applications of Biomass Energy, FAO Regional Wood Energy Development Programme in Asia, 1998.

McElligott, K.; Dumroese, D.; Coleman, M. *Bioenergy Production Systems and Biochar Application in Forests: Potential for Renewable Energy, Soil Enhancement, and Carbon Sequestration*; Res. Note RMRS-RN-46. Fort Collins, CO; US Department of Agriculture, Forest Service, Rocky Mountain Research Station, **2011;** Vol. 14; p 46.

Renewable Power Sources. *Biochemical Conversion of Biomass into Energy*; 2014. https://www.upsbatterycenter.com/blog/biochemical-conversion-biomass-energy/ (accessed Aug 18, 2018).

Victoria State Government. Environment, Land, Water and Planning. *Bioenergy: Sustainable Renewable Energy*. https://www.energy.vic.gov.au/renewable-energy bioenergy/bioenergy-sustainable-renewable-energy (accessed Aug 18, 2018).

CHAPTER 14

Third-Generation Biofuels: An Overview

JOÃO MOREIRA NETO[1,*], ANDREA KOMESU[2],
LUIZA HELENA DA SILVA MARTINS[3], VINICIUS O.O. GONÇALVES[4],
JOHNATT ALLAN ROCHA DE OLIVEIRA[5], and MAHENDRA RAI[6]

[1]Department of Engineering , Federal University of Lavras (UFLA), Lavras, MG, Brazil.

[2]Department of Marine Sciences, Federal University of São Paulo (UNIFESP), Santos, SP, Brazil.

[3]Department of Natural Sciences and Technology, State University of Pará (UEPA), Belém, PA, Brazil.

[4]Institute of Chemistry, Department of Physical Chemistry, Federal University of Rio de Janeiro, Rio de Janeiro, RJ, Brazil

[5]Institute of Health Sciences, Nutrition College, Federal University of Pará (UFPA), Belém, PA, Brazil.

[6]Nanobiotechnology Lab, Department of Biotechnology, SGB Amravati University, Amravati, Maharashtra, India.

[*]Corresponding author. E-mail: joao.neto@ufla.br

ABSTRACT

Algal cultures can be used to produce several products as vitamins, feed, biofuels, and other products on industrial scale. The growth in world energy demand has led to the development and analysis of efficient sources capable of producing fuels and chemicals. Thus, microalgae are seen as a promising alternative for the production of fuels and chemicals due to their high photosynthetic conversion efficiency. There are several processes to produce alternative energy by algal. Between them, we have biochemical conversion, thermochemical conversion, chemical reactions, and direct combustion; also, there is other alternative way to carry out the biomass treatments to produce high value chemicals as bioethanol. This chapter aimed to show some of the main technologies involved in algal transformation for commercial interests. Moreover, it is very important to study algal composition and apply the best technology to obtain the specific product of interest from this rich and potential feedstock.

14.1 INTRODUCTION

The rapid industrialization and high population growth are the two major factors that contribute to the global energy crisis (Jambo et al., 2016). Today, approximately 88% of worldwide energy consumption is a consequence of the combustion of fossil fuels (Chye et al., 2018). Unfortunately, the massive use of fossil fuels led to the problems such as depletion of its reserves, price fluctuation, negative environmental impacts, and climatic change (Jambo et al., 2016). Thereby, the alternative energy sources, which are readily available, facilely accessible, and greener in nature, are very much needed (Leong et al., 2018). Biofuels emerged as a promising solution to the fossil fuel resources.

The "biofuels" are referred to the energy-enriched chemicals generated through the biological processes or derived from the biomass of living organisms, such as microalgae, plants, and bacteria (Rodionova et al., 2017). The energy in biofuel is stabilized during a process of biological carbon fixation in which carbon dioxide (CO_2) is converted into sugar that is found only in living organisms and plants. In contrast with fossil fuel, biofuel is produced by, or derived from, living organisms in a relatively short period of time rather than being derived by the decomposition of organic matter over several million years (Alaswad et al., 2015).

Currently, three generations of biofuels have been flourished based on different feedstock. The "first generation" of biofuels, such as bioethanol, biobutanol, and biodiesels, are produced directly from food crops like corn, wheat, and soybeans. The biofuel is derived from the starch, sugar, and oil that these crops provide (Alaswad et al., 2015). The important characteristics of "first-generation" biofuels include their ability to be blended with petroleum-based fuels and their efficiency in internal combustion engines, as well as the compatibility with flexible fuel vehicles (Chye et al., 2018). The most controversial issue with "first-generation" biofuels is that it is necessary to choose one or other of the "fuel versus food" alternatives (Alaswad et al., 2015). Additionally, the crop feedstock requires large agricultural areas to produce sufficient quantities of biomass. In terms of environmental issues, this increased agriculture yield and subsequent harvesting may lead to enhanced land clearing, loss in biodiversity due to habitat destruction, water depletion, and air pollution (Chye et al., 2018).

The "second-generation" biofuels are made from nonfood crops, mainly lignocellulosic feedstock such as grass, wood, and other organic wastes (Alaswad et al., 2015). The advantages of these feedstocks are the ease of availability, which does not compete with food and thus eventually has a much lesser impact on the environment (Jambo et al., 2016). Nevertheless, the "second-generation" biofuels too face steep challenges in the form of technical difficulties during the pretreatment process and inefficient conversion of lignocellulosic materials due to their complex structures (Chye et al., 2018).

In a search for viable and cost-effective alternatives to fossil fuels, past studies have reported superior capabilities of algae-derived biomass for the production of an improved version: the "third-generation biofuels" (Chye et al., 2018).

Algae (alien to light green antenna entity) are very diverse micro- to macroplants and found almost everywhere on the planet Earth. They play important role in many ecosystems to provide food, feed, and vital nutrients and also supply up to 60% of needed oxygen to all living beings vital for their survival (Gajraj et al., 2018).

Lately, algal cultures have been used to produce vitamins, feed, and other products on an industrial scale. However, the growth in world energy demand has led to the development and analysis of efficient sources capable of producing fuels and chemicals. Thus, microalgae are seen as a promising alternative for the production of fuels and chemicals due to their high photosynthetic conversion efficiency. These organisms are unicellular and have carbon, hydrogen, oxygen, and nitrogen, and are categorized as aquatic biomass. The growth of microalgae depends heavily on the availability and intensity of light, availability of nutrients, like phosphorus and nitrogen, CO_2 and O_2 levels, temperature, and type of culture system used (Moncada et al., 2014).

Algae are a promising everlasting source of fuel and other valuable products (Gajraj et al., 2018). The algae have the ability to produce crude oil, which may then be easily processed to manufacture diesel and gasoline. The particular algae species, which may be categorized as either microalgae or macroalgae, may as well be genetically modified so that the carbon metabolic pathway facilitates the production of important end products like ethanol (Chye et al., 2018). In addition, algae have the ability to produce more energy per acre of land as compared with other conventional feedstock crops such as sugarcane bagasse and corn. For these reasons, microalgae are touted to be the biomass with the maximum potential to act as a substitute for petroleum-derived transport diesel without adversely affecting the food supply and other crop products (Chye et al., 2018).

Therefore, the main of the current review is to explore the potential of algae for the production of different biofuels. The physical–chemical composition of microalgae and types of treatments applied for the deconstruction of microalgae were also explored.

14.2 MICROALGAE POTENTIAL AS RAW MATERIAL FOR THE BIOFUELS PRODUCTION

Biofuels can be one of the alternatives to meet the energy needs generated by world population growth (Rodionova et al., 2017). This scenario has led to the development and analysis of efficient sources to produce fuels and chemicals (Sánchez-Tuirán et al., 2012). The emergence of third generation bioethanol provides more benefits as compared to the first and second generation. The third-generation bioethanol is focused on the use of marine organisms such as algae (Jambo et al., 2016).

Between that, different sources have been observed in many studies about the potential of the microalgae for producing bioenergy and chemicals (Bahadar and Khan, 2013; Sánchez-Tuirán et al., 2012).

Algae represents as a promising alternative feedstock due to its high lipid and carbohydrate contents, high proton conversion, easy cultivation in a wide variety of water environment, relatively low land usage, and high CO_2 absorption (Singh and Olsen, 2008).

In a search for viable and cost-effective alternatives to fossil fuels, past studies have reported superior capabilities of algae-derived biomass for the production of an improved version: the third-generation biofuels.

With the recent approaches involving the production of microbial biofuels, the cultivation of microalgae has been recommended as a possible strategy to obtain the biomass to be used in the production of fuel. This biomass could be obtained by culturing biofilms of microalgae or cyanobacteria (Demirbas, 2009; Heiman, 2016).

The production of biofuels can be achieved by several species of algae, but studies for the discovery of species richer in terms of carbohydrates, proteins, and lipids have been carried out, for their use in the production of biofuels (Razaghifard, 2013). According to Chisti (2007), algae have the ability to produce more bioenergy and biofuel energy per acre of land as compared with other conventional feedstock crops such as sugarcane bagasse and corn.

Microalgae are lauded to be a potential alternative feedstock for biofuel production as they are a sustainable energy source and do not compete with other edible feedstocks. Cultivation of microalgae also creates a carbon sink for greenhouse gasses, thus making it greener in nature (Leong et al., 2018).

Sugar is the basic molecular substrate for the production of bioethanol and biomethanol (Dias et al., 2009). The use of photosynthetic organisms as a source of biofuel is cheap and feasible, and it is based on the main component found in this type of material, the carbohydrate (Razaghifard, 2013).

Compared to other traditional biofuel crops, it is a better source for oil extraction and may be utilized to synthesize a multitude of diverse biofuels such as biomethane (via anaerobic digestion) and biodiesel (using microalgal oil), as well as biohydrogen (via photobiological synthesis) (Chisti, 2007).

The efficiency of their photosynthesis is higher than that of other plants, and some species are considered to be among the fastest growing plants in the world (Demirbas and Demirbas, 2010). It has been reported that photosynthetic efficiencies for algae range from 3% to 8%, compared with 0.5% for many terrestrial crops (Lardon et al., 2009).

Algae are simple plants ranging from microscopic (microalgae) to large seaweeds (macroalgae), from small green dots to larger cyanobacteria, and the tinsel diatoms. There are nine major groups of algae which are cyanobacteria (*Cyanophyceae*), green algae (*Chlorophyceae*), diatoms (*Bacillariophyceae*), yellow-green algae (*Xanthophyceae*), golden algae (*Chrysophyceae*), red algae (*Rhodophyceae*), brown algae (*Phaeophyceae*), dinoflagellates (*Dinophyceae*), and "picoplankton" (*Prasinophyceae* and *Eustigmatophyceae*) (Hu et al., 2008).

Microalgae like *Spirulina*, *Chlorella*, *Dunaliella*, and *Haematococcus* are currently cultivated commercially to produce photosynthetically grown biomass from a few tons to several hundred tons annually. They grow at an exceptional fast rate, 100 times faster than terrestrial plants, and they can double their biomass in less than 1 day (Tredici, 2010).

Harvesting microalga biomass is a major challenge because of their small size and their low concentration in the culture medium. In the overall production process, microalgae are initially grown to reach a maximum biomass concentration of 0.02–0.5%. From the culture medium, the biomass is concentrated to 15–20%, either in a single step or in a series of concentration steps, before they can be processed further via drying, extraction, or other downstream processing steps (Chisti, 2007; Bilad et al., 2014; Heasman et al., 2000).

14.3 PHYSICAL-CHEMICAL COMPOSITION OF MICROALGAE

The existence of nearly 300,000 species of algae, with varied oil contents and growth rates, were reported by Scott et al. (2010). Microalgal biomass is mainly constituted of lipids, carbohydrates, and proteins. A microscope is required to observe microalgae because of their size, being either unicellular or multicellular, but always smaller than 0.4 mm in diameter. The category includes groups such as the diatoms and cyanobacteria that have high growth rates (Alaswad et al., 2015). The three most important classes of microalgae in terms of abundance are the diatoms (*Bacillariophyceae*), the green algae (*Chlorophyceae*), and the golden algae (*Chrysophyceae*) (Alaswad et al., 2015).

Different microalgal species have significantly different compositions, where the average lipid oil content typically varies between 8% and 31% (Chye et al., 2018). The carbohydrate content was also found to be relatively high, which is up to 50% of dry weight for some species such as *Scenedesmus*, *Chlorella*, and *Chlamydomona* (Jambo et al., 2016). Table 14.1 summarizes the composition of protein, carbohydrate, and lipid of some species of microalgae on the dry mass basis.

Factors such as light, temperature, nutrient content, pH, O_2 and CO_2 level, salinity, and toxic chemicals influence the lipid and carbohydrate contents of microalgae (Jambo et al., 2016). The lipid content of microalgae varies in accordance with the culture conditions. It is possible to increase the lipid concentration by almost 80% above natural levels, by optimizing growth determining factors (Alaswad et al., 2015). Some of the common components in the cell wall of microalgae are cellulose, protein, lignin, pectin, hemicelluloses, and other carbohydrates, which can be converted to monomers through an acid or enzymatic hydrolysis to produce bioethanol (Jambo et al., 2016).

TABLE 14.1 Composition of Different Species of Microalgae on a Dry Matter Basis.

Microalgal species	Protein	Carbohydrate	Lipid
Anabaena cylindrica	43–56	25–30	4–7
Chlamydomonas rheinhardii	48	17	21
Chlorella pyrenoidosa	57	26	2
Chlorella vulgaris	51–58	12–17	14–22
Dunaliella bioculata	49	4	8
Porphyridium creuntum	28–39	40–57	9–14
Prymnesium parvum	28–45	25–33	22–39
Scenedesmus dimorphus	8–18	21–52	16–40
Scenedesmus obliquus	50–56	10–17	12–14
Spirogyra sp.	6–20	33–64	11–21
Spirulina maxima	60–71	13–16	6–7
Spirulina platensis	46–63	8–14	4–9
Synechoccus sp.	63	15	11
Tetraselmis maculate	52	15	3

Adapted from Um and Kim (2009).

14.4 GENERALITY USE OF BIOETHANOL

The greatest use of ethanol, no doubt, is utilized as motor fuel and gasoline additive. More than any other important country, Brazil relies on ethanol as a motor fuel for several motor vehicles. Gasoline sold in Brazil contains at least 25% anhydrous ethanol. Hydrated ethanol (about 95% ethanol and 5% water) can be used as fuel in more than 90% of new cars sold in the country. The United States uses mixtures of ethanol/gasoline gasohol (max 10% ethanol) and 85% ethanol (E85). Ethanol can also be used as rocket fuel and lightweight rocket-powered aircraft (Baeyens et al., 2015).

Australian law limits the use of pure ethanol from sugarcane residues to 10% in automobiles. It has been recommended that older cars, and also older cars designed to use a slower fuel, should have their valves updated or replaced (Baeyens et al., 2015).

Direct hydrated ethanol as an automotive fuel has been widely used in Brazil for pure ethanol vehicles and more recently for flexible vehicles: ethanol fuel is a distilled mixture of 95.63% by weight of ethanol and 4.73% in water. By the end of 2012, there were approximately 20 million flexible vehicles running on Brazilian highways. Hydrated ethanol imposes a limitation on the normal operation of the vehicle because the lower evaporative pressure of ethanol (compared to gasoline) causes problems when the engine starts to freeze at temperatures below 288 K. For this reason, both pure and flex fuels are built with an additional small gasoline reservoir inside the engine compartment to aid in starting the engine when cold, initially by injecting gasoline (Baeyens et al., 2015).

14.5 BIOETHANOL PRODUCTION

Bioethanol derived from sugar-based biomass in a fermentation process is subcategorized into first, second, and third generation of bioethanol. First-generation bioethanol is derived from fermentation of glucose contained in starch and sugar crops based on different feedstock such as corn and sugarcane. Second-generation bioethanol is produced by processes ligno-cellulosic feedstock and agricultural forest residues. Third-generation bioethanol is focused on the use of marine organisms such as algae (Kostas et al., 2016; Jambo et al., 2016).

In general, bioethanol production from biomass involves pretreatment, enzymatic hydrolysis, fermentation, and distillation (Borines et al., 2013).

The pretreatment of lignocellulosic feedstocks is a necessary step to unwind cellulose from hemicelluloses and lignin in which cellulose is embedded and makes cellulose more susceptible for easy access to enzymes during enzymatic hydrolysis and enhance the rate and yield of reducing sugars (Gupta and Verma, 2015; Kumar and Sharma, 2017).

Based on the type of the treatment process involved, lignocellulosic biomass pretreatment methods are broadly classified into different categories: physical (e.g., milling, microwave, ultrasound), chemical (e.g., dilute acid, organosolv, mild alkali, ionic liquid), physicochemical (e.g., steam explosion, liquid hot water, ammonia based, oxidative pretreatment), and biological (e.g., bacterial, brown, white, and soft-rot fungi) methods (Kumar and Sharma, 2017).

Such pretreatments, depending on the process conditions can generate a range of inhibitory by-products (e.g., hydroxymethylfurfural, furfural, phenolic compounds, or organic acids) which can impact on the viability of yeast in a subsequent fermentation process (Mosier et al., 2005; Kostas et al., 2016). Due to absence of lignin and low hemicellulose content, algae generally require relatively mild conditions for processing compared to lignocellulosic biomass (Alaswad et al., 2015).

The composition of carbohydrates in algae biomass may differ significantly from species to species (Jambo et al., 2016), thus is of great importance to select algae with high carbohydrate productivity as well as suitable sugar composition for bioethanol production. Although macroalgae has shown high potential as a sugar source, the majority of carbohydrate present in macroalgae biomass (e.g., brown macroalgae) is in the form of polysaccharides such as laminarin, mannitol, carrageenan, and alginate (Kostas et al., 2016). Degradation of these polysaccharides is slows and requires specific enzymes (Cho et al., 2013). The development of conversion methods to release monosaccharides from these polysaccharides and the choice of industrial microorganisms for bioethanol production from macroalgae are the bottlenecks to make feasible the utilizing seaweed as a feedstock (Kostas et al., 2016).

Wargacki et al. (2012) was capable of cofermenting alginate, mannitol, and glucan from the brown algal species (*Saccharina japonica*) using *Escherichia coli*, yielding 0.41 g ethanol/g of carbohydrate sugars.

Sunwoo et al. (2017) performed a fermentation with a mixed yeast of adapted *Saccharomyces cerevisiae* KCTC 1126 and *Pichia angophorae* KCTC 17574 to galactose and mannitol from waste seaweed obtained from Gwangalli Beach, Busan, Korea. The maximum ethanol concentration and ethanol yield obtained were of 13.54 g/L and 0.45, respectively.

Carbohydrates in microalgae biomass are mainly cellulose in the cell wall and starch in the plastids that can be readily converted into fermentable sugars, which is more suitable for ethanol production (Moreno-Garcia et al., 2017). Harvesting of microalgae is more expansive than macroalgae. Macroalgae can be harvested using nets, which require less energy, while microalgae are harvested by conventional processes, which include filtration, flocculation, centrifugation, sedimentation, froth floatation, and ultrasonic separation (Behera et al., 2015).

Several researches have been conducted on the utilization of different micro- and macroalgae for the bioethanol production such as *Chlorella vulgaris* (Kim et al., 2014), *Gracilaria verrucosa* (Kumar et al., 2013), *Gracilaria salicornia* (Wang et al., 2011), *Gelidium amansii* (Cho et al., 2014), *Sargassum* spp. (Borines et al., 2013), *Laminaria digitate* (Kostas et al., 2016), *Chondrus crispus* (Kostas et al., 2016), *Palmaria palmata* (Kostas et al., 2016), and *Kappaphycus alvarezii* (Khambhaty et al., 2012).

Saccharification is usually the rate-limiting step in biofuels production using lignocellulosic materials or microalgal biomass that contains a cellulose source (Chen et al., 2013). The saccharification or hydrolysis consists in the depolymerization of the polysaccharide contents (such as starch, cellulose, alginates, fucans, laminarin, and carrageenans) in algae cell walls into free monomer molecules, which can be fermented to bioethanol (Jambo et al., 2016). The absence or near absence of lignin and hemicelluloses content in algae in comparison to lignocellulosic biomass makes the enzymatic hydrolysis of algal cellulose easier (Saravanan et al., 2018).

The main methods that have been applied to produce sugars from microalgae can be categorized into two groups: enzymatic saccharification and acid saccharification (Chen et al., 2013).

In acid hydrolysis, a wide range of acids has been used in which sulfuric acid (H_2SO_4) is the most preferred. It has been found that the polysaccharides of the three classes of macroalgae (brown, red, and green) can be effectively hydrolyzed to monosaccharides by treatment with H_2SO_4 diluted at high temperature. The acidic role in the hydrolysis can be seen in its ability to break the bonds of long chains of polysaccharides. In the initial stage, the destruction of hydrogen bonds occurs to break up the chains of polysaccharides transforming them into molecules in the amorphous state. Polysaccharides are extremely susceptible to acid hydrolysis at this point. The acid will then serve as the catalyst where the cleavage of the polysaccharide will occur by the hydrolysis of the glycosidic bonds. At the end of the process, any addition or dilution with water at moderate temperature will provide complete hydrolysis and the hydrolyzate will be rich in monosaccharides (Jambo et al., 2016).

In enzymatic hydrolysis, cellulases are the enzymes that are employed to degrade polysaccharides and can be categorized into three major types, which include endoglucanases, exoglucanases, and β-glycosidase. The mechanisms possessed by endoglucanases have the ability to hydrolyze the complex sugars of the raw material, attacking the inner parts of the amorphous region of the cellulose. Exoglucanases degrade cellulose by the cleavage of cellobiose units (union of two glucose molecules) from the nonreducing end of a cellulose fiber that allows the enzymatic attack. With the combined efforts of β-glycosidase, the residues of cellobiase finally split into two units of glucose (Jambo et al., 2016).

Although the chemical hydrolysis usually gives a higher sugar production rate, this method requires harsh-reaction conditions and the hydrolysates may contain inhibitory by-products (such as hydroxymethylfurfural and furfural) for the subsequent bioethanol fermentation. On the other hand, enzymatic hydrolysis produces high amounts of fermentable sugars under mild conditions without producing sugar-degradation products or toxic by-products (Soliman et al., 2018).

To improve the production of high concentrations of fermentable sugars, acid and enzymatic saccharification can be combined. As red seaweeds contain galactan and glucans and galactan cannot be hydrolyzed by the same enzyme used for the hydrolysis of glucans, Yanagisawa et al. (2011) combined acid hydrolysis of galactan to produce galactose followed by the enzymatic hydrolysis of glucan to obtain glucose.

There are different groups of microorganisms like yeast, bacteria, and fungi, which can be used for the fermentation of saccharified algal biomass under anaerobic process for the production of bioethanol (Behera et al., 2015). *S. cerevisiae* (yeast) is the most commonly employed strain in fermentation of bioethanol due to its high selectivity, low accumulation of by-products, high ethanol yield as well as high rate of fermentation (Jambo et al., 2016).

To increase the production rate in the fermentation process, new process configurations were developed such as separated hydrolysis and fermentation (SHF) and simultaneous saccharification and fermentation (SSF).

In SHF process, the saccharification and fermentation of the biomass are performed into two distinct reactors. The saccharification will be conducted first to degrade the biomass into monomer sugars by utilization of enzyme. This process is followed by the fermentation reaction, which will utilize the sugars formed in hydrolysis stage. However, one major problem associated with the SHF process is the end-product inhibition by sugars, which are formed during the hydrolysis (Jambo et al., 2016).

In SSF, the saccharification and fermentation process is conducted simultaneously in a single reactor. During the reaction, the biomass, enzyme, and yeast are put together in an orderly manner so that the sugars released are rapidly converted into bioethanol. SSF can limit the end-product inhibition by removing the residual sugar (Jambo et al., 2016). Thus, SSF processes can achieve better ethanol yields than methods that apply separate hydrolysis of the fermentation (Gupta and Verma, 2015).

El-Dalatony et al. (2016) studied SHF and SSF processes for bioethanol production from microalgal (*Chlamydomonas mexicana*) biomass. SSF was selected as an efficient process to enhance the bioethanol yield through repeated batches using immobilized *S. cerevisiae* YPH499 cells. Combined sonication and enzymatic hydrolysis of *C. mexicana* generated 10.5 and 8.48 g/L of ethanol in SSF and SHF, respectively. Similarly, Tan and Lee (2014) found that SSF of seaweed solid wastes with *S. cerevisiae* was more effective as compared to SHF for bioethanol production. They were able to obtain a bioethanol yield of 55.9% for SHF process and a yield of 90.9% for SSF process.

14.6 FUEL PRODUCTION THROUGHOUT PYROLYSIS OF MICROALGAE

Thermochemical processes rely on the degradation of biomass following chemical reactions occurring at moderate–high temperatures. There are three very common thermochemical processes for converting biomass into biofuels and chemicals: pyrolysis, liquefaction, and gasification.

The main thermochemical route for biomass conversion is pyrolysis, which is conducted in the absence of oxygen and at temperatures between 300°C and 700°C. This process converts biomass into pyrolytic oil (or bio-oil), char (also so-called biochar), and light gases (H_2, CO, CO_2, and C1–C4 hydrocarbons). Biochar is similar to conventional coal but contains some oxygen. The pyrolytic bio-oil is a complex mixture containing several oxygenated compounds. Generally, pyrolysis bio-oils contain mainly water, aldehydes, carboxylic acids, carbohydrates, phenols, furfural, alcohols, and ketones. The water content in bio-oil is relatively high (up to 40 wt%), which very often creates an aqueous rich phase and an organic rich phase (Chew et al., 2017).

Pyrolysis operational conditions are varied and the process is mostly classified into fast, flash, slow pyrolysis, and catalytic pyrolysis. The operational conditions can be set to favor the production of biochar, bio-oil, or gas fraction. For example, fast pyrolysis and flash pyrolysis (low residence times) produce higher yields of bio-oil (up to 80%). In contrast, slow pyrolysis produces larger amounts of char due to increased degradation time of biomass components. Another possibility is to use microwave-assisted pyrolysis, which shows advantages over traditional processes such as rapid and uniform heating, which reduces residence time and accelerates chemical reactions. Regarding the use of fuels and chemical, bio-oil is preferable due to its variety of compounds in composition and flexibility to upgrade, being a platform for both renewable chemicals and fuels. In addition, the use of a catalyst is generally related to decrease the reaction temperature, remove oxygen content, and obtain more interesting fractions related to certain fuels.

Biofuels production is usually classified by its source. First-generation fuels come from sources like starch, sugar, animal fats, and vegetable oil producing, that is, bio-diesel and first-generation ethanol. Second-generation biofuels are related to fuels derived from nonfood biomass, including lignocellulosic biomass and woody crops. For third-generation biofuel, algae are used as raw material.

The dynamics of biomass production as well as where it grows are important factors for a sustainable chain of renewable fuel. Thus, the production of biofuel from algae is of great interest given the high growth rate of some species, the interesting ability to sequester CO_2, the robustness of growing in places that do not compete with arable lands, and noncompetition with human alimentation (Chew et al., 2017).

Algae are a diverse group of aquatic eukaryotic organisms that have the ability to conduct photosynthesis, existing in both unicellular and multicellular forms. They can be classified into (1) macroalgae and (2) microalgae. These photosynthetic organisms convert, through sunlight, CO_2, and water available in nature into algal biomass. Macroalgae are multicellular organisms that present some difficulties of cultivation/use in bioreactors. Because of that, macroalgae have not yet been as widely researched as microalgae. In contrast, microalgae are microorganisms that exist as individual cells or cell chains, living in saline or freshwater environments, with faster growth rates, higher oil contents, and less complex structures than macroalgae (Saber et al., 2016).

Differently than lignocellulosic biomass (composed by cellulose, hemicellulose, and lignin), algae are mainly formed by varied compositions of proteins (30–60 wt%), lipids (5–50 wt%), and carbohydrates (10–40 wt%), depending on the species. Interestingly, in mass basis, elemental analysis shows similar carbon and hydrogen content at present in algae and lignocellulosic materials. In contrast, algae have a much higher nitrogen content.

Promising aspects on the pyrolysis of many algae species, such as *Chlorella*, *Chlorella protothecoides*, *C. vulgaris*, *Dunaliella*, *Emilianiahuxleyi*, *Nannochloropsis* residue, *Plocamium*, *Sargassum*, *Spirulina*, *Synechococcus*, *Scenedesmus*, *Spirulina platensis*, *Tertiolecta*, and *Tetraselmis*, are reported. These results showed promising results that may envisage future industrial applications (Fermoso et al., 2018).

Since lignocellulosic biomass and algae are different in composition; algae bio-oils are quite different compared to lignocellulosic bio-oil. Pyrolysis of algae produces a bio-oil with considerably higher number of linear hydrocarbons, fatty oxygenates, and nitrogenated compounds, which are a result of the cleavage of bonds of lipids and proteins. In addition, since algae have no lignin in its composition, the concentration of phenolic compounds is very low. It is also reported that pyrolysis of algae produces less biochar and a more stable bio-oil with improved properties, which could be more feasibly upgraded (Li et al., 2017).

It is noteworthy to emphasize that the use of micro- and macroalgae as raw materials for biofuel production is in constant increasing. However, large-scale aspects concerning environmental impacts are yet considerations to consider for the near future. Moreover, bio-oil (either from algae or lignocellulose) direct application remains complicated due to its high oxygen content, which leads to several undesired properties such as poor heating values, corrosiveness, and high viscosity. In this context, bio-oil production throughout catalytic pyrolysis and bio-oil hydrotreatment upgrading are one of main research topics on the production of liquid fuels or chemicals in the biorefinery context. Concerning bio-oil complexity, upgrading algae bio-oil displays shows advantages compared to lignocellulosic bio-oil (Taylor, 2008).

14.7 PROTEIN BIOFUELS

Microalgae can offer great potential for large-scale and cost-effective production of recombinant proteins. Huo et al. (2011) studied the feasibility of the use of proteins as raw material for the bioreflex process through the application of a metabolic engineering strategy centralized in nitrogen. They integrate the merits of microorganisms, including the rapid growth and ease of cultivation with those from higher plants in posttranslational modification and photosynthesis (Xu et al., 2012; Gajraj et al., 2018).

Large-scale protein production using microalgae by natural selection under frequent or continuous harvesting conditions may favor robust and fast-growing microorganisms, which generally contain high protein content and are fully adapted to the local environment. Biofuels are currently produced from carbohydrates and lipids from this raw material. Proteins, in contrast, were not used to synthesize fuels due to the difficulties of hydrolysates of deaminating proteins. Thus, recombinant strategies have proven viable to produce large-scale enzymes that can be used to produce biofuels. Huo et al. (2011) as mentioned by Gajraj et al. (2018) applied metabolic engineering to generate *E. coli* that can deaminate protein hydrolysates, allowing cells to convert proteins into C4 and C5 alcohols, which present 56% of theoretical yield. They introduced three exogenous cycles of transamination and deamination, which provided an irreversible metabolic force to drive the deamination reactions to the conclusion (Gajraj et al., 2018).

Huo et al. (2011) demonstrated that *S. cerevisiae*, *E. coli*, *Bacillus subtilis*, and microalgae can be used as protein sources, producing up to 4035 mg/L of alcohols from biomass containing ~22 g/L amino acids. These results show the feasibility of using proteins for biorefineries, for which high protein microalgae could be used as feedstock, with the possibility of maximizing algal growth and total CO_2 fixation.

To test the viability of algae and bacteria proteins as raw materials, Huo et al. (2011) cultivated green algae *C. vulgaris*, red algae *Porphyridium purpureum*, blue-green algae *S. platensis*, and cyanobacterium *Synechococcus elongatus*, as well as *E. coli* and *B. subtilis*, in fluorine of 1- or 30-L tank. The biomass was harvested, digested with protease, and used as feedstock for the production of biofuel with the *E. coli* strain of engineering YH83. In all experiments, the protein concentration was adjusted to 21.6 g/L, which equivalent amount of protein in 4% of yeast extract (Gajraj et al., 2018).

14.8 DIRECT COMBUSTION

For this procedure, the algae biomass is burned in the presence of air in a furnace, boiler, or steam turbine, converting the chemical energy of the algal biomass into heat or electricity. Direct combustion of microalgae is only feasible with a moisture content of less than 50% of its dry weight. However, the process suffers from a limitation because it requires pretreatment of algae, for example, drying, which can lead to a decrease in energy efficiency. Instead, the production of algae and coal biomass is an effective way to generate electricity. During algal biomass aging, algae are cultivated and fed with the CO_2 recycling stream from the plant, and the cultured algae are fed to the plant together with charcoal (Milano et al., 2016; Chye et al., 2018).

14.9 ULTRASONICATION TO AID BIOFUEL YIELDS

According to Chye et al. (2018), the ultrasound process requires the incorporation of acoustic energy with frequencies in the range of 10 kHz to 20 MHz, which is applied to form microbubbles in constant expansion and shrinkage. The microbubbles destabilize as the acoustic energy exceeds a certain limit. When it resonates with the natural frequency of the microbubbles, it collapses, causing high-speed microdata of more than 100 m/s, which carry more than 100 MPa of pressure toward the surface of the biomass. This causes cavitation and the process improves biochemical reactions and may help maximize lipid yield during lipid extraction of microalgae or cause rupture of the cell wall to aid the hydrolysis step for bioethanol fermentation or anaerobic digestion. Suali and Sarbatly (2012) have demonstrated that ultrasound is able to reduce lipid extraction time by one-tenth of the conventional methods and reach 50% to more than 500% yield.

14.10 CONCLUDING REMARKS

In this chapter, micro- and macroalgae were presented as promising raw materials for the production of biofuels, and there are several technologies and studies aimed at improving the process of deconstruction of these algae. The technologies applied for the conversion of algae will depend on what one wants to obtain, as well as on the optimization recommendations for each process used. The characterization of algae biomass based on its lipid, carbohydrate, and protein contents is a crucial factor for obtaining products of interest in the chemical, food, and pharmaceutical industry. Some studies discussed in this chapter discuss the suitability and applicability of biofuels products, which may be an alternative to fossil-fuel shortages. Algae still deserve more focus, a more consolidated concept compared to biorefineries, as well as better future and commercial applications.

KEYWORDS

- **microalgae**
- **macroalgae**
- **biofuels**
- **bioethanol**
- **pyrolysis**

REFERENCES

Alaswad, A.; Dassisti, M.; Prescott, T.; Olabi, A. G. Technologies and Developments of Third Generation Biofuel Production. *Renew. Sustain. Energy Rev.* **2015,** *51*, 1446–1460.

Baeyens, J.; Kang, Q.; Appels, L.; Dewil, R.; Lv, Y.; Tan, T. Challenges and Opportunities in Improving the Production of Bio-Ethanol. *Progr. Energy Combust. Sci.* **2015,** *47*, 60–88.

Bahadar, A.; Khan, M. B. Progress in Energy from Microalgae: A Review. *Renew. Sustain. Energy Rev.* **2013,** *27*, 128–148.

Behera, S.; Singh, R.; Arora, R.; Sharma, N. K.; Shukla, M.; Kumar, S. Scope of Algae as Third Generation Biofuels. *Front. Bioeng. Biotechnol.* **2015,** *2* (90), 1–13.

Bilad, M. R.; Arafat, H. A.; Vankelecom, I. F. J. Membrane Technology in Microalgae Cultivation and Harvesting: A Review. *Biotechnol. Adv.* **2014,** *32*, 1283–1300.

Chisti, Y. Biodiesel from Microalgae. *Biotechnol. Adv.* **2007,** *25* (3), 294–306.

Borines, M. G.; Leon, R. L.; Cuello, J. L. Bioethanol Production from the Macroalgae *Sargassum* spp. *Bioresour. Technol.* **2013,** *138*, 22–29.

Chew, K. W.; Yap, J. Y.; Show, P. L, Suan, N. H.; Juan, J. C.; Ling, T. C.; Lee, D.-J.; Chang, J-S. Microalgae Biorefinery: High Value Products Perspectives. *Bioresour. Technol.* **2017,** *229*, 53–62.

Cho, H.; Ra, C. H.; Kim, S. K. Ethanol Production from the Seaweed *Gelidium amansii*, Using Specific Sugar Acclimated Yeasts. *J. Microbiol. Biotechnol.* **2014,** *24* (2), 264–269.

Cho, Y.; Kim, H.; Kim, S. Bioethanol Production from Brown Seaweed, *Undaria pinnatifida*, Using NaCl Acclimated Yeast. *Bioprocess Biosyst. Eng.* **2013,** *36*, 713–719.

Chye, J. T. T.; Jun, L. Y.; Yon, L. S.; Pan, S.; Danquah, M. K. Biofuel Production from Algal Biomass. In *Bioenergy and Biofuels*; Konur, O. Ed.; CRC Press (Taylor & Francis Group), Boca Raton, FL, **2018;** pp 87–117.

Demirbas, A. Political, Economic and Environmental Impacts of Biofuels: A Review. *Appl. Energy* **2009,** *86*, S108–S117.

Demirbas, A.; Demirbas, M. F. *Algae Energy. Algae as a New Source of Biodiesel.* Springer-Verlag London Limited: London, 2010.

Dias, M. O. S.; Ensinas, A. V.; Nebra, S. A, Filho, R. M.; Rossell, C. E. V.; Maciel, M. R. W. Production of Bioethanol and Other Bio-based Materials from Sugarcane Bagasse: Integration to Conventional Bioethanol Production Process. *Chem. Eng. Res. Des.* **2009,** *87*, 1206–1216.

El-Dalatony, M. M.; Kurade, M. B.; Abou-shanab, R.; Kim, H.; Salama, E.; Jeon, B. Long-Term Production of Bioethanol in Repeated-Batch Fermentation of Microalgal Biomass Using Immobilized *Saccharomyces cerevisiae*. *Bioresour. Technol.* **2016,** *219*, 98–105.

Fermoso, J.; Coronado, J.; Serrano, D.; Pizarro, P. Pyrolysis of Microalgae for Fuel Production. In *Microalgae-Based Biofuels and Bioproducts*; Elsevier: Amsterdam, **2018;** pp 259–281.

Gajraj, R. S.; Singh, G. P.; Kumar, A. Third-Generation Biofuel: Algal Biofuels as a Sustainable Energy Source. In *Biofuels: Greenhouse Gas Mitigation and Global Warming*; Kumar, A., et al., Eds.; Springer (India) Pvt. Ltd.: New Delhi, **2018;** pp 307–325.

Gupta, A.; Verma, J. P. Sustainable Bio-Ethanol Production from Agro-Residues: A Review. *Renew. Sustain. Energy Rev.* **2015,** *41*, 550–567.

Heiman, K. Novel Approaches to Microalgal and Cyanobacterial Cultivation for Bioenergy and Biofuel Production. *Curr. Opin. Biotechnol.* **2016,** *38*, 183–189.

Heasman, M.; Diemar, J.; O'connor, W.; Sushames, T.; Foulkes, L. Development of Extended Shelf-Life Microalgae Concentrate Diets Harvested by Centrifugation for Bivalve Molluscs—A Summary. *Aquacult. Res.* **2000,** *31*, 637–659.

Hu, Q.; Sommerfeld, M.; Jarvis, E.; Ghirardi, M.; Posewitz, M.; Seibert, M.; Darzins, A. Microalgal Triacylglycerols as Feedstocks for Biofuel Production: Perspectives and Advances. *Plant J.* **2008,** *54*, 621–639.

Huo, Y-X.; Cho, K. M.; Rivera, J. G. L.; Monte, E.; Shen, C. R.; Yan, Y.; Liao, J. C. Conversion of Proteins into Biofuels by Engineering Nitrogen Flux. *Nat. Biotechnol.* **2011,** *29* (4), 346–351.

Li, K.; Zhang, L.; Zhu, L.; Zhu, X. Comparative Study on Pyrolysis of Lignocellulosic and Algal Biomass Using Pyrolysis–Gas Chromatography/Mass Spectrometry. *Bioresour. Technol.* **2017,** *234*, 48–52.

Jambo, S. A.; Abdulla, R.; Azhar, S. H. M.; Marbawi, H.; Gansau, J. A.; Ravindra, P. A Review on Third Generation Bioethanol Feedstock. *Renew. Sustain. Energy Rev.* **2016,** *65*, 756–769.

Khambhaty, Y.; Mody, K.; Gandhi, M. R.; Thampy, S.; Maiti, P.; Brahmbhatt, H.; Eswaran, K.; Ghosh, P. K. *Kappaphycus alvarezii* as a Source of Bioethanol. *Bioresour. Technol.* **2012,** *103*, 180–185.

Kim, K. H.; Choi, I. S.; Kim, H. M.; Wi, S. G.; Bae, H. J. Bioethanol Production from the Nutrient Stress-Induced Microalga *Chlorella vulgaris* by Enzymatic Hydrolysis and Immobilized Yeast Fermentation. *Bioresour. Technol.* **2014,** *153*, 47–54.

Kostas, E. T.; White, D. A.; Du, C.; Cook, D. J. Selection of Yeast Strains for Bioethanol Production from UK Seaweeds. *J. Appl. Phycol.* **2016,** *28*, 1427–144.

Kumar, S.; Gupta, R.; Kumar, G.; Sahoo, D.; Huhad, R. C. Bioethanol Production from *Gracilaria verrucosa*, a Red Alga, in a Biorefinery Approach. *Bioresour. Technol.* **2013,** *135*, 150–156.

Kumar, A. K.; Sharma, S. Recent Updates on Different Methods of Pretreatment of Lignocellulosic Feedstocks: A Review. *Bioresour. Bioprocess.* **2017,** *4,* 7.

Lardon, L.; Hélias, A.; Sialve, B.; Steyer, J.-P.; Bernard, O. Life-Cycle Assessment of Biodiesel Production from Microalgae. *Environ. Sci. Technol.* **2009,** *43* (17), 6475–6481.

Leong, W.; Lim, J.; Lam, M.; Uemura, Y.; Ho, Y. Third Generation Biofuels: A Nutritional Perspective in Enhancing Microbial Lipid Production. *Renew. Sustain. Energy Rev.* **2018,** *91,* 950–961.

Milano, J.; Ong, H. C.; Masjuki, H. H.; Chong, W. T.; Lam, M. K.; Loh, P. K.; Vellayan, V. Microalgae Biofuels as an Alternative to Fossil Fuel for Power Generation. *Renew. Sustain. Energy Rev.* **2016,** *58,* 180–197.

Moncada, J.; Tamayo, J. A.; Cardona, C. A. Integrating First, Second, and Third Generation Biorefineries: Incorporating Microalgae into the Sugarcane Biorefinery. *Chem. Eng. Sci.* **2014,** *118,* 126–140.

Moreno-Garcia, L.; Adjallé, K.; Barnabé, S.; Raghavan, G. S. V. Microalgae Biomass Production for a Biorefinery System: Recent Advances and the Way towards Sustainability. *Renew. Sustain. Energy Rev.* **2017,** *76,* 493–506.

Mosier, N.; Wyman, C.; Dale, B.; Richard, E.; Lee, Y. Y.; Holtzapple, M.; Ladisch, M. Features of Promising Technologies for Pretreatment of Lignocellulosic Biomass. *Bioresour. Technol.* **2005,** *96,* 673–686.

Razaghifard, R. Algal Biofuels. *Photosynth. Res.* **2013,** *117,* 207–219.

Rodionova, M. V.; Poudyal, R. S.; Tiwari, I.; Voloshin, R. A.; Zharmukhamedov, S. K.; Nam, H. G.; Zayadan, B. K.; Bruce, B. D.; Hou, H. J. M.; Allakhverdiev, S. I. Biofuel Production: Challenges and Opportunities. *Int. J. Hydrogen Energy* **2017,** *42,* 8450–8461.

Saber, M.; Nakhshiniev, B.; Yoshikawa, K. A Review of Production and Upgrading of Algal Bio-Oil. *Renew. Sustain. Energy Rev.* **2016,** *58,* 918–930.

Sánchez-Tuirán, E.; El-Halwagi, M. M.; Kafarov, V. *Integrated Utilization of Algae Biomass in a Biorefinery Based on a Biochemical Processing Platform, Integrated Biorefineries*. CRC Press: Boca Raton, FL; pp 707–726.

Saravanan, K.; Duraisamy, S.; Ramasamy, G.; Kumarasamy, A.; Balakrishnan, S. Evaluation of the Saccharification and Fermentation Process of Two Different Seaweeds for an Ecofriendly Bioethanol Production. *Biocatal. Agric. Biotechnol.* **2018,** *14,* 444–449.

Scott, S. A. Davey, M. P.; Dennis, J. S.; Horst, I.; Howe, C. J.; Lea-Smith, D. J.; Smith, A. G. Biodiesel from Algae: Challenges and Prospects. *Curr. Opin. Biotechnol.* **2010,** *21,* 277–286.

Singh, A.; Olsen, S. I. A Critical Review of Biochemical Conversion, Sustainability and Life Cycle Assessment of Algal Biofuels. *Appl. Energy* **2011,** *88,* 3548–3555.

Soliman, R. M.; Younis, S. A.; El-Gendy, N. S.; Mostafa, S. S. M.; El-Temtamy, S. A.; Hashim, A. I. Batch Bioethanol Production via the Biological and Chemical Saccharification of Some Egyptian Marine Macroalgae. *J. Appl. Microbiol.* **2018,** *125* (2), 422–440.

Suali, E.; Sarbatly, R. Conversion of Microalgae to Biofuel. *Renew. Sustain. Energy Rev.* **2012,** *16* (6), 4316–4342.

Sunwoo, I. Y.; Kwon, J. E.; Nguyen, T. H.; Ra, C. H.; Jeong, G.; Kim, S. Bioethanol Production Using Waste Seaweed Obtained from Gwangalli Beach, Busan, Korea by

Co-culture of Yeasts with Adaptive Evolution. *Appl. Biochem. Biotechnol.* **2017,** *183*, 966–979.
Tan, I. S.; Lee, K. T. Enzymatic Hydrolysis and Fermentation of Seaweed Solidwastes for Bioethanol Production: An Optimization Study. *Energy* 2014, *78*, 53–62.
Taylor, G. Biofuels and the Biorefinery Concept. *Energy Policy* **2008,** *36*, 4406–4409.
Tredici, M. R. Photobiology of Microalgae Mass Cultures: Understanding the Tools for the Next Green Revolution. *Biofuels* **2010,** *1*, 143–162.
Um, B.-H.; Kim, Y-S. Review: A Chance for Korea to Advance Algal-Biodiesel Technology. *J. Ind. Eng. Chem.* **2009,** *15*, 1–7.
Xu, J.; Dolan, M. C.; Medrano, G.; Cramer, C. L.; Weathers, P. J. Green Factory: Plants as Bioproduction Platforms for Recombinant Proteins. *Biotechnol. Adv.* **2012,** *30* (5), 1171–1184.
Wang, X.; Liu, X.; Wang, G. Two-stage Hydrolysis of Invasive Algal Feedstock for Ethanol Fermentation. *J. Integr. Plant Biol.* **2011,** *53* (3), 246–252.
Wargacki, A. J.; Leonard, E.; Win, M. N.; Regitsky, D. D.; Santos, C. N. S.; Kim, P. B.; Cooper, S. R.; Raisner, R. M.; Herman, A.; Sivitz, A. B.; Lakshamanaswamy, A.; Kashiyama, Y.; Baker, D.; Yoshikuni, Y. An Engineered Microbial Platform for Direct Biofuel Production from Brown Macroalgae. *Science* **2012,** *335*, 308–313.
Yanagisawa, M.; Nakamura, K.; Ariga, O.; Nakasaki, K. Production of High Concentrations of Bioethanol from Seaweeds That Contain Easily Hydrolyzable Polysaccharides. *Process Biochem.* **2011,** *46*, 2111–2116.

CHAPTER 15

Bioethanol Production from Different Lignocellulosic Biomass

SAURABH SINGH[1] and JAY PRAKASH VERMA[1,2,*]

[1]*Institute of Environment and Sustainable Development, Banaras Hindu University, Varanasi 221005, Uttar Pradesh, India*

[2]*Hawkesbury Institute for the Environment, Hawkesbury Campus, Western Sydney University, Penrith, NSW 2750, Sydney, Australia*

[*]*Corresponding author.*
E-mail: jpv.iesd@bhu.ac.in, verma_bhu@yahoo.co.in, j.verma@westernsydney.edu.au

ABSTRACT

Lignocellulosic biomass is the most abundant organic material present on the surface of the earth. It has the capacity to provide almost 13% of the worldwide energy consumption . Having the production potential to be around 220 billion tones, it is not fully exploited of its potential. Technologies for the production of bioethanol from lignocellulosic biomass have evolved quite a lot since the introduction of the concept of biofuel. The major obstacles which arise are the presence of lignin in the complex web-like structure with cellulose and the inefficiency of the microbes identified to produce all three set of enzymes of the cellulase system (endoglucanase, exoglucanase, and β-1,4-glucosidase) efficiently. Other issues, which arise at the pretreatment step, are the cost-effectiveness and also some environmental issues. Apart from this at the hydrolysis step, the feedback inhibition by the products creates a major hindrance. Bioethanol production from lignocellulosic biomass gives us an alluring insight to the future of renewable energy development.

15.1 INTRODUCTION

Biofuels have become a hot topic of late owing to their eco-friendly properties. They are known to reduce the emissions resulting from the use of fossil fuels. These are gaining prevalence because of the sustainable development goals, which have come into existence as a need to develop a sustainable future on Earth. The need to reduce the emissions resulting from the fossil fuels is urged because of the fact that the mean global temperature is increasing, due to excessive greenhouse effect, which is known as global warming. The natural mean average temperature of the Earth is nearly 15°C, for which the mainly responsible gases are H_2O and CO_2. This natural mean average temperature of the Earth has been affected in the recent past because of the emissions resulting from fossil-fuel combustion and other anthropogenic activities. The emissions have resulted in the abundance of gases, such as N_2O, excessive CO_2, CFCs, CH_4, and SF_6. The result of emissions is such that the concentration of CO_2 in the preindustrial level, which was 280 ppm, has been increased to 410 ppm to the present time. In addition, as a result of this increase in the concentration of these greenhouse gases in the atmosphere, they have led to an increase in the global average temperature of the Earth, by absorbing the terrestrial radiations of the atmospheric window. The presence of the anthropogenically generated greenhouse gases in the atmosphere leads to the absorbance of the terrestrial radiations of 8–14 μm wavelengths. Apart from CFCs, all other anthropogenically generated greenhouse gases are a result of mainly fossil-fuel combustion. The fossil-fuel combustion leads to the formation of nitric oxide, which upon oxidation in the presence of oxygen leads to the formation of nitrous oxide. NOx, SOx, and hydrocarbons are the main constituents of the vehicular emissions. To reduce these emissions from the atmosphere, there has been an urge in the nations to curtail the use of fossil fuel with a vision of sustainable Earth. This has been logically presented with a solution of biofuels. The major biofuel, which is in existence at present, is bioethanol. Bioethanol is the leading biofuel mainly because of its multiple eco-friendly properties. Bioethanol production falls under the category of second-generation biofuels, which utilizes waste materials for the formation of fuels. It provides us with double benefits; first, it reduces the emissions resulting from the waste itself, and second, it minimizes the emissions, which will generate if the same amount of fossil fuel is burnt. Nowadays, bioethanol is generated mainly from the agricultural waste. Agricultural waste is lignocellulosic in nature. The lignocellulosic waste when acted upon by the microbes leads to its degradation, and upon employment of several techniques, ethanol can be generated from it. The ethanol produced from it can directly be used as a replacement for the fossil fuels, either completely or partly depending upon the composition of the engine. This chapter discusses in detail the production of bioethanol from lignocellulosic biomass, along with the composition of lignocellulosic biomass and its prevalence, and also the processes and current status of bioethanol production in today's world. This chapter will make the readers familiar about the process involved in the production of bioethanol from the lignocellulosic biomass. Further, it will also inform them about the recent advances and current technological aids that are explored and focused for the bioethanol production from lignocellulosic biomass. At the end, it will tell the readers about the current status of bioethanol production in the world.

15.2 LIGNOCELLULOSIC BIOMASS

Plant material is composed of three basic components: cellulose, hemicellulose, and lignin. Together, these three form lignocellulosic complex. Cellulose and hemicellulose are the carbohydrate polymers, whereas the lignin is the aromatic polymer. Lignocellulosic biomass is the most abundant biomass present on the Earth's surface. All the plant parts in trees, shrubs, herbs, or grasses have all these components in varying amounts. Cellulosic biomass, sometimes called lignocellulosic biomass, is a heterogeneous complex of carbohydrate polymers and lignin, a complex polymer of phenylpropanoid units. Lignocellulosic biomass typically contains 55–75% carbohydrates by dry weight. Cellulose is a polymer of glucose and favors the formation of tightly packed structures in the form of tightly held chains. This property of cellulose provides them the crystalline structure and also enables them to be resistant against depolymerization. Cellulose is the polymer, which is the main constituent of the plant cell as well as bacteria, fungi, and algae. The ratio of different constituents in the plants varies with the type, age, species, stage of growth, and several other conditions as well. The dry weight composition of different lignocellulosic biomass feedstock is presented in Table 15.1. Chemically, cellulose is a homopolymer of β-D-glucopyranose moieties which are linked through β-1,4-glycosidic bonds. The degree of polymerization of cellulose can reach up to 15,000 or more. Cellulose is the most abundant polymer present on the surface of the Earth. The second most abundant polymer is hemicellulose, which comprises about 20–50% of the total lignocellulosic biomass. Hemicelluloses are those compounds that are the branched polymers of glucose or xylose substituted with arabinose, xylose, galactose, fucose, mannose, glucose, or glucuronic acid. The presence of the hydrogen bonding between the hemicellulose compounds and the cellulosic compounds provides the structural backbone to the plant. Its composition also varies with type, age, species, stage of growth, and several other conditions. Basically, they are composed of xylan in agricultural biomass, whereas in the softwood hemicelluloses, glucomannan is dominant. In many plants, xylans in the form of heteropolysachharides with backbone chains of 1,4-linked β-D-xylopyranose is present. Apart from xylose, xylans may contain arabinose, glucuronic acid, or its 4-*O*-ethyl ester, acetic acid, ferulic, and *p*-coumaric acids. Xylans are easy to extract in comparison to glucomannan, which requires stronger alkaline environment than the former. Hemicellulose is thermochemically sensitive. Hemicellulose coats the cellulose fibrils in the plant structure and it has been predicted that the 50% of hemicellulose should be removed to effectively cause degradation of the cellulose. But then, it should be carefully done so as to not allow the product formation from lignin degradation, which is reported to inhibit fermentation (Palmqvist and Hahn-Hägerdal, 2000). Lignin, another major component of the plant material, is formed of aromatic polymers and thus makes them recalcitrant. The presence of lignin in the plant structures provides them strength as well as resistance to external pathogens. It is the major obstacle in the enzymatic hydrolysis as it is not easily degradable. Apart from the major hindrance of its digestibility, it also sticks to the hydrolytic enzymes of cellulose hydrolysis. This creates a site competition for the enzymes and thus makes the hydrolysis of cellulose slower.

Different types of biomass are prevalent in the different regions. The main feedstock for the preparation of bioethanol in the United States is corn, while that in Brazil is sugarcane. Apart from the production feedstocks, high amounts of lignocellulosic agricultural wastes are generated in the form of wheat straw and rice straw in the northern part of India, cassava in Indonesia, and likewise many others. The production of these types of biomass is region-specific and should be utilized in the same.

TABLE 15.1 Dry Weight Composition in Percentage of Lignocellulosic Feedstocks (Mosier et al., 2005; Gupta and Verma, 2015).

Feedstock	Cellulose	Hemicellulose	Lignin	References
Corn stover	37.5	22.4	17.6	Mosier et al. (2005)
Corn fiber	14.28	16.8	8.4	Mosier et al. (2005)
Pine wood	46.4	8.8	29.4	Wiselogel et al. (2018), Wyman (1996)
Popular	49.9	17.4	18.1	Wiselogel et al. (2018), Wyman (1996)
Wheat straw	38.2	21.2	23.4	Wiselogel et al. (2018), Wyman (1996)
Switch grass	31.0	20.4	17.6	Wiselogel et al. (2018), Wyman (1996)
Cereal straws	35–40	26	15–20	Schell et al. (2004), Shenoy et al. (2011)
Rice straw	32.6	27.3	18.4	Zheng et al. (2012)
Sugarcane bagasse	65	–	18.4	Sarkar et al. (2012)
Cotton seed hairs	80–85	5–20	0	Abbasi and Abbasi (2010)
Oil palm frond	49.8	83.5	20.5	Khalil et al. (2007)
Coconut	44.2	56.3	32.8	Khalil et al. (2007)
Pineapple leaf	73.4	80.5	10.5	Khalil et al. (2007)
Banana stem	63.9	65.2	18.6	Khalil et al. (2007)
Softwood	40–50	25–30	25–35	Dwivedi et al. (2009)
Big blustem (whole plant)	29–37	21–25	17–24	Wei et al. (2009)
Switchgrass (whole plant)	31–35	24–28	17–23	Wei et al. (2009)
Jatropha waste	56.31	17.47	23.91	Sricharoenchaikul et al. (2007), Khalil et al. (2013)
Paper	85–99	0	0–15	Abbasi and Abbasi (2010)
Poplar aspen	42.3	31.0	16.2	Isahak et al. (2012), Scott et al. (1985)

The lignocellulosic biomass contains on an average 50–55% of carbohydrate in the form of cellulose and hemicellulose. Thus, the production of bioethanol focuses on the carbohydrate portion, mainly cellulose. The nonprioritization of hemicellulose is because of the fact that its degradation causes the formation of by-products which are inhibitory in nature to the hydrolytic enzymes. The production of bioethanol from lignocellulosic biomass can be divided into three basic furcation, that is, pretreatment, enzymatic hydrolysis, and fermentation.

15.3 PRETREATMENT

It is the process of loosening of cellulose from hemicellulose and lignin. The lignin, which is present in the lignocellulosic biomass, offers strength and hardness to the structure of the biomass. Hemicellulose also acts in the same manner to provide firmness to the structure of the lignocellulosic biomass. The process of partial degradation of the lignin and the hemicellulose to free the cellulose chains is called pretreatment. Based on the type of method used for the loosening of the biomass, pretreatment process has been divided into three basic types: physical pretreatment, chemical pretreatment, and biological pretreatment.

15.3.1 PHYSICAL PRETREATMENT

The use of different techniques such as milling, grinding, or chipping to reduce the crystallinity of the biomass can be categorized under physical pretreatment (Sun and Cheng, 2002). Different types of milling can be wet milling, dry milling, vibratory ball milling, and compression milling (Gupta and Verma, 2015). Physical pretreatments function by minimizing the size of the biomass and thus allowing more surface area for the enzymatic hydrolysis. The strength of the biomass is directly related to the amount of power that needs to be applied for mechanical comminution. Other forms of milling can be compression milling, ball milling, cryo-milling, and attrition milling and steam treatment using poplar, wheat straw, newspaper, oat straw, and so on. Overall, there are six types of physical pretreatment: wet treatment, dry treatment, compression treatment, attrition treatment, cryo-treatment, and steam explosion. Wet treatment uses the moisture or water to swell the adjoining layers and thus making the inner layers of the biomass accessible. Dry treatment refers to the treatment of the biomass in zero-moisture conditions to bring about the change. Compression pretreatment refers to the use of mechanical pressure to bring about the breaking of the biomass into smaller pieces, and presenting more surface area. Attrition treatment is the use of wet and dry treatment alternatively to bring about the change. Cryo-treatment uses very low temperatures for the purpose of degradation. In cryo-treatment technology, liquid nitrogen has also been reported to be used. Uncatalyzed steam treatment is also considered under physical pretreatment process.

15.3.2 CHEMICAL PRETREATMENT

Chemical pretreatment is the use of chemicals such as acids or alkalis for the partial breakdown of the lignocellulosic biomass. Lignin in the agricultural residues, which is mainly in the form of xylose, is comparatively easy to break with the use of mild acids or alkali, whereas the lignin from softwood, which contains mainly glucomannan, is broken down by the use of strong alkali. The alkali used in the chemical treatment is mainly NaOH (Gupta and Verma, 2015). Acids used for the purpose are $HClO_4$, HCH_3COOH, H_2SO_4, and HCOOH. There are certain organic solvents used for the chemical pretreatment, such as propylamine, ethylene diamine, and *n*-butylamine (Martínez et al., 2005). The use of chemical pretreatment in the paper industry to produce good-quality paper products has been in existence since long. Some ionic liquids too significantly impact the structure of the lignocellulosic biomass (Fengel and Wegner, 1984, Swatloski et al., 2002). High-energy pretreatment can also be categorized under this category.

15.3.3 BIOLOGICAL PRETREATMENT

Biological pretreatment refers to the treatment of the lignocellulosic waste with microorganisms, especially fungus for the liberation of cellulose from the lignocellulosic complex. In biological pretreatment process, various microorganisms like brown rot, white rot, and soft rot fungi can be used to degrade the lignocellulosic biomass. White rot fungus has been found to be the most effective amongst all the abovementioned microorganisms. Brown rot fungus acts by attacking on the cellulose fibers, whereas the white rot fungus functions by attacking both cellulose and lignin. The function of fungus hydrolysis is through the release of hydrolytic enzymes and also the formation of hyphal structures.

15.3.4 PHYSICOCHEMICAL PRETREATMENT

Physicochemical pretreatments are the type of pretreatments that combine both the physical and chemical process. There are many processes that have been categorized into physicochemical pretreatment, such as steam explosion which takes place in the presence of chemicals, namely, SO_2, ammonia, carbon dioxide, and many more. Autohydrolysis, which is also called steam explosion, is the most commonly used type of physicochemical pretreatment. It removes most of the hemicellulose, thereby improving enzymatic digestion. In this process, there is a sudden decrease in the pressure, due to which the materials undergo an explosive decompression. And then, high pressure and temperature conditions (160–260°C) are given for a specified period of time, ranging between few seconds and several minutes (Boussaid et al., 1999; Varga et al., 2004; Sun et al., 2004). The temperature of the treatment is in direct correlation with the effective release of hemicellulose sugars up in the mentioned range. However, if the temperature is further increased, it leads to the loss of sugars, thereby results in a decrease in total sugar recovery (Ruiz et al., 2008). Apart from this, steam explosion with the addition of SO_2 is done to improve the recovery of both cellulose and hemicellulose. Treatment takes place with the addition of 1–4% SO_2 (w/w substrate) at temperatures ranging between 160 and 230°C (Taherzadeh and Karimi, 2008). The maximum glucose yield of 95% was obtained when 1% of SO_2 treatment was applied on willow at 200°C (Eklund et al., 1995).

Ammonia fiber explosion is an alkaline physicochemical process. In this, liquid ammonia is used at temperatures of around 100°C for a certain period of time (Taherzadeh and Karimi, 2008). This process mainly acts to remove the lignin content from the lignocellulosic biomass, leaving hemicellulose and cellulose intact. The biggest advantage presented by this type of pretreatment is that no inhibitory by-product formation takes place, which forms in other types of pretreatment processes.

CO_2 explosion process involves the use supercritical carbon dioxide. About 100% glucose yield has been reported by Park et al. (2001) on the application of supercritical CO_2 along with enzymatic hydrolysis simultaneously. Apart from this, supercritical carbon dioxide is readily available at low cost, nontoxic, nonflammable, easy to recover after extraction, and is environmentally acceptable (Zheng and Tsao, 1996).

Liquid hot water pretreatment involves the use of very little or no chemicals in the pretreatment unit. In some cases, there is an addition of small amount of acid or base to adjust the pH of the process. The process takes place at temperatures between 100°C and 250°C (Taherzadeh and Karimi, 2008). This process mainly removes hemicellulose and can be used in combination with delignification pretreatment processes.

Microwave chemical pretreatment is a physicochemical pretreatment, which involves the use of microwaves along with the use of chemicals such as acids and bases. Microwave-chemical pretreatment results in more effective pretreatment process (Taherzadeh and Karimi, 2008; Zhu et al., 2005, 2006). Apart from the use of acids and alkali, H_2O_2 is also used in combination for the pretreatment purpose. In a study, it was found that microwave–acid–alkali–H_2O_2 pretreatment had the highest hydrolysis rates and also maximum glucose content in the hydrolysate.

15.4 ENZYMATIC HYDROLYSIS OR SACCHARIFICATION

The pretreated substrate when ready for hydrolysis is acted upon by the hydrolytic enzymes produced by the microbial strains. The microbial strains such as *Clostridium*, *Cellulomonas*, *Thermonospora*, *Bacillus*, *Bacteriodes*, *Ruminococcus*, *Erwinia*, *Acetovibrio*, *Microbispora*, *Streptomyces*, and other fungi such as *Trichoderma*, *Penicillium*, *Fusarium*, *Phanerochaete*, *Humicola*, and *Schizophillum* sp., which possess the ability of degrading cellulose secrete enzyme called cellulase (Gupta and Verma, 2015). These enzymes have ability to convert the cellulose to glucose or galactose monomer. Cellulase is a group of enzymes, which act synergistically to bring about the degradation of the cellulose. The cellulase enzyme is composed of three different enzymes, namely, endoglucanases, exoglucanases, and β-glucosidases. The degradation of cellulosic biomass is accomplished by the most prominent form of associated enzymes, that is, cellulases (Hasunuma et al., 2013). The complex form of cellulase consists of endoglucanases (1,4-β-D-glucanohydrolases) and exoglucanases that also contain cellodextrinases (1,4-β-D-glucan glucanohydrolases), cellobiohydrolases (β-D-glucan cellobiohydrolases), and β-glucosidases (β-glucoside glucohydrolases). The mechanism of action of cellulases is such that initially the packed structure of the cellulose needs to be broken down, which is in the form of D-glucopyranose units linked to each other by β-1,4-glycosidic linkages. According to study, it has been found that 90–95% of the cellulose degrading strains of bacteria and fungus are aerobic while the remaining 10% account for the anaerobic microbes.

The end product of the cellulose hydrolysis gives mainly glucose as other components in the lignocellulosic biomass also degrade to form some or the other components. In a study by Palmqvist and Hahn-Hägerdal (2000), it is shown that the hydrolysis of cellulose gives glucose, but the hydrolysis of hemicellulose may yield xylose, mannose, galactose, or glucose depending upon the composition of the hemicellulose. The further breakdown of xylose yields furfural, which then leads to the formation of formic acid. Likewise, the formation of hydroxymethylfurfural from the mannose, glactose, and glucose can lead to the formation of formic acid and levulinic acid (Palmqvist and Hahn-Hägerdal, 2000). The formation of these acids is enhanced by the presence of high amounts of hemicellulose in the biomass. The presence of these acids in the reaction chamber leads to the inhibition of the process and the desired yield is reduced.

15.4.1 CELLODEXTRINASES OR ENDOGLUCANASES

These are also called endoglucanases or 1,4-β-D-glucan glucanohydrolase. These act by randomly incising the random amorphous sites in the cellulose polysaccharide chain. In other words, they attack the region of high crystallinity and produce oligosaccharides of various lengths. The break in the structure is brought about by the insertion of water molecule in place of 1,4-β bond. Chemically, they break β-1,4-glucosidic linkages randomly and break cellulose with a high degree of polymerization. These also create new reducing and nonreducing ends in the cellulose chains to allow cellobiohydrolases to act upon them.

15.4.2 CELLOBIOHYDROLASES OR EXOGLUCANASES

These are also called exoglucanases or β-D-glucan cellobiohydrolases. These act upon the end product of endoglucanses or cellodextrins to form cellobiose. They bring about the hydrolysis by catalyzing the reducing and nonreducing ends of cellulose, releasing cellobiose molecules as the main product, the substrate for β-glucosidases. They are almost 40–70% of the total composition of the cellulase enzyme system. This means that most of the hydrolytic activity is done by the cellobiohydrolases in the cellulase system. Their main function is the breakdown of oligomers into dimers. Cellobiose is the dimer of glucose, that is, it is a repeating unit of two units of glucose.

15.4.3 β-GLUCOSIDASES

They bring about the hydrolysis of cellobiose and remaining cellodextrins, producing glucose. These are also called 1,4-β-D-glucosidase glucanohydrolase. They bring about the hydrolysis of cellobiose and the remaining cellodextrins to form glucose. They can bring about the degradation of cellodextrins with a degree of polymerization up to six. But the matter of fact remains that the rate of hydrolysis decreases significantly as the degree of polymerization increases. Temperature plays an important role in the hydrolytic action of β-glucosidases and is favored at high temperatures. In a study, it was found that some soil microbes showed highest enzyme activity at temperature of 70°C incubation temperature (Eivazi and Tabatabai, 1988).

15.5 FERMENTATION

Fermentation is the process by which the saccharification products are converted into ethanol and other by-products in the absence of oxygen. It is carried out under the presence of fermentative microbes, which are mostly anaerobic. There are many microbes which can carry out the function of fermentation, such as *Zymomonas mobilis* and *Saccharomyces cerevisiae* (Gunasekaran and Raj, 1999). The hydrolysis temperature, time, and acid concentration influence the generation of fermentation inhibitors. The severity of different pretreatment conditions can be compared by calculating a severity parameter, where the reaction temperature, T (°C), and residence time, t (min), are combined into a single reaction ordinate. Fermentation is also governed by several factors, such as temperature, by-products in the chamber, presence of acids, and many others. The presence of weak acids in the reaction chamber is one of the biggest factors that inhibit the formation of the ethanol from sugars. Weak acids have a high dissociation constant and thus are not easy to break and flow inside the cytosol, creating an inhibitory effect on the growth of the cells.

15.5.1 DIFFERENT TYPES OF FERMENTATION

Fermentation for bioethanol production can be divided into two basic types: simultaneous saccharification and fermentation, and separate hydrolysis and fermentation. In separate hydrolysis and fermentation, the pretreated agrowastes are subjected to enzymatic hydrolysis for the saccharification and then are allowed to ferment separately by the use of fermentative microbes such as *S. cerevisiae*, rumen microbes such as *Bacteroides succinogenes*, *Ruminococcus flavefaciens*, *Ruminococcus albus*, *Butyrivibrio fibrisolvens*, *Phycomycets*, *Bacteroides ruminicota*, *Bacteroides amylophilus*, and so on (Lin et al., 1985). Separate hydrolysis and fermentation involve step-by-step processing and hence is not cost-effective. This avoided by the use of simultaneous saccharification and fermentation. Simultaneous saccharification and fermentation involve the use of microbes to obtain products of hydrolysis and fermentation from a single reactor. In a study by Öhgren et al. (2007), it was concluded that ethanol yield with simultaneous saccharification and fermentation was 78.2%, whereas with separate hydrolysis and fermentation, it was 64.1%. This suggests that ethanol yield increases with the use of simultaneous saccharification and fermentation technology.

With advancement in the technology for the production of bioethanol, a new one, which has come into existence, is solid-state fermentation. Solid-state fermentation is the fermentation which involves the use of solid material in almost negligible water, but enough moisture to allow the growth of microbes involved (Pandey, 2003). Solid-state fermentation is a better and efficient technology as it has less energy requirements, produces less waste water, and is eco-friendly as well. Initially, it was thought that the simultaneous saccharification fermentation (SSF) is suitable only for fungus as the bacterial cultures require high amounts of water for their activity, but later with the emergence of new bacterial strains and experiments, it was found that bacterial cultures can also perform solid-state fermentation (Selvakumar and Pandey, 1999; Pandey et al., 1998). SSF is now seen as process, which generates higher yields than the submerged fermentation technology, even though the exact reasons behind the same still remain unclear (Pandey, 2003; Pandey et al., 1998).

15.6 GENETICALLY MODIFIED MICROBES FOR BIOETHANOL PRODUCTION

Bioethanol production with the use of naturally occurring microbes has not been very much efficient. So, researchers are trying to produce new techniques that can provide them with enhanced yields of bioethanol. One of the best techniques is through the modification of the microbe being used at the genetic level so as to produce higher amounts of hydrolytic enzymes. To get economically feasible cellulose conversion of the lignocellulosic biomass, the genetic improvements in the microbial strains are required. Most of the research activity that has been carried out with respect to the genetic modifications in the microbial strains is in the fungal strains. Much of the effort has been put in *Trichoderma reesei* by mutating them and selecting highly efficient mutants for the purpose of degradation of cellulose. In *T. reesei* strains, mutant strains have been obtained which are resistant to catabolite repression, hence achieving the extracellular protein production, most of which is cellulase, more than 35 g/L (Béguin et al., 1987). Apart from catabolite repression, two major obstacles, which form hindrances in the feasible degradation of cellulose, are the slow growth of this filamentous fungi and the presence of the secreted extracellular protein, affecting the metabolism of the organism. Engineering of the microbes for improvement in cellulase activity, which includes enhanced catalytic activity as well as increased thermostability of the enzyme, has become an important aspect for the commercialization of the lignocelluloses biorefinery. Presently, the study on cellulose degrading bacteria is very much less than the cellulose degrading fungi. In general, the cellulase activity that has been reported in cellulose degrading bacteria has been very much less compared to cellulose degrading fungi. As the cellulase activity is known to perform better at high temperatures, that is, near 50°C, high-temperature stable enzymes can work more efficiently. The use of thermophilic microbes is an useful option as its enzymes will have high thermostability. Many genetically modified microbes have been used for the purpose of cellulose degradation. Some of the recent studies are *Bacillus amyloliquefaciens* S1, in which a gene encoding for a type of cellulase was cloned (Sun et al., 2017), *Cellulomonas biazotea* (Chan et al., 2018), in which two novel β-glucosidase genes encoding isoenzymes, were cloned.

15.6.1 CELLULOSE DEGRADATION IN ANAEROBIC BACTERIA

In anaerobic bacteria, the cellulose degradation is performed with the help of special units called cellulosomes (Béguin et al., 1987). Cellulosomes were first discovered as large protuberances, consisting of a large saffoldin protein unit with enzymatic subunits attached to it. They were first observed in *Clostridium thermocellum* (Bayer et al., 1983; Lamed et al., 1983). Cellulosome units have cellulases as their enzymatic subunits as well as hemicellulases in some of the bacterial strains. First, scaffoldin protein was sequenced in 1991 in *Clostridium cellulovorans* (Shoseyov et al., 1992). The cellulosome assembly is in itself efficient enough to cause the degradation of cellulose, which in other cases is the result of the synergistic mechanism of the different cellulases (Bayer et al., 1983). Cellulosome are encoded by genes, which may exceed that required for the coding 14–18 polypeptides (Lamed et al., 1983; Béguin et al., 1987). In short, cellulosomes are a collection of free cellulases, which are independently capable of cellulolysis.

15.6.2 GENETICALLY MODIFIED FUNGUS FOR BIOETHANOL PRODUCTION

Genetically modified organisms have always presented us with better strains in comparison to the wild types. Among the naturally existing organisms, the fungal strains show better activity in degrading the lignocellulosic waste. Thus, genetic engineering on fungal strains is more in prevalence with enhanced activity results. In a recent work by Ramani et al. (2015), recombinant BGL (β-glucosidase) in *Pichia pastoris* showed optimal activity at a decreased pH of 5.0 and temperature 60°C with cellobiose conversion rate 2.083 μmol min^{-1} mg^{-1}. It also showed glucose tolerance levels up to 400 mM. Another such study by Chen et al. (2012) obtained results of 34.5 U/mg for cellobiose degradation. The results were better than obtained in saccharification with Novozyme 188. Similarly, Yan et al. (2012) obtained results of 274.4 U/mL through genetic modification in *Paecilomyces thermophila*. Shen et al. (2008), Murray et al. (2004), and Yao et al. (2015) obtained results of 1.02 IU/mg, 512 IU/mg V_{max}, and 150 U/mL of BGL activity, respectively. *T. reesei* is one of the major fungal strains in bioethanol production from lignocellulosic biomass. Several studies have been shown to increase their activities tremendously through the use of genetic modifications (Dashtban and Qin, 2012; Ma et al., 2011; Nakazawa et al., 2012). Ma et al. obtained results of 6–8-folds higher BGL activity in comparison to the native strain, whereas Wang and Xia (2011) and Dashtban and Qin (2012) obtained 106 times and 10.5 times higher BGL activity, respectively. Nakazawa et al. (2012) obtained results of 10 U/mg for BGL activity. In general, fungal strains have shown better results with respect to bioethanol production from lignocellulosic biomass.

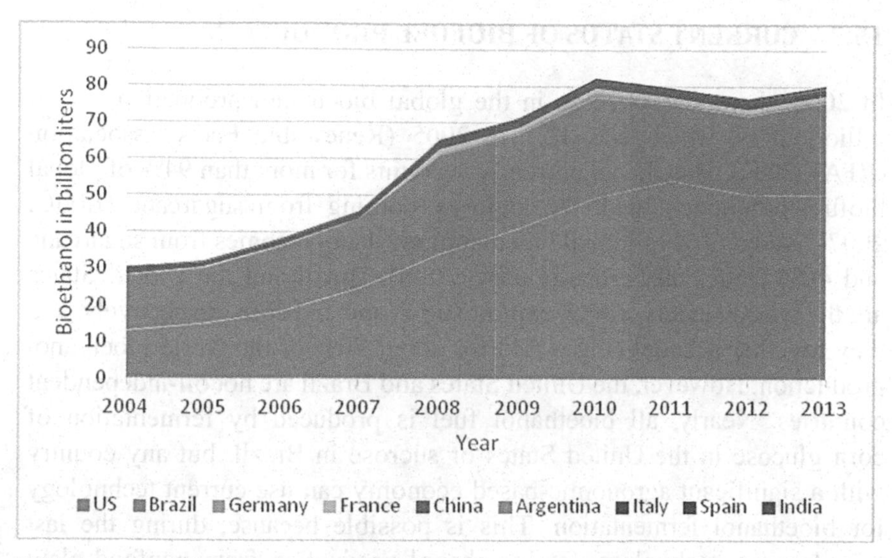

FIGURE 15.1 Bioethanol production by some major countries in billion liters (Gupta and Verma, 2015).

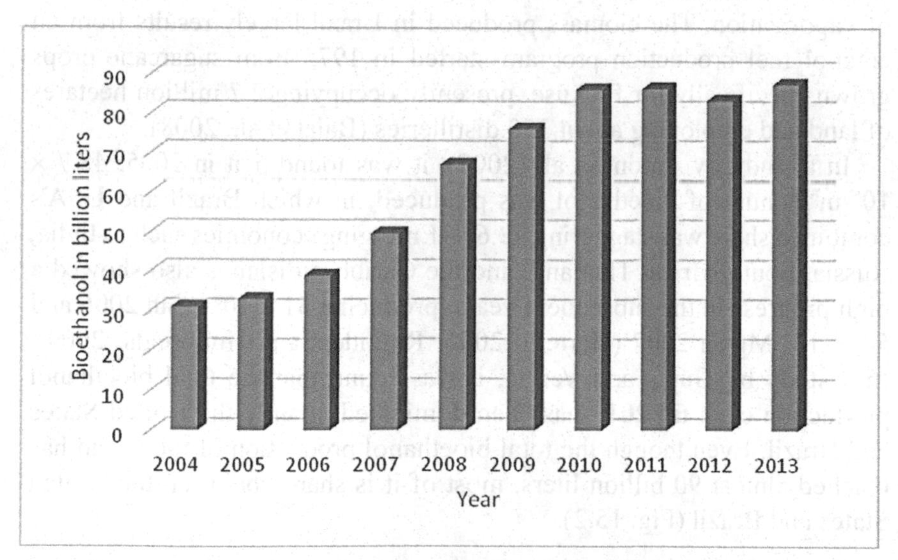

FIGURE 15.2 Total bioethanol production of the world in billion liters. Source: Reprinted with permission from Gupta and Verma, 2015. © 2015 Elsevier.

15.7 CURRENT STATUS OF BIOFUEL PRODUCTION

In 2006, there was increase in the global bioethanol production by 1.4 billion liters, which was 12.1 in 2005 (Renewable Fuels Association (RFA), 2007). Bioethanol currently accounts for more than 94% of global biofuel production, with the majority coming from sugarcane (IRGC, 2007). About 60% of global bioethanol production comes from sugarcane and 40% from other crops (Dufey, 2006). Brazil and the United States are the world leaders, which exploit sugarcane and corn, respectively, and they together account (Fig. 15.1) for about 70% of the world bioethanol production. However, the United States and Brazil are not oil-independent countries. Nearly, all bioethanol fuel is produced by fermentation of corn glucose in the United States or sucrose in Brazil, but any country with a significant agronomic-based economy can use current technology for bioethanol fermentation. This is possible because, during the last two decades, technology for bioethanol production from nonfood-plant sources has been developed to the point at which large-scale production will be a reality in the next few years (Lin and Tanaka, 2006). Ethanol is an oxygenated fuel that contains 35% oxygen, which reduces particulate and NOx emissions from combustion. Ethanol has a higher octane number (108), broader flammability limits, higher flame speeds, and higher heats of vaporization. The biomass produced in Brazil largely results from an ethanol fuel production program started in 1975 from sugarcane crops grown specifically for fuel use, presently occupying 2.7 million hectares of land and employing about 350 distilleries (Balat et al., 2008).

In a study by Antoni et al. (2007), it was found that in 2005, 48.7×10^6 m^3/annum of bioethanol was produced, in which Brazil and USA's combined share was staggering 72.6%. Emerging economies such as India, Russia, South Africa, Thailand, and the Caribbean islands also showed a high progress in the subsequent years, producing 51×10^6 m^3 in 2006 and 54×10^6 Mg in 2007 (Herrera, 2006; Reijnders and Huijbregts, 2009). In a study by Gupta and Verma, it was found that the total bioethanol production even till 2013 has been dominated, mainly the United States and Brazil. Even though the total bioethanol production of the world has reached almost 90 billion liters, most of it is shared between the United States and Brazil (Fig. 15.2).

KEYWORDS

- **biofuels**
- **fossil fuels**
- **global warming**
- **bioethanol**
- **lignocellulosic**

ACKNOWLEDGMENTS

Authors are highly grateful to Council of Scientific and Industrial Research for providing the funds to carry out the research work related to bioethanol production through lignocellulosic materials. Authors have taken idea from CSIR project.

REFERENCES

Abbasi, T.; Abbasi, S. A. Biomass Energy and the Environmental Impacts Associated with Its Production and Utilization. *Renew. Sustain. Energy Rev.* **2010,** *14* (3), 919–937.

Antoni, D.; Zverlov, V. V.; Schwarz, W. H. Biofuels from Microbes. *Appl. Microbiol. Biotechnol.* **2007,** *77* (1), 23–35.

Balat, M.; Balat, H.; Öz, C. Progress in Bioethanol Processing. *Progress Energy Combust Sci.* **2008,** *34* (5), 551–573.

Bayer, E. A.; Kenig, R.; Lamed, R. Adherence of *Clostridium thermocellum* to Cellulose. *J. Bacteriol.* **1983,** *156* (2), 818–827.

Béguin, P.; Gilkes, N. R.; Kilburn, D. G.; Miller, R. C.; O'neill, G. P.; Warren, R. A. J. Cloning of Cellulase Genes. *Crit. Rev. Biotechnol.* **1987,** *6* (2), 129–162.

Boussaid, A.; Robinson, J.; Cai, Y. J.; Gregg, D. J.; Saddler, J. N. Fermentability of the Hemicellulose-Derived Sugars from Steam-Exploded Softwood (Douglas Fir). *Biotechnol. Bioeng.* **1999,** *64* (3), 284–289.

Chan, A. K.; Ng, A. K.; Ng, K. K.; Wong, W. K. R. Cloning and Characterization of Two Novel β-Glucosidase Genes Encoding Isoenzymes of the Cellobiase Complex from *Cellulomonas biazotea*. *Gene* **2018,** *642*, 367–375.

Chen, H. L.; Chen, Y. C.; Lu, M. Y. J.; Chang, J. J.; Wang, H. T. C.; Ke, H. M.; Wang, T. Y.; Ruan, S. K.; Wang, T. Y.; Hung, K. Y.; Cho, H. Y. A Highly Efficient β-Glucosidase from the Buffalo Rumen Fungus *Neocallimastix patriciarum* W5. *Biotechnol. Biofuels* **2012,** *5* (1), 24.

Dashtban, M.; Qin, W. Overexpression of an Exotic Thermotolerant β-Glucosidase in *Trichoderma reesei* and Its Significant Increase in Cellulolytic Activity and Saccharification of Barley Straw. *Microb. Cell Factories* **2012,** *11* (1), 63.

Dufey, A. *Biofuels Production, Trade and Sustainable Development: Emerging Issues (No. 2)*; International Institute for Environment and Development: London, **2006;** pp 1–62.

Dwivedi, P.; Alavalapati, J. R.; Lal, P. Cellulosic Ethanol Production in the United States: Conversion Technologies, Current Production Status, Economics, and Emerging Developments. *Energy Sustain. Dev.* **2009,** *13* (3), 174–182.

Eivazi, F.; Tabatabai, M. A. Glucosidases and Galactosidases in Soils. *Soil Biol. Biochem.* **1988,** *20* (5), 601–606.

Eklund, R.; Galbe, M.; Zacchi, G. The Influence of SO_2 and H_2SO_4 Impregnation of Willow Prior to Steam Pretreatment. *Bioresour. Technol.* **1995,** *52* (3), 225–229.

Gunasekaran, P.; Raj, K. C. Ethanol Fermentation Technology—*Zymomonas mobilis*. *Curr. Sci.* **1999,** *77*, 56–68.

Gupta, A.; Verma, J. P. Sustainable Bio-Ethanol Production from Agro-Residues: A Review. *Renew. Sustain. Energy Rev.* **2015,** *41*, 550–567.

Hasunuma, T.; Okazaki, F.; Okai, N.; Hara, K. Y.; Ishii, J.; Kondo, A. A Review of Enzymes and Microbes for Lignocellulosic Biorefinery and the Possibility of Their Application to Consolidated Bioprocessing Technology. *Bioresour. Technol.* **2013,** *135*, 513–522.

Herrera, S. Bonkers about Biofuels. *Nat. Biotechnol.* **2006,** *24* (7), 755.

Isahak, W. N. R. W.; Hisham, M. W.; Yarmo, M. A.; Hin, T. Y. Y. A Review on Bio-Oil Production from Biomass by Using Pyrolysis Method. *Renew. Sustain. Energy Rev.* **2012,** *16* (8), 5910–5923.

Khalil, H. A.; Aprilia, N. S.; Bhat, A. H.; Jawaid, M.; Paridah, M. T.; Rudi, D. A Jatropha Biomass as Renewable Materials for Biocomposites and Its Applications. *Renew. Sustain. Energy Rev.* **2013,** *22*, 667–685.

Khalil, H. S. A.; Alwani, M. S.; Omar, A. K. M. Chemical Composition, Anatomy, Lignin Distribution, and Cell Wall Structure of Malaysian Plant Waste Fibers. *BioResources* **2007,** *1* (2), 220–232.

Lamed, R.; Setter, E.; Bayer, E. A. Characterization of a Cellulose-Binding, Cellulase-Containing Complex in *Clostridium thermocellum*. *J. Bacteriol.* **1983,** *156* (2), 828–836.

Lin, K. W.; Patterson, J. A.; Ladisch, M. R. Anaerobic Fermentation: Microbes from Ruminants. *Enzyme Microb. Technol.* **1985,** *7* (3), 98–107.

Lin, Y.; Tanaka, S. Ethanol Fermentation from Biomass Resources: Current State and Prospects. *Appl. Microbiol. Biotechnol.* **2006,** *69* (6), 627–642.

Ma, L.; Zhang, J.; Zou, G.; Wang, C.; Zhou, Z. Improvement of Cellulase Activity in *Trichoderma reesei* by Heterologous Expression of a Beta-Glucosidase Gene from *Penicillium decumbens*. *Enzyme Microb. Technol.* **2011,** *49* (4), 366–371.

Martínez, Á. T.; Speranza, M.; Ruiz-Dueñas, F. J.; Ferreira, P.; Camarero, S.; Guillén, F.; Martínez, M. J.; Gutiérrez Suárez, A.; Río Andrade, J. C. D. Biodegradation of Lignocellulosics: Microbial, Chemical, and Enzymatic Aspects of the Fungal Attack of Lignin. *Int. Microbiol.* **2005,** *8* (3), 195–204.

Mosier, N.; Wyman, C.; Dale, B.; Elander, R.; Lee, Y. Y.; Holtzapple, M.; Ladisch, M. Features of promising technologies for pretreatment of lignocellulosic biomass. *Bioresour. Technol.* **2005,** *96* (6), 673–686.

Murray, P.; Aro, N.; Collins, C.; Grassick, A.; Penttilä, M.; Saloheimo, M.; Tuohy, M. Expression in *Trichoderma reesei* and Characterisation of a Thermostable Family 3 β-Glucosidase from the Moderately Thermophilic Fungus *Talaromyces emersonii*. *Prot. Express. Purif.* **2004,** *38* (2), 248–257.

Nakazawa, H.; Kawai, T.; Ida, N.; Shida, Y.; Kobayashi, Y.; Okada, H.; Tani, S.; Sumitani, J. I.; Kawaguchi, T.; Morikawa, Y.; Ogasawara, W. Construction of a Recombinant *Trichoderma reesei* Strain Expressing *Aspergillus aculeatus* β-Glucosidase 1 for Efficient Biomass Conversion. *Biotechnol. Bioeng.* **2012,** *109* (1), 92–99.

Öhgren, K.; Bura, R.; Lesnicki, G.; Saddler, J.; Zacchi, G. A Comparison between Simultaneous Saccharification and Fermentation and Separate Hydrolysis and Fermentation Using Steam-Pretreated Corn Stover. *Process Biochem.* **2007,** *42* (5), 834–839.

Palmqvist, E.; Hahn-Hägerdal, B. Fermentation of Lignocellulosic Hydrolysates. I: Inhibition and Detoxification. *Bioresour. Technol.* **2000,** *74* (1), 17–24.

Pandey, A. Solid-State Fermentation. *Biochem. Eng. J.* **2003,** *13* (2–3), 81–84.

Pandey, A.; Kavita, P.; Selvakumar, P. Culture Conditions for Production of 2-1-β-d-Fructan-Fructanohydrolase in Solid Culturing on Chicory (*Cichorium intybus*) Roots. *Braz. Arch. Biol. Technol.* 1998, *41* (2), https://dx.doi.org/10.1590/S1516-89131998000200010.

Park, C. Y.; Ryu, Y. W.; Kim, C. Kinetics and Rate of Enzymatic Hydrolysis of Cellulose in Supercritical Carbon Dioxide. *Korean J. Chem. Eng.* **2001,** *18* (4), 475–478.

Ramani, G.; Meera, B.; Vanitha, C.; Rajendhran, J.; Gunasekaran, P. Molecular Cloning and Expression of Thermostable Glucose-Tolerant β-Glucosidase of *Penicillium funiculosum* NCL1 in *Pichia pastoris* and Its Characterization. *J. Ind. Microbiol. Biotechnol.* **2015,** *42* (4), 553–565.

Reijnders, L.; Huijbregts, M. Transport Biofuels: Their Characteristics, Production and Costs. *Biofuels for Road Transport: A Seed to Wheel Perspective*; Springer-Verlag: London, **2009;** pp 1–48.

Renewable Fuels Association (RFA). *Ethanol Industry Statistics*; Renewable Fuels Association: Washington, DC, **2007** [available from: www.ethanolrfa.org].

Ruiz, E.; Cara, C.; Manzanares, P.; Ballesteros, M.; Castro, E. Evaluation of Steam Explosion Pre-treatment for Enzymatic Hydrolysis of Sunflower Stalks. *Enzyme Microb. Technol.* **2008,** *42* (2), 160–166.

Sarkar, N.; Ghosh, S. K.; Bannerjee, S.; Aikat, K. Bioethanol Production from Agricultural Wastes: An Overview. *Renew. Energy* **2012,** *37* (1), 19–27.

Schell, D. J.; Riley, C. J.; Dowe, N.; Farmer, J.; Ibsen, K. N.; Ruth, M. F.; Toon, S. T.; Lumpkin, R. E. A Bioethanol Process Development Unit: Initial Operating Experiences and Results with a Corn Fiber Feedstock. *Bioresour. Technol.* **2004,** *91* (2), 179–188.

Scott, D. S.; Piskorz, J.; Radlein, D. Liquid Products from the Continuous Flash Pyrolysis of Biomass. *Ind. Eng. Chem. Process Des. Dev.* **1985,** *24* (3), 581–588.

Selvakumar, P.; Pandey, A. Solid State Fermentation for the Synthesis of Inulinase from *Staphylococcus* sp. and *Kluyveromyces marxianus*. *Process Biochem.* **1999,** *34* (8), 851–855.

Shen, Y.; Zhang, Y.; Ma, T.; Bao, X.; Du, F.; Zhuang, G.; Qu, Y. Simultaneous Saccharification and Fermentation of Acid-Pretreated Corncobs with a Recombinant *Saccharomyces cerevisiae* Expressing β-Glucosidase. *Bioresour. Technol.* **2008,** *99* (11), 5099–5103.

Shenoy, D.; Pai, A.; Vikas, R. K.; Neeraja, H. S.; Deeksha, J. S.; Nayak, C.; Rao, C. V. A Study on Bioethanol Production from Cashew Apple Pulp and Coffee Pulp Waste. *Biomass Bioenergy* **2011,** *35* (10), 4107–4111.

Shoseyov, O.; Takagi, M.; Goldstein, M. A.; Doi, R. H. Primary Sequence Analysis of *Clostridium cellulovorans* Cellulose Binding Protein A. *Proc. Nat. Acad. Sci.* **1992,** *89* (8), 3483–3487.

Sricharoenchaikul, V.; Marukatat, C.; Atong, D. Fuel Production from Physic Nut (*Jatropha curcas* L.) Waste by Fixed-Bed Pyrolysis Process. *Thai. J.* **2007,** *3*, 23–25.

Sun, L.; Cao, J.; Liu, Y.; Wang, J.; Guo, P.; Wang, Z. Gene Cloning and Expression of Cellulase of *Bacillus amyloliquefaciens* Isolated from the Cecum of Goose. *Anim. Biotechnol.* **2017,** *28* (1), 74–82.

Sun, X. F.; Xu, F.; Sun, R. C.; Wang, Y. X.; Fowler, P.; Baird, M. S. Characteristics of Degraded Lignins Obtained from Steam Exploded Wheat Straw. *Polym. Degrad. Stab.* **2004,** *86* (2), 245–256.

Sun, Y.; Cheng, J. Hydrolysis of Lignocellulosic Materials for Ethanol Production: A Review. *Bioresour. Technol.* **2002,** *83* (1), 1–11.

Swatloski, R. P.; Spear, S. K.; Holbrey, J. D.; Rogers, R. D. Ionic liquids: New Solvents for Non-derivitized Cellulose Dissolution. In *Abstracts of Papers of the American Chemical Society*; American Chemical Society: Washington, DC, August **2002;** Vol 224, p U622.

Taherzadeh, M. J.; Karimi, K. Pretreatment of Lignocellulosic Wastes to Improve Ethanol and Biogas Production: A Review. *Int. J. Mol. Sci.* **2008,** *9* (9), 1621–1651.

Varga, E.; Réczey, K.; Zacchi, G. Optimization of Steam Pretreatment of Corn Stover to Enhance Enzymatic Digestibility. In *Proceedings of the Twenty-Fifth Symposium on Biotechnology for Fuels and Chemicals Held in May 4–7, 2003, in Breckenridge, CO*; Humana Press: Totowa, NJ, **2004;** pp 509–523.

Wang, B.; Xia, L. High Efficient Expression of Cellobiase Gene from *Aspergillus niger* in the Cells of *Trichoderma reesei*. *Bioresour. Technol.* **2011,** *102* (6), 4568–4572.

Wei, H.; Xu, Q.; Taylor II, L. E.; Baker, J. O.; Tucker, M. P.; Ding, S. Y. Natural Paradigms of Plant Cell Wall Degradation. *Curr. Opin. Biotechnol.* **2009,** *20* (3), 330–338.

Wiselogel, A.; Tyson, S.; Johnson, D. Biomass Feedstock Resources and Composition. In *Handbook on Bioethanol*; Routledge: Abingdon, **2018;** pp 105–118.

Wyman, C. *Handbook on Bioethanol: Production and Utilization*; CRC Press: Boca Raton, FL, 1996.

Yan, Q.; Hua, C.; Yang, S.; Li, Y.; Jiang, Z. High Level Expression of Extracellular Secretion of a β-Glucosidase Gene (PtBglu3) from *Paecilomyces thermophila* in *Pichia pastoris*. *Prot. Expr. Purif.* **2012,** *84* (1), 64–72.

Yao, G.; Li, Z.; Gao, L.; Wu, R.; Kan, Q.; Liu, G.; Qu, Y. Redesigning the Regulatory Pathway to Enhance Cellulase Production in *Penicillium oxalicum*. *Biotechnol. Biofuels* **2015,** *8* (1), 71.

Zheng, L.; Hou, Y.; Li, W.; Yang, S.; Li, Q.; Yu, Z. Biodiesel Production from Rice Straw and Restaurant Waste Employing Black Soldier Fly Assisted by Microbes. *Energy* **2012,** *47* (1), 225–229.

Zheng, Y.; Tsao, G. T. Avicel Hydrolysis by Cellulase Enzyme in Supercritical CO_2. *Biotechnol. Lett.* **1996,** *18* (4), 451–454.

Zhu, S.; Wu, Y.; Yu, Z.; Liao, J.; Zhang, Y. Pretreatment by Microwave/Alkali of Rice Straw and its Enzymic Hydrolysis. *Process Biochem.* **2005,** *40* (9), 3082–3086.

Zhu, S.; Wu, Y.; Yu, Z.; Wang, C.; Yu, F.; Jin, S.; Ding, Y.; Chi, R. A.; Liao, J.; Zhang, Y. Comparison of Three Microwave/Chemical Pretreatment Processes for Enzymatic Hydrolysis of Rice Straw. *Biosyst. Eng.* **2006,** *93* (3), 279–283.

CHAPTER 16

Algal Biomass and Biodiesel Production

SAMAKSHI VERMA and ARINDAM KUILA*

Department of Bioscience and Biotechnology, Banasthali Vidyapith, Rajasthan 304022, India

**Corresponding author. E-mail: arindammcb@gmail.com*

ABSTRACT

To enable large-scale production of algae biomass, technologies should be developed so that biodiesel production will have significant effect on renewable fuels standards. Cultivation of algae is followed by harvesting, and then, processing of microalgae for producing biodiesel reviewed in this chapter. Microalgae species, which are commonly used to produce biodiesel, are presented and are compared with the other available biodiesel sources having various lipid contents. Triacylglycerol is basically a fatty acid, which is being utilized for producing biodiesel, and it is a main source of energy reserves in microalgae. In this chapter, it is shown that due to the lack of nutrients, triacylglycerol will be aggregated in different species of microalgae. Recently, it has been experienced by bioenergy industry that microalgae can be used as a potential biomass feedstock for the production of biodiesel, bio-hydrogen, and methane. Microalgae can also be used as a viable source for producing high value-added products.

16.1 INTRODUCTION

Energy crisis is one of the major problems that the world is facing today which makes our environment polluted and nonpeaceful. The fossil fuels requirement is increasing day by day due to the increasing population because of which available resources start decreasing very fast, and in future, they will be finished. Because of this condition, renewable energy sources gain more attention toward themselves. Due to large exploitation of fossil fuels (unsustainable), CO_2 level is increased and greenhouse gas (GHG) emissions get accumulated results in making environment polluted or unsafe. To keep environment clean and to maintain sustainability, renewable and eco-friendly fuels are need to be produced (Schenk et al., 2008). Biodiesel is synthesized from algae (Dunahay et al., 1996; Roessler et al., 1994; Sawayama et al., 1995; Sheehan et al., 1998), biobutanol (Durre, 1997), *Jatropha curcas* (Becker and Makkar, 2008), and vegetable oils (Shay, 1993). Brazil, the European Union, and the United States are all well known as world's largest producers of biodiesel (Balat, 2007).

Energy content of algae is 80% which is occupied by petroleum (Chisti, 2007, 2013). Other sources of biodiesel which includes soybeans and palm oils (Kligerman and Bouwer, 2015; Lam and Lee, 2011) do not have as much lipid content than that of algae cells which comprises 30% lipid content (Lam and Lee, 2012). Dry weight of lipid content is 30–40% in microalgae and it kept on increasing until it reached up to 85%. Cultivation of micro- and macroalgae can be done on large scale within a short time period. Due to the heterotrophic and photosynthetic nature of microalgae, they are capable of growing as energy crops and can produce some economically significant compounds like oils and fats (Pittman et al., 2011; Rawat et al., 2011). After combustion, there will be no harm to environment because algal biofuels do not possess any toxic chemicals. Blending method is utilized for extraction of fatty acids and separation of biodiesel on experimental or small scale. This simple method comprises the following steps which have been shown in Figure 16.1.

According to recent research, it is clear that microalgae are used as an additional renewable source. Microalgae are the prominent source for the production of third-generation fuels, mainly if they are linked with the utilization of other high-value compounds, which are being produced by microalgae (Harun et al., 2010).

16.2 MICROALGAE IS UTILIZED FOR PRODUCING BIODIESEL

Prokaryotic or eukaryotic thallophytes that contain chlorophyll *a* (photosynthetic pigment) and do not have sterile layer of cells around the reproductive cells are called microalgae (Lee, 1980). More than 50,000 species of microalgae are found within the wide range of environments (Richmond, 2004). Prokaryotic microalgae (cyanobacteria, blue-green

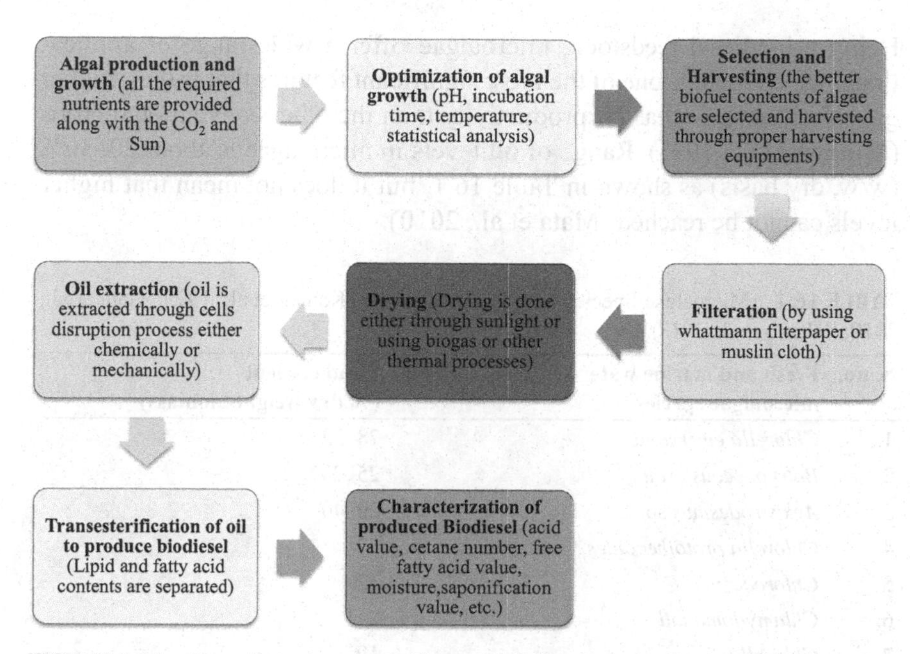

FIGURE 16.1 By using microalgae, biomass biodiesel production is taking place for experimental purpose and for small-scale production.

algae) lack membrane-bound organelles and are more similar to bacteria than to algae. Some specific organelles are found in eukaryotic microalgae, which regulate the functioning of their cells (Brennan and Owende, 2010). There are several classes of eukaryotic microalgae depending upon their basic cellular structure, life cycle, and pigmentation (Khan et al., 2009). Diatoms, green algae, and red algae are some of the most significant classes of eukaryotic microalgae. Algal biomass is divided into three types: (1) autotrophic, (2) heterotrophic, and (3) mixotrophic (Lee, 1980). During unfavorable conditions, that is, when nitrogen content is exhausted or excessive amount of carbon is accumulated, then triacylglycerols (TAGs) (main form of lipids in microalgae) are synthesized and stored by microalgae. TAGs are stored as a carbon energy source within the cells of microalgae which will no longer divide (Meng et al., 2009). The process of TAG biosynthesis completes in following three steps:

1. Acetyl-CoA is converted into malonyl-CoA
2. The carbon chain of fatty acids are elongated and desaturated
3. TAGs are synthesized (Huang et al., 2010)

Being a biodiesel feedstock, microalgae offer a wide range of applications, which include one of the most significant features that is their higher growth rates as well as oil productivity than the other conventional crops (Minowa et al., 1995). Range of oil levels in microalgae is about 20–50% (w/w, dry basis) as shown in Table 16.1, but it does not mean that higher levels cannot be reached (Mata et al., 2010).

TABLE 16.1 Microalgae Species with Lipid Contents (Kanani et al., 1999; Mata et al., 2010; Salama et al., 2013).

S. no.	Fresh and marine water microalgae species	Lipid content (% dry weight biomass)
1.	*Chlorella emersonii*	28–32
2.	*Botryococcus braunii*	25–75
3.	*Ankistrodesmus* sp.	29–40
4.	*Chlorella protothecoides*	57
5.	*Chlorocccum*	12
6.	*Chlamydomonad*	2
7.	*Chlorella*	19
8.	*C. Mexicana*	15
9.	*Tetraselmis maculate*	3
10.	*Chlorella pyrenoidosa*	2
11.	*Spirulina maxima*	6–7
12.	*Scenedesmus obliquus*	12–14

During exponential growth phase, multiplication of microalgae cells is faster in every 3.5 h because of which microalgae are capable of increasing (doubling) their biomass within 24 h (Chisti, 2007). Microalgae comprises 58,700 L ha^{-1} year^{-1} oil yield within 30% oil, on comparing with 636 L ha^{-1} year^{-1} oil yield of soybean and 5366 L ha^{-1} year^{-1} oil yield of palm oil (as given in Table 16.2). If microalgae are having 70% oil, then 136,900 L oil ha^{-1} year^{-1} can be produced by microalgae (Mata et al., 2010).

In contrast with cultivation area, microalgae that possess higher biomass and oil productivity are more significant than the land plants because microalgae do not compete for land with those crops which are utilized for fodder, food, and other products (Huang et al., 2010). The basic procedure to obtain algal biomass and extract oil/lipids is shown in Figure 16.2. Brackish, fresh, salty water, or nonarable lands are some examples of environments that are unfavorable for growing other crops

TABLE 16.2 Comparison of Different Biodiesel Feedstocks.

S. no.	Plant source	Oil content (% dry weight biomass)	Oil productivity (L oil ha^{-1})	Utilization of land (m^2 kg^{-1} biodiesel)
1.	Microalgae (oil content is high)	70	136,900	0.1
2.	Microalgae (oil content is medium)	50	97,800	0.1
3.	Microalgae (oil content is low)	30	58,700	0.2
4.	*Jatropha curcas* L.	28	1892	15
5.	*Helianthus annuus* L.	40	1070	11
6.	*Brassica napus* L.	41	1190	12
7.	*Camelina sativa* L.	42	915	12
8.	*Elaeis guineensis*	36	5950	2
9.	*Cannabis sativa* L.	33	363	31
10.	*Zea mays* L.	44	172	66
11.	*Ricinus communis*	48	1307	9
12.	*Glycine max* L.	18	446	18
13.	Rapeseed	50.7	1190	

except microalgae (Patil et al., 2008). Autotrophic growth of microalgae is only possible in the presence of carbon dioxide, which is supplied by industrial facilities where the amount of carbon dioxide in emitted gases of boilers and power plants may reach up to 15% (v/v) (Salih, 2011; Zhao et al., 2011). Microalgae can also be utilized to generate an environmental significance, to remove nitrogen and phosphorus efficiently, and to treat wastewater (Mallick, 2002).

16.3 LIPIDS FROM MICROALGAE

There are several species of microalgae and they all possess different lipid yields, that is, lipid content is directly proportional to microalgae species (Nacimento et al., 2013). Lipid production is taking place throughout the cell cycle, but in the beginning, it is focused toward structural synthesis of lipids for 1° metabolism and then toward the formation of TAGs. Energy is reserved in the form of fatty-acid class called TAGs within the microalgae, and the composition of these TAGs is not only associated with the genetics of microalgae species but also with the cellular stress and environmental factors because of which microalgae are submitted. Kyoto Encyclopedia of Genes and Genomes is basically utilized for analyzing the illustrations about TAGs biosynthesis metabolism in microalgae (Chiu et al., 2008; Pereira et al., 2013).

Among different microalgae species and at different stages of growth, the composition and concentration of oils produced by microalgae can vary depending on application. In case of green algae, polyunsaturated fatty acids and polar lipids are produced in higher concentrations during the early stages of growth, but green algae produce higher neutral lipids during their stationary phase (Li et al., 2014; Sanchez-Saavedra et al., 2005). Major estimation of lipids and fatty-acid methyl is given in Table 16.3.

Unsaponifiable fraction of lipids denotes the proportion of lipids in this method. Availability of unsaponifiable fractions (carotenoids and chlorophylls) is excluded, but methyl esters were used. The values given in the brackets correlate with the same cultures that are found in the stationary growth phase. According to recent studies, this hypothesis supports that fatty acids of microalgae show relative stability and taxonomic specificity compared with the vascular plants. It is a fascinating task to regulate the transfer of materials between fresh or marine water food chains and to measure a mono-specific marking between the dominant groups in the environment.

TABLE 16.3 Under Continuous Cultivation in Tubular Photobioreactor, Composition of Polar and Neutral Lipids of *Isochrysis galbana*, *Phaeodactylum tricornutum*, and *Porphyridium cruentum* Is Studied (Cartens et al., 1996; Gimenez Gimenez et al., 1998; Molina Grima et al., 2003).

S. no.	Lipids	Percentage of lipids		
		Isochrysis galbana	*Phaeodactylum tricornutum*	*Porphyridium cruentum*
1.	Glycolipids	55–60 (37.0)	47–51 (35.0)	43–47 (43.0)
2.	Neutral	20–28 (43.0)	22–25 (51.0)	37–41 (47.0)
3.	Phospholipids	11–15 (20.0)	25–29 (14.0)	13–19 (10.0)
4.	Polar	70–75 (57.0)	73–76 (49.0)	58–62 (53.0)

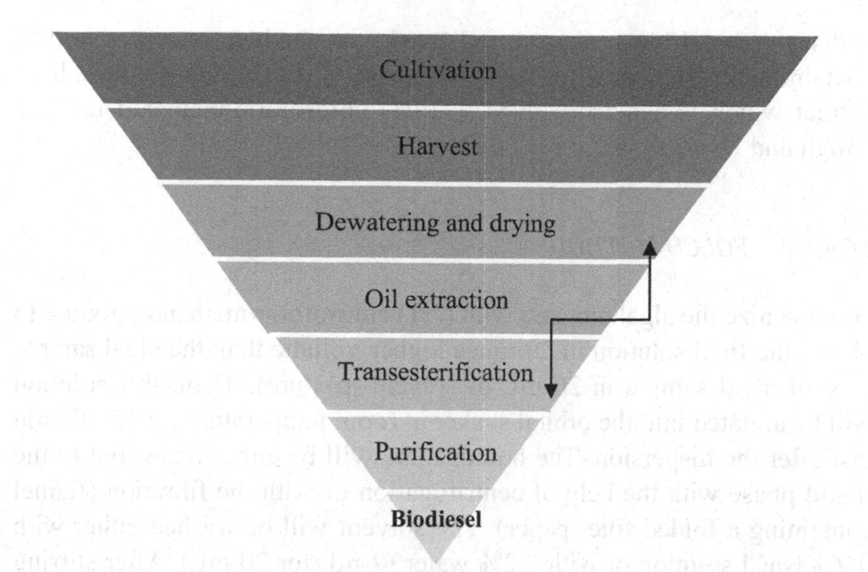

FIGURE 16.2 Pyramid diagram to show the steps followed for biodiesel production via microalgae. In some cases, the oil-extraction step before transesterification is bypassed (shown by lined arrow).

16.4 PURIFICATION OF BIODIESEL FROM MICROALGAE

Biodiesel is purified from microalgae within two major following processes: (1) Extraction of oil/lipids and (2) Transesterification of extracted oil.

16.4.1 EXTRACTION OF OIL

Two–three extraction solvent systems were used for the extraction of oil so that the oil content in each case can be compared and the most suitable solvent system can be selected for the highest biodiesel production.

16.4.1.1 BLIGH AND DYER METHOD

About 1.0 g of microalgae is added with 3.75 mL of chloroform/methanol (1/2) mixture and shaken for 10–15 min, then again addition of 1.25 mL of chloroform and 1.25 mL of water with continuous mixing for 1 min for each. After that centrifugation of the content, upper phase of solution is discarded and lower phase is collected with the help of a protein disk containing a Pasteur pipette. The lower phase of the solution contains lipid extract which is again dissolved in (2/1) chloroform/methanol mixture (Bligh and Dyer, 1959).

16.4.1.2 FOLCH METHOD

Homogenize the algal biomass with (2/1) chloroform/methanol mixture to obtain the final solution of 20 times higher volume than the algal sample (1 g of algal sample in 20 mL of solvent mixture). Then, this solution will be agitated into the orbital shaker at room temperature for 15–20 min just after the dispersion. The homogenate will be either recovered to the liquid phase with the help of centrifugation or with the filtration (funnel containing a folded filter paper). The solvent will be washed either with 0.9% NaCl solution or with 0.2% water (4 mL for 20 mL). After stirring the solution, it will be centrifuged at 2000 rpm (low speed) for the separation of two phases. The upper phase is purified by siphoning and is kept to measure the contents of small organic polar molecules or gangliosides. If it is mandatory to remove the labeled molecules, then the interface of solution will be washed once or twice with (1/1) methanol/water without mixing the whole solution. After centrifugation and siphoning of the upper phase, the lower chloroform phase containing lipids will be evaporated under vacuum in a rotary evaporator or under a nitrogen stream if the volume is under 2–3 mL.

16.4.1.3 SOLVENT EXTRACTION METHOD

For the separation of oil, 20 g of grinded algal biomass will be treated with 20 mL of solvent (di-ethyl ether and *n*-hexane). These solvents can also be utilized separately as well as a blend for oil extraction. This mixture of solvents and algal biomass will be kept at room temperature for 24 h. An oily layer is formed on the solvent surface, which will be extracted in a rotary evaporator. To evaporate di-ethyl ether, the evaporator will be initially kept at 34°C for 15 min; after that, the temperature will be raised up to 69°C for the removal of *n*-hexane, and hence, solvent-free oil will be left behind in the evaporation flask of solvent extractor.

16.4.2 TRANSESTERIFICATION AND BIODIESEL PRODUCTION

The solvent mixture solution is released by evaporating the extracted oil at 40–45°C under vacuum with the help of rotary evaporator. Then, each algal species will produce oil which will be added to a mixture of 24 mL methanol and catalyst (0.25 g NaOH); this whole process is known as transesterification (as shown in Fig. 16.3), with proper stirring of 20 min. The above mixture is kept in electric shaker at 3000 rpm for 3 h (National Biodiesel Board, 2002). Then, the solution is kept for settling the biodiesel and to sediment layers visibly for 16 h. By using flask separator, biodiesel is purified from solvent solution via sedimentation. Amount of sediments (glycerin, pigments, etc.) is also estimated carefully. Purified biodiesel is washed repeatedly with 5% H_2O until it appears clear; after that, biodiesel is dried with the help of dryer, and then, it is kept under the running fan for 12 h. By using measuring cylinder, the amount of produced biodiesel is analyzed and pH is recorded and stored for further analysis.

16.5 BIODIESEL CHARACTERIZATION

The produced biodiesel will be washed and dried by the following standard method (Indhumathi et al., 2014). The dried biodiesel will be used for characterization of different parameters (acid value, ash content, calorific value, cetane number, cloud point, fire point, flash point, free fatty-acid value, moisture, pH, pour point, saponification value, specific gravity density, and viscosity). All parameters will be measured by following the method of Indhumathi et al. (2014). Comparison of all the characteristic parameters of petroleum diesel, microalgae diesel, and standard of ASTM is given below in the Table 16.4.

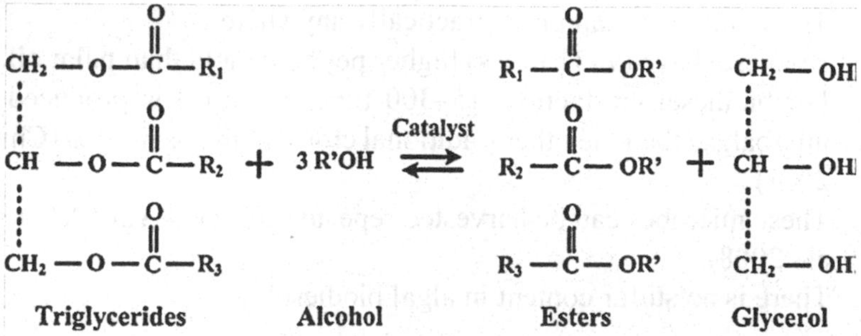

$$
\begin{array}{l} CH_2-O-C(=O)-R_1 \\ CH-O-C(=O)-R_2 \\ CH_2-O-C(=O)-R_3 \end{array}
+ 3\,R'OH \xrightleftharpoons{\text{Catalyst}}
\begin{array}{l} R_1-C(=O)-OR' \\ R_2-C(=O)-OR' \\ R_3-C(=O)-OR' \end{array}
+ \begin{array}{l} CH_2-OH \\ CH-OH \\ CH_2-OH \end{array}
$$

Triglycerides **Alcohol** **Esters** **Glycerol**

FIGURE 16.3 Transesterification of triglycerides.

16.6 PROPERTIES OF BIODIESEL

- Renewable source of energy is biodiesel which will be sustainably available to someone as the reserves of petroleum will be degraded within 50 years due to their increased exploitation (Sheehan et al., 1998).
- Biodiesel possesses various suitable environmental properties due to which release of CO_2 does not increase and also decrease in the concentration of sulfur (Antolin et al., 2002). If biodiesel is used as an energy source, then the release of sulfur content is 30% and the release of CO_2 is 10% decreased. Biodiesel contains higher oxygen content because of the gas reduction that is produced during combustion and also due to decrease in CO_2 content. Biodiesel does not have any harmful substances like chemical substances and aromatic compounds. According to the recent research, the use of biodiesel in place of common diesel source can reduce 95% of cancers and 90% of air toxicity.
- In terms of biodegradability and flash point, biodiesel is preferred over petroleum diesel (Ma and Hanna, 1999).

16.7 IMPORTANCE OF USING MICROALGAE FOR PRODUCING BIODIESEL

Among several algae, microalgae are the one which appears to be promising because:

- Microalgae shows rapid rate of growth, for example, doubling in 24 h (Rittmann, 2008).
- These microbes can grow practically anywhere.
- Biodiesel have (7–31 times) higher per-acre yield than palm oil.
- For biodiesel production, 15–300 times more oil is produced by microalgae than the other traditional crops on the basis area (Chisti, 2007).
- These microbes can be harvested repeatedly over a year (Schenk et al., 2008).
- There is no sulfur content in algal biodiesel.

- Algal biodiesel is nontoxic in nature and are highly biodegradable (Schenk et al., 2008).
- Oil extracts of algae can be utilized as livestock source and further can be transformed into ethanol.
- Higher levels of polyunsaturated fatty acids are found in algal biodiesel which is favorable for cold weather climates.
- Depending on the place where it is grown, it can reduce carbon emissions.
- Salt rich or waste water can also be utilized (Schenk et al., 2008).
- Carbon source for growing microalgae is atmospheric CO_2 (Schenk et al., 2008).
- By modifying the constituents of growth media, the lipid productivity of microalgae could be adjusted (Naik et al., 2006).

16.8 CHALLENGES, PROSPECTS, AND CONCLUSION

Most useful limitation of biodiesel is its crop oils cost, for example, net operating value of canola oil is 80% which is utilized in biodiesel production (Demirbas, 2007). Oil crop accessibility is limited for the biodiesel production (Chisti, 2008). So, there is a need of searching new acceptable feedstock that does not burden the supply of edible vegetable oil for producing biodiesel. Algae comprises lipids that are appropriate for the transesterification; so, that is why it is an alternative for all other oil crops.

Economic viability is the major challenge for commercializing microalgal biodiesel. Several other obstacles have to be solved until the energy required for cultivation, harvesting, oil extraction, and conversion of oil becomes higher than the released energy. The requirements of water, nutrients, and carbon are also considered. From a technical point of view, biodiesel production from microalgae is not yet feasible but with possibilities for improvements (Carioca et al., 2009; Lardon et al., 2009).

In the course of fatty-acid production, there are other useable products such as proteins, carbohydrates, essential fatty acids, and other nutrient contents. The encouragement of the total use of biomass improves the industrial value of microalgae biomass and becomes economically attractive microalgae biodiesel production (Carioca et al., 2009; Stuart et al., 2010).

TABLE 16.4 According to Khalil (2006) Comparison of Characteristics Between the Microalgae Biodiesel, Petroleum Diesel, and Standard of the American Society for Testing and Materials (ASTM).

S. no.	Characteristics	Microalgal biodiesel	Petroleum diesel	Standard of ASTM
1.	Acid value (mg KOH g^{-1})	0.374	Max. 0.5	–
2.	Calorific value (MJ kg^{-1})	41	40–45	–
3.	Density (kg L^{-1})	0.864	0.838	0.86–0.9
4.	Flash point (°C)	115	75	Min. 100
5.	Freezing point (°C)	−12	−50–10	–
6.	Pour point (°C)	−11	−3.0 (max. −6)	Summer max. = 0; Winter <−15
7.	Proportion of H/C	1.81	1.81	–
8.	Viscosity (mm^2 s^{-1} CST 40°C)	5.2	1.9–4.1	3.5–5.0

As part of the integrated biorefineries concept, it is recommended that the efficient use of natural resources and recovery of materials, energy, and nutrients are contained in the by-products or generated from other processes. The utilization of domestic wastewater in microalgae ponds has demonstrated a possibility for the integrated process to produce oil for biodiesel. These ponds can be used in secondary or tertiary treatment and are advantageous in terms of cost, energy requirements, and GHG mitigation (Sydney et al., 2011). Some countries, such as China, Taiwan, Israel, India, Germany, Canada, and the United States, have announced initiatives for the commercial production of microalgae for biodiesel production purposes. It is clear about the advance of knowledge about this issue. However, it is also observed that this effort takes place in an isolated form, with few cross-actions and exchange of know-how (Franco et al., 2013). The Brazilian market is faced with the prospect of a significant increase in biodiesel demand by the evolution of biodiesel addition to the diesel blend. A lot of work is still needed before the potential offered by microalgae as source of biodiesel is fully exploited. The literature cited throughout this chapter shows that most of these studies were performed with strains isolated from temperate regions and therefore environmental conditions different than those observed in Brazil. Another point is that these studies are not effective in elucidating what would be the best growing conditions in both laboratory and scale production, which enables the use of microalgae as the raw material for the production of biodiesel.

It reiterates the need to obtain Brazilian strains that can withstand the climatic conditions to which they will be exposed to in our country, as well as to enhance the design and development of technologies that can reduce costs while increasing yields.

KEYWORDS

- **biodiesel**
- **microalgae**
- **transesterification**
- **high lipid content**
- **oil extraction**

REFERENCES

Antolin, G.; Tinaut, F. V.; Briceno, Y. Optimisation of Biodiesel Production by Sunflower Oil Transesterification. *Bioresour. Technol.* **2002,** *83*, 111–114.

Balat, M. An Overview of Biofuels and Policies in the European Union. *Energy Sour. B: Econ. Plann.* **2007,** *2* (2), 167–181.

Becker, K.; Makkar, H. P. S. *Jatropha curcas*: A Potential Source for Tomorrow's Oil and Biodiesel. *Lipid Technol.* **2008,** *20* (5), 104–107.

Bligh, E. G.; Dayer, W. J. A Rapid Method for Total Lipid Extraction and Purification. *Can. J. Biochem. Physiol.* **1959,** *37*, 911–917.

Brennan, L.; Owende, P. Biofuels from Microalgae—A Review of Technologies for Production, Processing, and Extractions of Biofuels and Co-products. *Renew. Sustain. Energy. Rev.* **2010,** *14*, 557–577.

Carioca, J. O. B.; Hiluy Filho, J. J.; Leal, M. R. L. V.; Macambira, F. S. The Hard Choice for Alternative Biofuels to Diesel in Brazil. *Biotechnol. Adv.* 2009, *27*, 1043–1060.

Cartens, M.; Molina, E.; Robles, A.; Gimenez, A.; Ibenez, M. J. Eicosapentaenoic Acid (20:4n-3) from the Marine Microalga *Phaeodactylum tricornutum*. *J. Am. Oil Chem. Soc.* **1996,** *73*,1025–1031.

Chisti, Y. Constraints to Commercialization of Algal Fuels. *J. Biotechnol.* **2013,** *167* (3), 201–214.

Chisti, Y. Biodiesel from Microalgae Beats Bioethanol. *Cell Press* **2008,** *26*, 126–131.

Chisti, Y. Biodiesel from Microalgae. *Biotechnol. Adv.* **2007,** *25*, 294–306.

Chiu, S. Y.; Kao, C. Y.; Tsai, M. T.; Ong, S. C.; Chen, C. H.; Lin, C. S. Lipid Accumulation and CO_2 Utilization of *Nannochloropsis oculata* in Response to CO_2 Aeration. *Bioresour. Technol.* **2008,** *100*, 833–838.

Demirbas, A. Importance of Biodiesel as Transportation Fuel. *Energy Policy* **2007,** *35*, 4661–4670.

Dunahay, T. G.; Jarvis, E. E.; Dais, S. S.; Roessler, P. G. Manipulation of Microalgal Lipid Production Using Genetic Engineering. *Appl. Biochem. Biotechnol.* **1996,** *57–58*, 223–231.

Durre, P. Biobutanol: An Attractive Biofuel. *Biotechnol. J.* **1997,** *2* (12), 1525–1534.

Franco, A. L. C.; Lobo, I. P.; Cruz, R. S. Biodiesel de Microalgas: Avanços e desafios. *Quim. Nova* **2013,** *36*, 437–448.

Gimenez Gimenez, A.; Ibanez, M. J.; Robles, A.; Molina, E.; Garcia, S.; Esteban, L. Downstream Processing and Purification of Eicosapentaenoic (20:5n-3) and Arachidonic Acids (20:4n-6) from the Microalga *Porphyridium cruentum*. *Bioseparation* **1998,** *7*, 89–99.

Harun, R.; Singh, M.; Forde, G. M.; Danquah, M. K. Bioprocess Engineering of Microalgae to Produce a Variety of Consumer Products. *Renew. Sustain. Energy Rev.* **2010,** *14*, 1037–1047.

Huang, G.; Chen, F.; Wei, D.; Zhang, X.; Chen, G. Biodiesel Production by Microalgal Biotechnology. *Appl. Energy* **2010,** *87*, 38–46.

Indhumathi, P.; Syed-Shabudeen, P. S.; Shoba, U. S. A Method for Production and Characterization of Biodiesel from Green Micro Algae. *Int. J. Biosci. Biotechnol.* **2014,** *6*, 111–122.

Khalil, C. N. As Tecnologias de Produçao de Biodiesel. *Coletanea de Artigos—O Futuro da Industria: Biodiesel*; Ministerio do Desenvolvimento, Industria e Comercio Exterior: Brazil, 2006.

Khan, S. A.; Rashmi, Hussain, M. Z.; Prasad, S.; Banerjee, U. C. Prospects of Biodiesel Production from Microalgae in India. *Renew. Sustain. Energy Rev.* 2009, *13*, 2361–2372.

Kligerman, D. C.; Bouwer, E. J. Prospects for Biodiesel Production from Algae Based Wastewater Treatment in Brazil: A Review. *Renew. Sustain. Energy Rev.* **2015,** *52*, 1834–1846.

Lam, M. K.; Lee, K. T. Renewable and Sustainable Bioenergies Production from Palm Oil Mill Effluent (POME): Win-Win Strategies toward Better Environmental Protection. *Biotechnol. Adv.* **2011,** *29* (1), 124–141.

Lam, M. K.; Lee, K. T. Microalgae Biofuels: A Critical Review of Issues, Problems and the Way Forward. *Biotechnol. Adv.* **2012,** *30* (3), 673–690.

Lardon, L.; Helias, A.; Sialve, B.; Steyer, J. P.; Bernard, O. Life-Cycle Assessment of Biodiesel Production from a Microalgae. *Environ. Sci. Technol.* **2009,** *43*, 6475–6481.

Lee, R. E. *Phycology*; Cambridge University Press: New York, 1980.

Li, S.; Xu, J. L.; Chen, J.; Chen, J. J.; Zhou, C. X.; Yan, X. J. The Major Lipid Changes of Some Important Diet Microalgae during the Entire Growth Phase. *Aqua* **2014,** *428*, 104–110.

Ma, F.; Hanna, M. A. Biodiesel Production. *Bioresour. Technol.* 1999, *70*, 1–15.

Mallick, N. Biotechnological Potential of Immobilized Algae for Wastewater N, P and Metal Removal: A Review. *Biometals* **2002,** *15*, 377–390.

Mata, T. M.; Martins, A. A.; Caetano, N. S. Microalgae for Biodiesel Production and Other Applications: A Review. *Renew. Sustain. Energy Rev.* **2010,** *14*, 217–232.

Meng, X.; Yang, J.; Xu, X.; Zhang, L.; Nie, Q.; Xian, M. Biodiesel Production from Oleaginous Microorganisms. *Renew. Energy* **2009,** *34*, 1–5.

Minowa, T.; Yokoyama, S.; Kishimoto, M.; Okakura, T. Oil Production from Algal Cells of *Dunaliella tertiolecta* by Direct Thermochemical Liquefaction. *Fuel* **1995,** *74,* 1735–1738.

Molina Grima, E.; Belarbi, E.-H.; Acien Fernandez, F. G.; Robles Medina, A.; Chisti, Y. Recovery of Microalgal Biomass and Metabolites: Process Options and Economics. *Biotechnol. Adv.* **2003,** *20,* 491–515.

Nacimento, I. A.; Marques, S. S. I.; Cabanelas, I. T. D.; Pereira, S. A.; Duzian, J. I.; Souza, C. O.; Vich, D. V.; Carvalho, G. C.; Nacimento, M. A. Screening Microalgae Strains for Biodiesel Production: Lipid Productivity and Estimation of Fuel Quality Based on Fatty Acids Profiles as Selective Criteria. *Bioenergy Res.* **2013,** *6,* 1–13.

Naik, S. N.; Meher, L. C.; Sagar, D. V. Technical Aspects of Biodiesel Production by Transesterification—A Review. *Renew. Sustain. Energy Rev.* **2006,** *10,* 248–268.

National Biodiesel Board. National Biodiesel Board: USA, 2002. Available at www.biodiesel.org/.

Patil, V.; Tran, K.-Q.; Giselrod, H. R. Towards Sustainable Production of Biofuels from Microalgae. *J. Mol. Sci.* **2008,** *9,* 1188–1195.

Pereira, C. M. P.; Hobuss, C. B.; Maciel, J. V.; Ferreira, L. R.; Del Pino, F. B.; Mesko, M. F. Biodiesel Renovavel Derivado de Microalgas: Avanço e Perspective Astecnologicas. *Quim. Nova* **2013,** *35,* 2013–2018.

Pittman, J. K.; Dean, A. P.; Osundeko, O. The Potential of Sustainable Algal Biofuel Production Using Wastewater Resources. *Bioresour. Technol.* **2011,** *102* (1), 17–25.

Rawat, I.; Kumar, R. R.; Mutanda, T.; Bux, F. Dual Role of Microalgae: Phycoremediation of Domestic Wastewater and Biomass Production for Sustainable Biofuels Production. *Appl. Energy* **2011,** *88* (10), 3411–3424.

Richmond, A. *Handbook of Microalgal Culture: Biotechnology and Applied Phycology*; Blackwell Science Ltd.: Oxford, 2004.

Rittmann, B. E. Opportunities for Renewable Bioenergy Using Microorganisms. *Biotechnol. Bioeng.* **2008,** *100,* 203–212.

Roessler, P. G.; Brown, L. M.; Dunahay, T. G.; Heacox, D. A.; Jarvis, E. E.; Schneider, J. C.; et al. Genetic-Engineering Approaches for Enhanced Production of Biodiesel Fuel from Microalgae. *ACS Symp. Ser.* **1994,** *566,* 255–270.

Salih, F. M. Microalgae Tolerance to High Concentrations of Carbon Dioxide: A Review. *J. Environ. Prot.* **2011,** *2,* 648–654.

Sanchez-Saavedra, M. P.; Votolina, D. The Growth Rate, Biomass Production and Composition of *Chaetoceros* sp. Grown with Different Light Sources. *Aqua Eng.* **2005,** *35,* 161–165.

Sawayama, S.; Inoue, S.; Dote, Y.; Yokoyama, S.-Y. CO_2 Fixation and Oil Production through Microalga. *Energy Convers. Manage.* **1995,** *36,* 729–731.

Schenk, P. M.; Thomas-Hall, S. R.; Stephens, E.; Marx, U. C.; Mussgnug, J. H.; Posten, C.; Kruse, O.; Hankamer, B. Second Generation Biofuels: High-Efficiency Microalgae for Biodiesel Production. *Bioenergy Res.* **2008,** *1,* 20–43.

Shay, E. G. Diesel Fuel from Vegetable Oils: Status and Opportunities. *Biomass Bioenergy* **1993,** *4* (4), 227–242.

Sheehan, J.; Dunahay, T.; Benemann, J.; Roessler, P. *A Look Back at the U.S. Department of Energy's Aquatic Species Program Biodiesel from Algae*; National Renewable Energy Laboratory: Golden, CO, 1998.

Stuart, A. S.; Atthew, P. D.; John, S. D.; Irmtraud, H.; Christopher, J. H.; David, J. L. S.; Alison, G. S. Biodiesel from Microalgae: Challenges and Prospects. *Curr. Opin. Biotechnol.* **2010,** *21*, 277–286.

Sydney, E. B.; da Silva, T. E.; Tokasrki, A.; Novak, A. C.; de Carvalho, J. C.; Woiciecohwski, A. L.; Larroche, C.; Soccol, C. R. Screening of Microalgae with Potential for Biodiesel Production and Nutrient Removal from Treated Domestic Sewage. *Appl. Energy* **2011,** *88*, 3291–3294.

Zhao, B.; Zhang, Y.; Xiong, K.; Zhang, Z.; Hao, X.; Liu, T. Effect of Cultivation Mode on Microalgal Growth and CO_2 Fixation. *Chem. Eng. Res. Des.* **2011,** *89*, 1758–1762.

CHAPTER 17

Bioenergy: Sources, Research, and Advances

SUNANDA JOSHI[1], MONIKA CHOUDHARY[1],
SAMEER SURESH BHAGYAWANT[2], and NIDHI SRIVASTAVA[1,*]

[1]*Department of Bioscience and Biotechnology, Banasthali Vidyapith, Rajasthan, India*

[2]*School of Studies in Biotechnology, Jiwaji University, Gwalior, Madhya Pradesh, India*

[*]*Corresponding author. E-mail: nidhiscientist@gmail.com*

ABSTRACT

Presently, the energy demand of the world is continuously increasing; this increasing demand is the reason for environmental pollution and thus results in energy disaster. Bioenergy is evolving and availability of sustainable production system for going forward is helpful for marginal land resources. It is important to reduce food versus fuel competition for main croplands. Biomass as an organic substance, such as agricultural residues, wood waste, farming manure, and wood, is one of the most extensive energy resources worldwide. Biomass has a potential to be a significant part of future energy sources; the energy derived from it is long term and stable, thus can be supplied in almost every area of the world. Bioenergy is renewable, biomass-derived energy, highly available, and geographically dispersed. For the production of bioenergy, wide range of renovations in technologies exists at present, and these techniques are ranging from research stage to commercialization. Today the dominant biomass conversion technology consists of the combustion of biomass as fuel wood, as field and forest residues.

17.1 INTRODUCTION

Bioenergy is produced by using organic matter or natural resources with the purpose of avoiding geological formations (fossils). Biomass energy is produced from the present living organisms. Original form of biomass can be used as fuel. Biomass can be eminent to different procedures like liquid, solid, or gaseous biofuels. These kinds of fuels can be used for transport, industrial processes, and production of heating, cooling, and electricity (Varvel et al., 2008).

Fuels, heat, and electrical power are forms in which bioenergy is available. Farmers are well equipped and knowledgeable about the production of biomass. Thus in the energy production, it can be used in a better way. Farmers are consumers of bioenergy as they can produce and use it at the same time (Zhen, 2013).

Production of bioenergy can reduce the harmful effects of residue, waste, and disposal, and it is beneficial for agricultural residues. People are considering bioenergy, bioethanol, biodiesel, biomass gasification, biomethanation, and biomass cookstove for regular usage (Fig. 17.1). Biofuels are the solution for shortage of supply of electricity, increasing price of petroleum products, and domestic atmosphere pollution caused by heating with cooking.

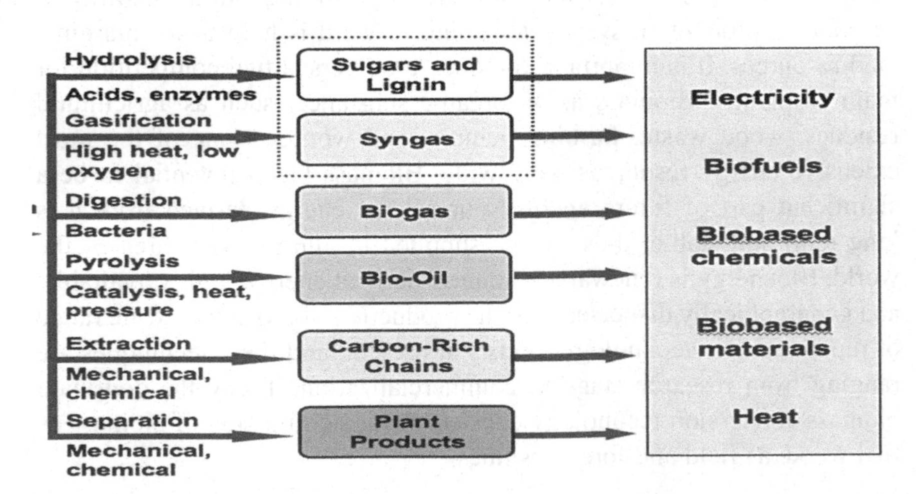

FIGURE 17.1 (See color insert.) Bioenergy formation process.

People are considering bioenergy, bioethanol, biomass gasification biodiesel, biomass cookstove, etc., for daily use purpose. Biofuels are the answer of problems like shortage of electricity supply, increasing prices of petroleum product,s and domestic atmosphere pollution caused by heating with cooking. In developing countries, approximately 10% of primary energy is made up of biofuels. Developing countries use it inadequately that affect millions of children and women while cooking and heating daily and are the reasons for atmospheric pollution. Cleaner fuels for cooking, electricity for water pumping, and increasing incomes through enhanced employment opportunities are some of the essential advantages for bioenergy production.

Previously, "bioenergy" is principally produced and used in the form of heat. Basic methodology is used from sustainability of biomass to biofuel and bioproduct synthesis (Fig. 17.2). Direct combustion is used for burning the biomass or products. Practical bioenergy technologies are used for some selective types of farming operations, anaerobic digestion, and gasification of biomass (ianswer4u.com, 2012). Bioenergy and biochemicals are produced from industrial and agricultural wastewater while treating these residues, such as methanogenic anaerobic digestion, biological hydrogen production, microbial fuel cells, and fermentation for production of valuable products. However, there are some scientific and technical barriers behind the implementation of these strategies (Angenent et al., 2004).

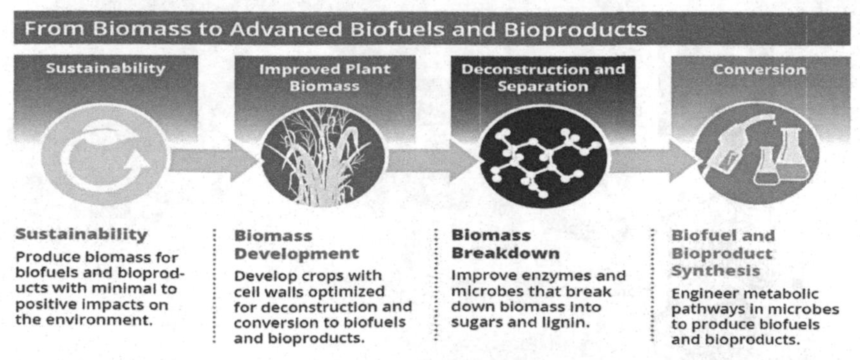

FIGURE 17.2 (See color insert.) Basic methodology of biofuel and bioproduct synthesis.

17.2 SOURCES OF BIOMASS

17.2.1 BIOFUELS

Biomass can be used in favor of gathering several means of transport fuels; in this way, biomass serves to recover the emissions of greenhouse gas (GHG) of the haulage region as well as lightens the petroleum products' burden.

Currently, for the biodiesel production, rapeseed and soy are used and sugarcane and corn are used for the production of ethanol. But in market, several companies are moving forward insistently.

Second generation of biofuels is advanced in which nonfood feedstocks, such as algae, wood chips, municipal waste, and perennial grasses, are used (Fig. 17.3). These fuels contain methanol, biobutanol, cellulosic ethanol, and various other synthetic diesel/petrol equivalents (Adler et al., 2009).

17.2.2 BIOBASED PRODUCTS

For the energy production, biomass can be alternatively used for fossil fuels and it is capable of producing a renewable replacement. There are

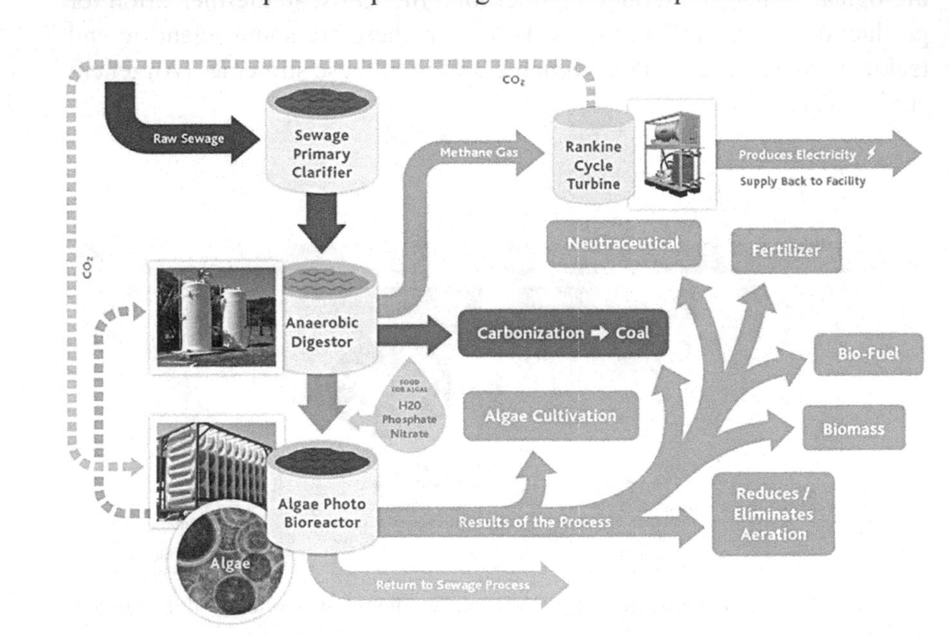

FIGURE 17.3 Algae biofuel production.

a lot of industrialized products and resources that are made up of natural gas or petroleum such as fertilizers, biobased foams, industrial chemicals, plastics, and lubricants.

17.2.3 BIOMASS ENERGY

The energy stored in biomass is used for production. Biomass is capable for manufacturing renewable heat or electricity. Incineration or gasification of dried biogas or biomass is used for biopower generation through proscribed anaerobic incorporation. Thermal energy (cooling and heating) is produced by direct burning of timber chips as well as its pellets and further sources of dry biomass. Low cost of biomass and fossil fuels (usually coal) is the way of improving cost effectiveness, dropping GHG emissions, and declining air pollutants in accessible power plants (http://www.epa.gov/chp/basic/efficiency.html). Agriculture and forestry crops and residues, municipal waste, industrialized residue, animal residues, etc. are basic source of biomass and it produces biomass energy (Fig. 17.3).

Process of production from biomass can be completed at anywhere if animals and plants are exiting. An extensive diversity of biomass feedstock is available. Moreover, heat, biobased products, liquid fuels, and electric power are used for most feedstocks (Biomass Energy Resource Center (BERC), n.d.).

17.2.4 AGRICULTURAL AND FOREST RESIDUES

By-products from crops are agricultural residues, that is, wheat, corn, rice straw, etc. Slurry, compost, seed husks, and wheat straw as well as many other are agricultural by-products. In various facets, the formation of energy from crop-quality biomass is an uncertain approach (Fig. 17.4). On this Earth's imperfect land, there is lesser amount of agricultural and forest residues to generate biofuel than food. On the other hand, in developing countries, many natives use woody compost and crop residues left after sorting as a fuel source. Although they might give the impression of a feasible energy resource, forest and crop residues contain high concentrations of nutrients.

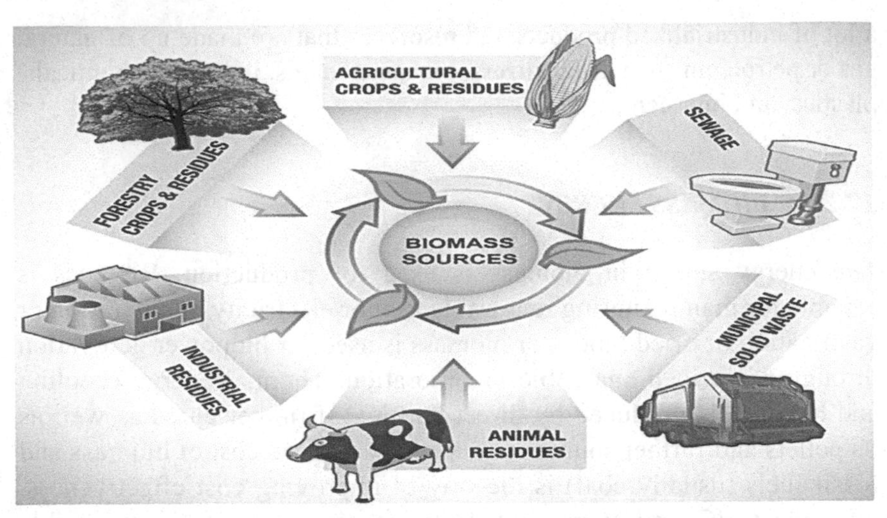

FIGURE 17.4 Sources of biomass energy.

17.2.5 *MUNICIPAL SOLID WASTE*

It is a method to generate electrical power by inexpensive fuel. A large part of metropolitan waste is organic, so incinerated in some cities equally as a resource of keeping the waste out of landfills. On an average, the burning of 1 t of typical municipal solid waste produces as much heat as one barrel of oil and more than 70 waste-to-energy plants are now installed in the United States, and for that many others are being planned.

17.2.6 *INDUSTRIAL WASTE*

Some industrial wastes are good sources of bioenergy, like wood waste obtained from construction, biogases, a by-product of milling sugar, and destruction firms, and this is possible if their use involves a cogeneration capability.

Cogeneration is the use of heat or material left over from manufacturing, often in combination with a conventional fuel such as oil, to produce electricity, and it can be used to run the factory, by putting any power left into the regional transmission grid.

17.2.7 BIOENERGY CROPS

The grass biomass is able to pack as one into dense pellets and resourcefully transported and burned in this type. The rising of annual crops such as corn may not be efficient enough in the long run for this purpose; too much labor and energy are required to grow this annual plant. On the other hand, perennial grasses such as switch grass can be more competently cultivated for bioenergy.

As bioenergy sources, wood and other high-biomass crops can be grown up entirely. Woody crops, such as high-yield hybrid poplar trees, can be grown in plantations with the biomass harvested on a 3–10-year cycle and then regenerated from stump sprouts. New trees have to be replanted every 15–20 years.

17.2.8 FOOD WASTE

Waste is produced by food-processing waste. Food (and fodder) crops are the edible parts of sugar, starch, and oil plants traditionally developed and grown to produce food for humans and animals. Various progression used for fuel (biotechnol, biodiesel, etc.) production includes wheat, maize, soya, palm oil, and sugar cane as food waste (Fig. 17.5).

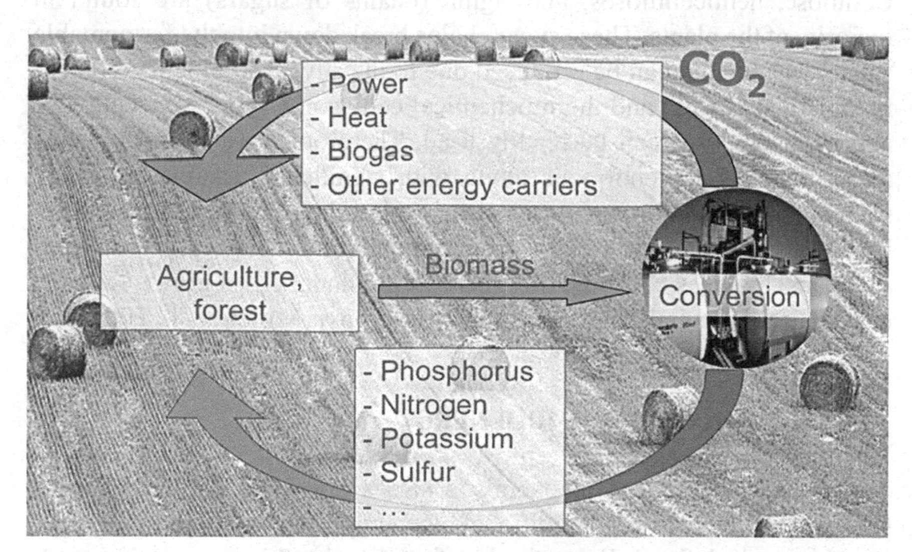

FIGURE 17.5 Agricultural and forest residues in bioconversion.

Forestry materials—Forestry and forest residues denote woody material from existing forests (which may or may not be managed) plus residues from sawmills, forest floors, tree pruning, logging residues, forest thinning, etc.

17.2.9 ANIMAL BY-PRODUCTS

These by-products are tallow, fish oil, manure, etc.

17.2.10 ENERGY CROPS

These are switch grass, *Miscanthus*, hybrid poplar, willow, algae, etc.

In the United States, forest-based woods are used as majority of biomass for the production of energy; similarly in the national energy portfolio, grassland- and agriculture-based biomass materials play a crucial and better role. Agricultural systems—grasslands, including lands in row-crop production—are the source of four types of biomass resources: starches, sugars, non-woody lignocellulosic materials, and woody lignocellulosic materials.

Small portion of plant matter is made through starches and sugars. Cellulose, hemicelluloses, and lignin (chains of sugars) are found in majority of the plants. These sugar chains break down into the fermentable sugars and ethanol can be produced due to the advanced bioethanol technology. For thermal and thermochemical conversion processes, cellulose and hemicelluloses can be readily used. Therefore, agricultural systems and grasslands are enormous along with significant feasible feedstock sources for production of bioenergy as well as their product. Agriculture-based biomass materials come from annual commodity crop (corn and soybeans) residues collected after harvest of annual crops grown for food or feed and perennial crops (grass and tree crops) (Miller et al., 2006).

17.3 ADVANCEMENT IN BIOENERGY

Traditional and modern bioenergy is also a classified part of biomass source. The line between "traditional bioenergy" and "modern bioenergy" is not so well defined. From the last few decades, bioenergy (firewood) is used for heating or cooking. It is still a relatively large energy source, which is used at a large scale.

In the production of various biofuels, usage of microorganisms has been steadily increasing day by day (Liao et al., 2016), particularly because of the metabolic diversity of different microorganisms that enable the production of biofuels from various substrates (Liao et al., 2016).

Electricity production can be produced by solid, liquid, or gaseous biofuels, with the biggest fraction of biopower, and it is divided into three broad categories (Fig. 17.6).

17.3.1 MODERN BIOMASS—BIOMASS TO ELECTRICITY AND HEAT

Modern biomass technologies include producing heat in boilers, used to power automobiles, biorefineries used in generating electricity, and liquid biofuels and pellet heating systems (Fig. 17.7). Biomass using modern technology differs from traditional biomass in two key characteristics; the source of organic matter should be sustainable, and second, the technology used to obtain the energy should limit or mitigate emissions of flue gases

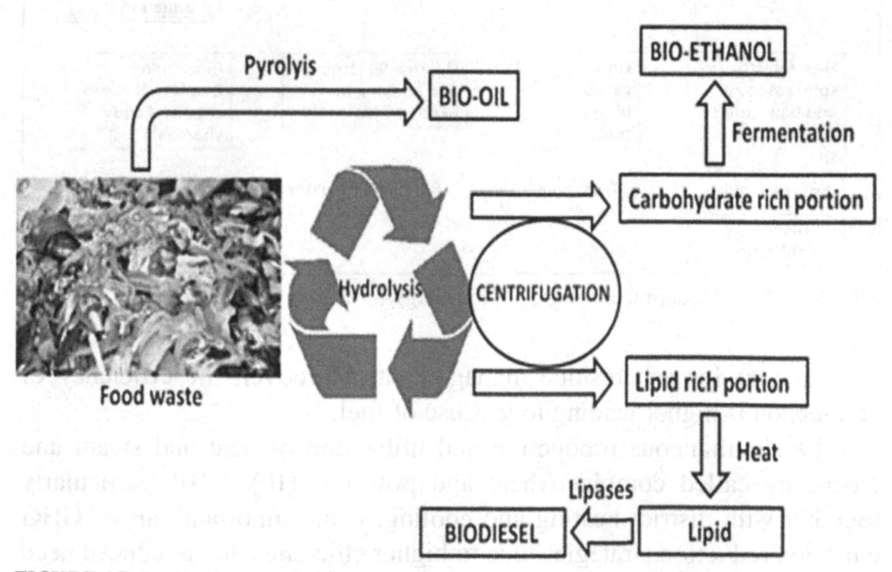

FIGURE 17.6 Food waste in bioconversion.

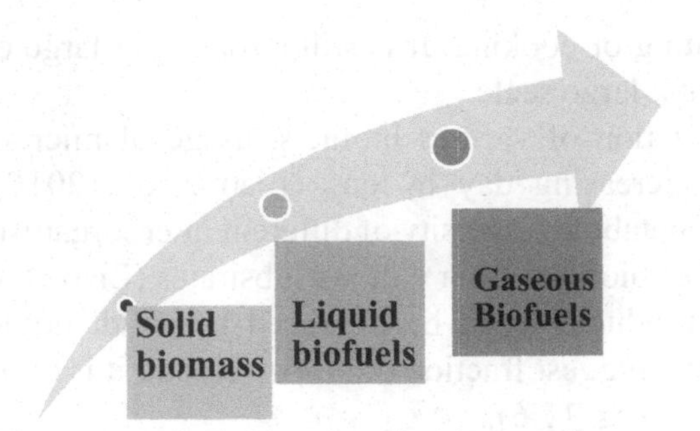

FIGURE 17.7 Bioenergy is divided into three broad categories.

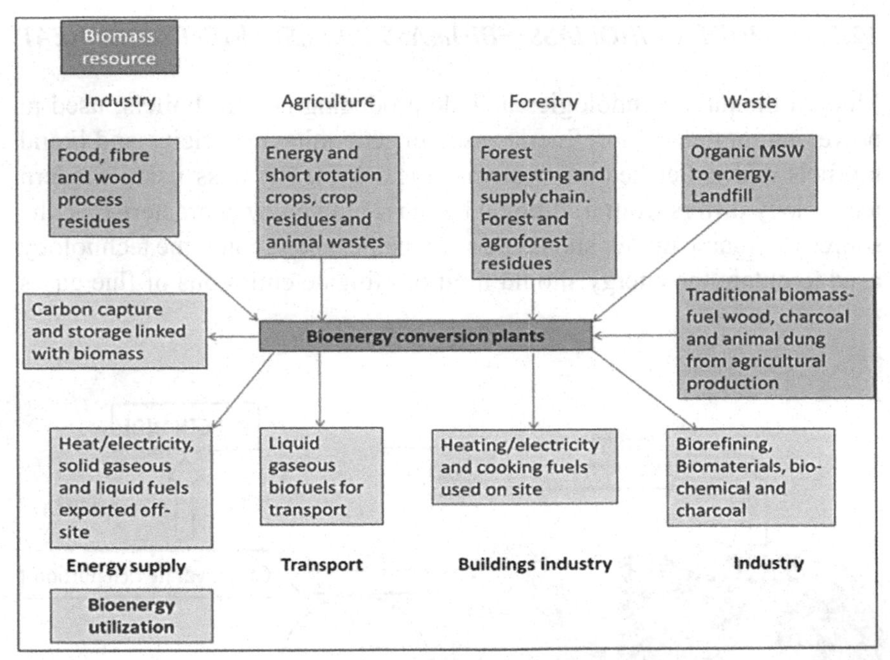

FIGURE 17.8 Modern technologies of bioenergy formation.

and account for ash residue management. Moreover, the efficiency of conversion is higher leading to less use of fuel.

The simultaneous production and utilization of heat and steam and electricity called combined heat and power (CHP). CHP, particularly together with district heating and cooling, is an important part of GHG emission reduction strategies, due to higher efficiency and a reduced need for fuels in comparison to stand-alone systems.

17.3.2 BIOENERGY FEEDSTOCKS AND TECHNOLOGIES

Food and agricultural residues have given energy crops, fodder crops, waste, and forestry to be the key of feedstocks of bioenergy (Fig. 17.8). Presently, the vast majority of modern bioenergy (excluding traditional biomass, that is, small-scale uses for heating, lighting, and cooking) comes from food and fodder crops, through a range of established conversion processes are used to produce conventional liquid biofuels.

In last decade, central as well as some state governments and international development agencies in terms of policymakers have shown considerable interest in bioenergy (Kracke et al., 2015). Instability of oil prices (and in oil producing regions), surging energy demand in developing countries, and greater awareness about climate change threats due to fossil fuel usage have primarily contributed to this renewed interest in bioenergy (Helsel and Specca, 2009) (Fig. 17.9).

17.3.3 ANNUAL BIOENERGY CROPS

Currently, in the United States for the production of liquid transportation fuel (ethanol and biodiesel, respectively), mainly grain, soybeans, and corn are used as crucial feedstocks. As a substitution, biodiesel is used,

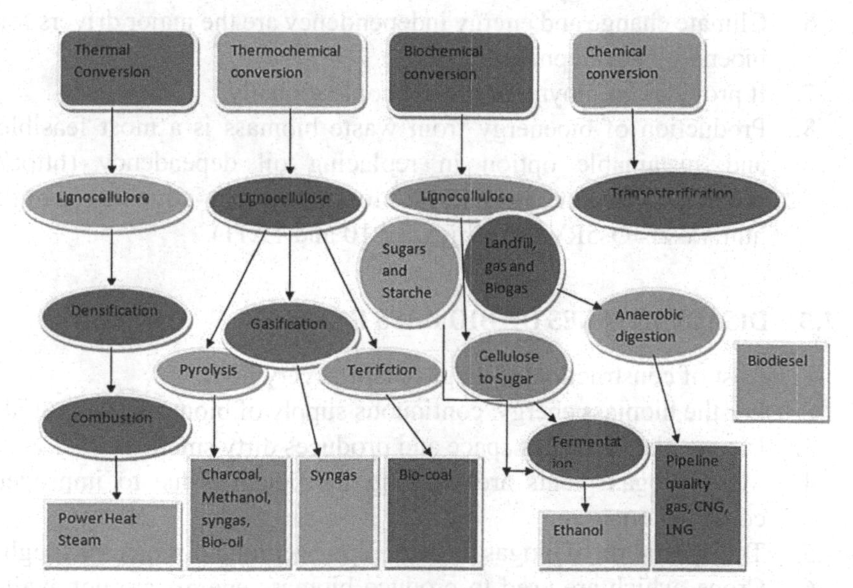

FIGURE 17.9 Bioenergy feedstocks and conversions.

and for the biodiesel formation, soybean and vegetable oils are mainly used. Biodiesel production during a distillation progression called "transesterification."

Annual crops are used for biofuel production; so, each year, these crops are replanted. Replantation of these crops depends on them. At the time of harvesting or before it, seeds fall on the ground to germinate. Annual crops, like corn, are dedicated bioenergy crop only when they are grown specifically for bioenergy, for example, when a farmer grows the crop under contract with a bioenergy company such as an ethanol producer (Brumfield and Helsel, 2011).

17.4 ADVANTAGES OF BIOMASS ENERGY

1. It is becoming an expensive service as oil necessities are getting exhausted day by day.
2. The heat energy obtained from biogas is 3.5 times higher than the usual burning wood.
3. Bioenergy is an inventive energy source.
4. Bioenergy is modern and globally traded commodity.
5. Countries with high share of renewable energy also have high share of bioenergy.
6. Climate change and energy independency are the major drivers for bioenergy development.
7. It provides employment to the people globally.
8. Production of bioenergy from waste biomass is a most feasible and sustainable option in replacing oil dependency (http://www.ianswer4u.com/2012/02/biomass-energy-advantages-and.html#axzz3Qs5RV0Sb) (Figs. 17.10 and 17.11).

17.5 DISADVANTAGES OF BIOMASS ENERGY

1. Cost of construction of biogas plant is very high.
2. For the biomass energy, continuous supply of biomass is required.
3. Biogas plant requires space and produces dirty smell.
4. Many biogas plants are working inefficiently due to improper construction.
5. Transportation of biogas through pipe over long distances is tough.
6. Crops, which are used to produce biomass energy, are not available for the entire year.

Provides manure for the agriculture and gardens and biomass can be converted into solid, liquid and gaseous fuels
Relatively cheaper and reliable
Time required for cooking is lesser
In cleanliness in villages and cities
Generated from human and animal wastes, left-over
Pressure on the surroundin g forest and scrubs can be reduced
Reduces pollution and spread of diseases.

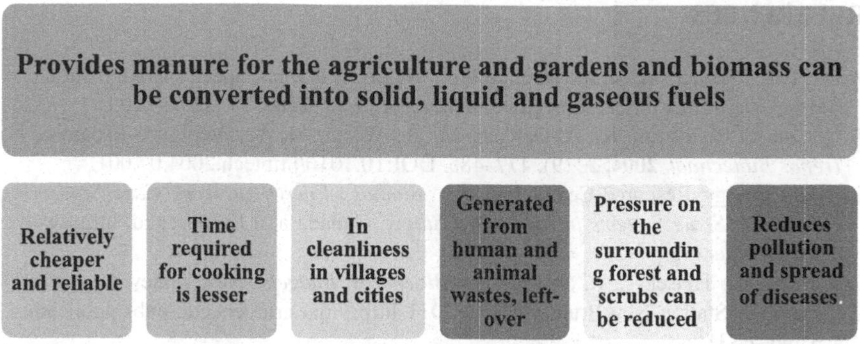

FIGURE 17.10 Advantages of biomass energy.

Bio- Energy's Future Aspects
ADVANTAGES
Fine way of recycling wastes
May be less economical then energy derived from fossil fuel
Positive impact on surroundings.
DISADVANTAGES
Tough to remain a enormous quantity of waste at all time.
Burning of methane causes greenhouse gases
Devoid of appropriate running bioenergy plants can pollute the environment
Potential strategy involving bio-energy includes drilling in to landfills and capturing methane instead of letting it escape to the atmosphere.

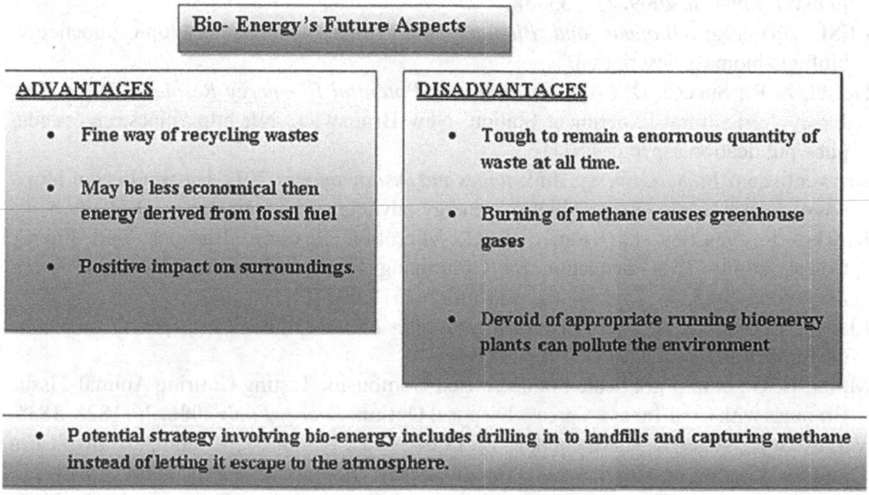

FIGURE 17.11 Bioenergy future aspects including advantages and disadvantages.

KEYWORDS

- **bioenergy**
- **biomass**
- **agriculture**
- **forestry**
- **production**
- **resources**
- **biofuel**

REFERENCES

Adler, P. R.; Sanderson, M. A.; Weimer, P. J.; Vogel, K. P. Plant Species Composition and Biofuel Yields of Conservation Grasslands. *Ecol. Appl.* **2009,** *19*, 2202–2209.

Angenent, L. T.; Karim, K.; Al-Dahhan, M. H.; Wrenn, B. A.; Domíguez-Espinosa, R. *Trends Biotechnol.* **2004,** *22* (9), 477–485. DOI:10.1016/j.tibtech.2004.07.001.

Biomass Energy Resource Center (BERC). *Biomass Energy at Work Case Studies of Community-Scale Systems in the United States*; Canada and Europe, n.d. http://www.biomasscenter.org.

Brumfield, R.; Helsel, Z. R. *Switchgrass Bioenergy Budgets*. New Jersey Agricultural Experiment Station, New Brunswick, NJ, 2011. http://njaes.rutgers.edu/pubs/publication.asp?pid=E331.

DeHaan, L. R.; Weisberg, S.; Tilman, D.; Fornara, D. Agricultural and Biofuel Implications of a Species Diversity Experiment with Native Perennial Grassland Plants. *Agric. Ecosyst. Environ.* **2009,** *137*, 33–38.

EESI. *Bioenergy, Biofuels and Biomass*. n.d. https://www.eesi.org/topics/bioenergy-biofuels-biomass/description.

Helsel, Z. R.; Specca, D. *Crop Residues as a Potential Bioenergy Resource*, 2009, New Jersey Agricultural Experiment Station: New Brunswick, NJ. http://njaes.rutgers.edu/pubs/publication.asp?pid=FS1116.

ianswer4u.com. *Biomass Energy: Advantages and Disadvantages*. 2012, ianswer4u.com. http://www.ianswer4u.com/2012/02/biomass-energy-advantages-and.html#axzz3Qs5RV0Sb.

Kracke, F.; Vassilev, I.; Krömer, J. O. Microbial Electron Transport and Energy Conservation—The Foundation for Optimizing Bioelectrochemical Systems. *Front. Microbiol.* **2015,** *6*, 575. DOI:10.3389/fmicb.2015.00575.

Liao, I. T.; Shan, H.; Xu, G.; Zhang, R. *Bridging Evolution and Development in Plants*; November 2016. DOI:10.1111/nph.14294.

Miller, B. G.; et al. Pilot Scale Fluidized Bed Combustor Testing Cofiring Animal-Tissue Biomass with Coal Tar as a Carcass Disposal Option. *Energy Fuels* **2006,** *20*, 1828–1835.

Varvel, G. E.; Vogel, K. P.; Mitchell, R. B.; Follet, R. F.; Kimble, J. M. Comparison of Corn and Switchgrass on Marginal Soils for Bioenergy. *Biomass Bioenergy* **2008,** *32*, 18–21.

Zhen, F. *Biofuels—Economy, Environment and Sustainability*. February, 2013. ISBN 978-953-51-0950-1. DOI:10.5772/50478.

Index

A

Algal biomass
  energy crisis, 301
  microalgae
    biodiesel characterization, 309–310
    biodiesel production, 309
    Bligh and dyer method, 307–308
    extraction of oil, 307–308
    Folch method, 308
    lipids from, 305–306
    producing biodiesel, importance of, 310–311
    solvent extraction method, 308
    transesterification, 309
    utilized for producing biodiesel, 302–305
Anaerobic digestion (AD), 255–256
  large-scale implementations, implication
    biofertilizer, 238
    biomethane, 227–228
    biomethane production, agrowastes for, 235–236
    carbon to nitrogen ratio, variation in, 234–235
    four stages, 226–227
    inhibitory action of VFA, 233–234
    kinetic modeling of, 236–238
    microbial communities, 231–232
    mixing in anaerobic digester, 230–231
    organic loading rate (OLR), 231
    organic wastes, 228–229
    pH value, 233
    plant growth promoter, 238–240
    rural family-size biogas plants, barriers in, 229–230
    schematic illustrations, 226
    volatile fatty acids (VFAs), 232–233
Aquatic energy crops
  perspective of liquid and gaseous fuel production from
    dedicated aquatic energy crops, 170–179
    economic growth, 167
    feedstocks, 169
    weeds as energy crops, 169–170

B

Biodiesel production
  energy crisis, 301
  microalgae
    biodiesel characterization, 309–310
    biodiesel production, 309
    Bligh and dyer method, 307–308
    extraction of oil, 307–308
    Folch method, 308
    lipids from, 305–306
    producing biodiesel, importance of, 310–311
    solvent extraction method, 308
    transesterification, 309
    utilized for producing biodiesel, 302–305
Bioenergy
  research and advances
    advantages of, 328
    agricultural and forest residues, 321
    animal by-products, 324
    annual bioenergy crops, 327–328
    biobased products, 320–321
    bioenergy, 319
    bioenergy crops, 323
    bioenergy feedstocks and technologies, 327
    biofuels, 320
    biomass energy, 321
    biomass to electricity and heat, 325–326
    combined heat and power (CHP), 326
    developing countries, 319
    disadvantages of, 328
    energy crops, 324

food-processing waste, 323–324
industrial wastes, 322
municipal solid waste, 322
production of, 318
sources
advantages of, 328
agricultural and forest residues, 321
animal by-products, 324
annual bioenergy crops, 327–328
biobased products, 320–321
bioenergy, 319
bioenergy crops, 323
bioenergy feedstocks and technologies, 327
biofuels, 320
biomass energy, 321
biomass to electricity and heat, 325–326
combined heat and power (CHP), 326
developing countries, 319
disadvantages of, 328
energy crops, 324
food-processing waste, 323–324
industrial wastes, 322
municipal solid waste, 322
production of, 318
Bioenergy production
biomass sources and applications, 246
agricultural residue, 248
anaerobic digestion (AD), 255–256
animal manure biomass, 248
biochemical conversion, 254–256
bioenergy, conversion of, 248–249
cellulosic plants, 247–248
combustion method, 250–252
fermentation, 256
major barriers in, 257
oil and sugar biomass, 247
production of bioenergy, 249–250
pyrolysis and torrefaction, 252–253
thermal conversion, 250–253
thermochemical conversion, 253–254
torrefaction, 253
transesterification, 257
Bioethanol production
feedstock
advantages of, 219–220
biomass derived, 217–218
E. crassipes, 216
ethanol, 214
L. camara, 215
limitation of fossil fuel, 213
limitations, 220–221
P. hysterophorus, 215–216
steps involved in, 219
lignocellulosic biomass
agricultural waste, 283
anaerobic bacteria, cellulose degradation in, 293–294
β-glucosidases, 291
bioethanol production, genetically modified fungus for, 294
biofuel production, current status of, 296
biological pretreatment, 287
cellobiohydrolases or exoglucanases, 290
cellodextrinases or endoglucanases, 290
cellulose, 283–284
chemical pretreatment, 287
dry weight composition in, 285
enzymatic hydrolysis or saccharification, 288–290
fermentation, 291–292
genetically modified microbes, 292–293
hemicelluloses, 284
mean average temperature, 282
nonprioritization of hemicellulose, 285
physical pretreatment, 286–287
physicochemical pretreatments, 287–289
Biofuel cells (BFCs), 95–96
conversion of, 109–110
industrial applications, 110–112
stimulant for production of, 107–109
types of
advancements in, 105
chemical oxygen demand (COD), 104
EFCs, 99–101
mediator or co-substrate, 105–107
MFCs, 98–99

producer of, 101–102
proton exchange membrane (PEM), 104
substrate for, 103–105
Biofuel production
algal biomass
fossil fuels, 121
fossil hydrocarbons, utilization of, 120
microalgae, 119
studies, 120
biomass resources and bioenergy prospective
biotechnology, application of, 14–15
brown grease, use, 13–14
conventional agricultural products, 12–13
dedicated energy crops, 7–10
direct use and blending, 10–11
first-generation feedstock biofuels, 5–7
inedible feedstock for, 13
lignocellulosic products, 13
microalgae for, 14
microemulsion process, 11
pyrolysis, 11–12
sustainability issues of, 7
transesterification method, 12
lipase and its biotechnological application
biodiesels, 2–3
biomass resources and bioenergy prospective, 5–7
classification of, 5
feedstocks and consumption, 4–5
main types of biofuels, 3–4
sources and applicable methods, 4
microalgae and macroalgae, 121, 123
anaerobic digestion, 133
biodiesel from, 137
bioethanol, 136
biogas and biohydrogen, 137–138
closed photobioreactor systems, 125
cultivation & harvesting technologies, 129–130
direct combustion, 135
fermentation, 133–134
gasification is, 131
harvesting technologies, 127–128
industrial and commercial prospects, 138–139
lipid induction technique, 126–127
liquefaction, 132–133
open-air systems, 124
production of, 122
pyrolysis, 131–132
second-generation, 122
thermochemical conversion, 130–131
transesterification, 134–135
two-stage hybrid systems, 125–126
ultrasonification to aid biofuel yield, 135–136
second-generation ethanol production in Brazil
challenges in, 161–162
Cocoa, 160–161
lignocellulosic materials, 148–160
Biohydrogen production
bio-syngas for biofuels and chemicals
conventional fuels, 74
conventional production of, 79–80
entrained flow gasification, 82–84
fertilizer industry, ammonia production for, 88
fluidized bed gasification technique, 80–82
FT synthesis, 87
hydrogen production in refineries, 88
market, 88–89
methanol, 87–88
power generation, 86–87
production of, 77–78
purification and conditioning, 84–86
second generation biofuels, 75–76
synthetic natural gas (SNG), 75
torrefaction, 84
dark fermentation (DF), lignocellulosic biomass
commercialization, 31, 33
coupling of, 32
effluents, bioenergy recovery, 30–31
fossil fuels, 19
inhibition of, 26–28
organic fraction municipal solid wastes (OFMSW), 24

pretreatment, 28–30
principles of, 20–21
process parameters, 21–24
yield, 25–26
in plants, tailoring triacylglycerol biosynthetic pathway
biodiesel *via* transesterification, 55–56
energy densification in vegetative tissues, metabolic engineering, 53–55
transcription factors (TFs), 49–53
triacylglycerol (TAG), 43–49
industrial technology, lignocellulosic biomass
biological conversion, 63
cost-effective and fast production, 63
economic perspective, 68–70
first-generation biofuels, 65–66
fourth-generation biofuels, 67
global scenario, 64–65
International Energy Agency (IEA), 62
plant biomass, 64
second-generation biofuels, 66
technologies and processes, 67–68
third-generation biofuels, 66–67
Biomass resources and bioenergy prospective
first-generation feedstock biofuels, 5
bioethanol production, sugar and sugar crops, 6
oil crops, 6–7

C

Chemical oxygen demand (COD), 104
Combined heat and power (CHP), 326

D

Dark fermentation (DF), lignocellulosic biomass
commercialization, 31, 33
coupling of, 32
effluents, bioenergy recovery, 30–31
fossil fuels, 19
inhibition of, 26–28
metal ions concentrations, 27
mixed microflora, 26–27
soluble metabolites, concentration of, 27–28
organic fraction municipal solid wastes (OFMSW), 24
pretreatment, 28–30
principles of, 20–21
process parameters, 21–24
hydraulic retention time (HRT), 23
microflora, 21–22
organic loading rate (OLR), 24
pH, 22
temperature, 22–23
yield, 25–26
Dedicated aquatic energy crops, 170
Azolla, 174–176
marine macrophytes (sea weeds), 171–173
pistia stratiotes (water lettuce), 174
water hyacinth (*Eichhornia crassipes* Martius), 176–179
Dedicated energy crops
forestry and agricultural residues, 8–9
municipal solid waste (MSW), 9
perennial grasses, 7–8
short rotation wood crops
Jatropha, 8
third-generation feedstocks, 9–10

E

Enzymatic fuel cells (EFCs), 97

L

Large-scale implementations
implication
biofertilizer, 238
biomethane, 227–228
biomethane production, agrowastes for, 235–236
carbon to nitrogen ratio, variation in, 234–235
four stages, 226–227
inhibitory action of VFA, 233–234
kinetic modeling of, 236–238
microbial communities, 231–232

mixing in anaerobic digester, 230–231
organic loading rate (OLR), 231
organic wastes, 228–229
pH value, 233
plant growth promoter, 238–240
rural family-size biogas plants, barriers in, 229–230
schematic illustrations, 226
volatile fatty acids (VFAs), 232–233
Lignocellulosic biomass
agricultural waste, 283
anaerobic bacteria, cellulose degradation in, 293–294
β-glucosidases, 291
bioethanol production, genetically modified fungus for, 294
biofuel production, current status of, 296
biological pretreatment, 287
cellobiohydrolases or exoglucanases, 290
cellodextrinases or endoglucanases, 290
cellulose, 283–284
chemical pretreatment, 287
dry weight composition in, 285
enzymatic hydrolysis or saccharification, 288–290
fermentation, 291–292
genetically modified microbes, 292–293
hemicelluloses, 284
mean average temperature, 282
nonprioritization of hemicellulose, 285
physical pretreatment, 286–287
physicochemical pretreatments, 287–289

M

Metabolic engineering for liquid biofuels generations
lignocellulosic biomass
accessibility, 187–190
composition, 185–187
improving liquid, 195–201
organic residues, 184
stultification of environment and carbon imbalance, 184
synergistic approach for, 202
yield and decomposition, 190–195
Microalgae
biodiesel
characterization, 309–310
production, 309
Bligh and dyer method, 307–308
extraction of oil, 307–308
Folch method, 308
lipids from, 305–306
producing biodiesel
importance of, 310–311
solvent extraction method, 308
transesterification, 309
utilized for producing biodiesel, 302–305
Microbial fuel cells (MFCs), 97
Municipal solid waste (MSW), 9

O

Organic loading rate (OLR), 231

P

Proton exchange membrane (PEM), 104

S

Second-generation ethanol production in Brazil
lignocellulosic materials, 148
biomass structure, 149
cassava residue, 153–155
leaves, 150–153
Palm Oil industry, 156–160
peach palm (bactris gasipaes), 156
straw, 150–153
sugarcane bagasse, 150–153
tucumã (astrocaryum aculeatum), 156

T

Third-generation biofuels
aid biofuel yields
ultrasonication to, 276
bioethanol
generality use of, 267–268
bioethanol production, 268–271
biofuels production
microalgae potential as raw material, 264–266

direct combustion, 275–276
high population growth, 262
microalgae
fuel production throughout pyrolysis of, 272–274
physical-chemical composition of, 266–267
protein biofuels, 274–275
rapid industrialization, 262
Triacylglycerol (TAG), 43
biosynthesis, 44
acyl-CoA, 46
DAG in, 46
lysophosphatidic acid acyltransferase enzyme (LPAAT), 45
polyunsaturated fatty acids, 45
mobilization, 46–49

V

Volatile fatty acids (VFAs), 232–233

W

Weed biomass
bioethanol production, feedstock
advantages of, 219–220
biomass derived, 217–218
E. crassipes, 216
ethanol, 214
L. camara, 215
limitation of fossil fuel, 213
limitations, 220–221
P. hysterophorus, 215–216
steps involved in, 219

Printed in the United States
by Baker & Taylor Publisher Services

Printed in the United States
by Baker & Taylor Publisher Services